AIR POLLUTION SAMPLING & ANALYSIS DESKBOOK

by

Paul N. Cheremisinoff

Angelo C. Morresi

ANN ARBOR SCIENCE
PUBLISHERS INC
P.O. BOX 1425 • ANN ARBOR, MICH. 48106

Copyright © 1978 by Ann Arbor Science Publishers, Inc.
230 Collingwood, P. O. Box 1425, Ann Arbor, Michigan 48106

Library of Congress Catalog Card No. 77-093385
ISBN 0-250-40234-3

Manufactured in the United States of America

PREFACE

During the late 1960s, after years of neglect, society began to focus on the need for pollution control. It became apparent that trade-offs between techno-logical progress, economics and the environment are vital to man's survival. We have seen this recognition implemented through both environmental legislation and a desire for greater knowledge of the effects pollution has on public health and welfare.

Environmental contaminants manifest themselves in minute quantities—parts per million, parts per billion, parts per trillion. It has become necessary to identify and measure these contaminants so that their effects can be assessed. One confusing aspect to engineers, managers and technicians in-volved in sampling and analysis has been that many methods and techniques do essentially the same thing. While many methods supposedly achieve the same results, often there are differences, as well as advantages and disad-vantages, among procedures. This book is an attempt to address this problem for the air pollution sector. Sampling and analytical methods for point sources, fugitive emissions and ambient conditions are identified. References are provided for those requiring more detailed information.

Having experienced the cost and time involved in an environmental pollution investigation, we present sampling and analytical methods for use in screening studies, such as general qualitative investigations, and for more extensive studies. Environmental assessments and impact statements are now a way of life for industry. Sampling and analytical data may be re-quired for designing control equipment, for process control, for submission to government enforcement agencies to comply with source measurement requirements, as legal evidence of compliance, and for protection of health and property for both workers and the public at large.

We are grateful to the many people who have contributed in making this book possible. Particular acknowledgment is due Dr. Gordon Lewandowski for his contribution of Chapter 3 on stack sampling and William B. Rossnagel, President of Rossnagel Associates, for Chapter 13 on comparison of source emission limits.

Paul N. Cheremisinoff
Angelo Morresi

CONTENTS

v

COMMON TERMS AND THEIR ABBREVIATIONS
FOR AIR POLLUTION SAMPLING AND ANALYSIS

AAS	Atomic Absorption Spectroscopy
DL	Detection Limit
FAAS	Flameless Atomic Absorption Spectroscopy
FID	Flame Ionization Detector
FPD	Flame Photometric Detector
GC	Gas Chromatography
GCMS	Gas Chromatography/Mass Spectroscopy
HE	Hydride Evolution Technique
HPLC	High-Performance Liquid Chromatography
HRMS	High-Resolution Mass Spectrometry
IR	Infrared Spectroscopy
LRMS	Low-Resolution Mass Spectrometry
MS	Mass Spectrometry
NDIR	Nondispersive Infrared Spectroscopy
NMR	Nuclear Magnetic Resonance Spectroscopy
SASS	Source Assessment Sampling System
SIE	Specific Ion Electrode
SSMS	Spark-Source Mass Spectrometry
TCD	Thermal Conductivity Detector
TLC	Thin-Layer Chromatography
UV	Ultraviolet Visible Spectroscopy
XAD-2 Module	SASS Component

SAMPLING AND ANALYSIS:
METHODS AND PROCEDURES

INTRODUCTION

This section will outline the various sampling and analysis methods and procedures for air pollution. The first portion is devoted primarily to the available sampling techniques; the latter portions deal with analysis of these samples. In most cases, references have been provided for further information.

PROGRAM FOR SAMPLING AND ANALYSIS

A sampling and analysis program can be relatively simple when the analyst knows, in general terms, the compound and order of magnitude of concentration for which he is sampling. This might be the case in an examination of the emissions from a methylene chloride drying operation. In this instance, the compound is known, and through calculation, the amount of methylene chloride given off can be determined. The sampling and analysis is done to verify the calculation and material balance.

However, for example, in the case of an industrial waste incinerator or a chemical operation, a qualitative and quantitative analysis of a single sample can become rather cumbersome and costly because of the almost infinite possibilities of contaminants. Further, depending on how well equipped the laboratory, a comprehensive pollutant investigation could take several weeks, which can be discouraging to a company trying to prove that the odor causing community complaints is effected by a compound it does not use.

By contrast, a quick, simple, incomplete analysis could result in omissions, which could be dangerous and legally controversial should the missing data concern toxic material.

The procedures outlined in this book provide for a logical order of sample analysis in a manner that is least costly but does not limit results. Initially, a general analysis is done on the sample. This involves determination of a material's presence, its approximate concentration and whether it is organic or inorganic. It is, in effect, a screening study known as a *Level 1 investigation.*

From the results of a Level 1 investigation, a more comprehensive effort can be given to those areas with defined potential. In this phase of the analysis a complete quantitative and qualitative identification can be made based on the guidelines set forth in the Level 1 investigation. This postplanning effort is known as a *Level 2 investigation.*

Figures 1-1, 1-2 and 1-3 give the general characteristics of Level 1 and 2 investigations, outlining procedures, sampling techniques and analytical techniques. All pollutant classes from sources are screened in a Level 1 investigation to be followed by a study of selected pollutants and selected sources in the Level 2 investigation.

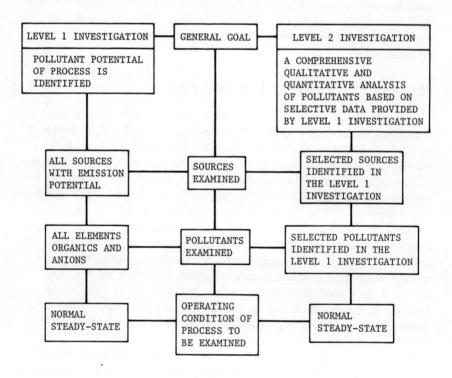

Figure 1-1. General characteristics of Level 1 and Level 2 investigations.

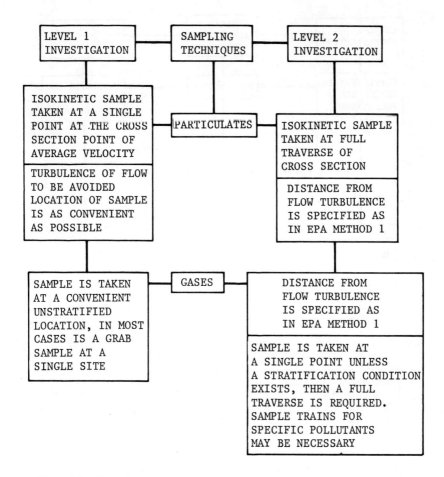

Figure 1-2. General sampling techniques for Level 1 and Level 2 investigations.

Step-by-step procedures for a gas and particulate analysis of an emission source are shown in Figures 1-4 and 1-5. Note that all pollutants are considered potentially existent; only through the screening study of the Level 1 investigation, are they eliminated from consideration.

Sampling and Analytical Procedures

Before sampling, information pertaining to the process or equipment to be sampled must be gathered. This would include a site survey for adequacy and accessibility of sampling location, personnel expertise, literature searches and a sampling plan. Other types of information necessary are:

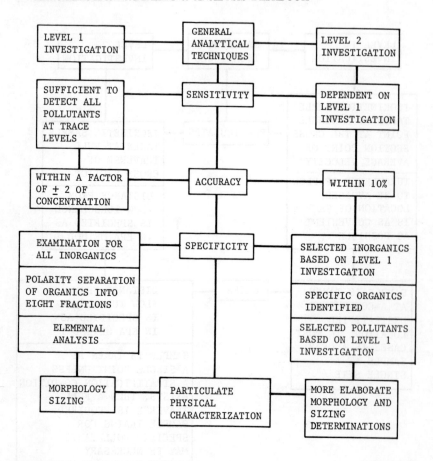

Figure 1-3. General analytical techniques for Level 1 and Level 2 investigations.

1. process flow diagrams;
2. equipment layouts and specifications;
3. equipment operating procedures; and
4. operating parameters (temperature, flow rate, etc.).

Sample Size

Sample size depends on the analyst's level of investigation. Level 2 investigation sample size requirements depend on the output of the Level 1 investigation because the amount of analytical work to be done is not known previously. Minimum sample requirements for typical Level 1 investigations are as follows:

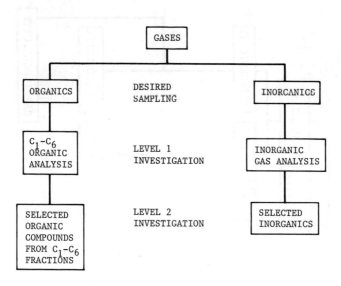

Figure 1-4. Gas sample analysis program.

Sample Type	Sample Size
Ambient Air (fugitive emissions)	480 m^3 (STP)
Stack Gases	3 liters
Stack Particulates	30 m^3

Special Sampling Handling Requirements

Precautions listed below to be taken for sample handling are necessary because of the potential of extremely low pollutant levels:

1. Sampling equipment must be inert, that is, it must not in any manner change the sample constituents.
2. Sample preservation techniques should be utilized.
3. Sampling equipment must be closely cared for to avoid contamination. Clean equipment and work environments are required. Seals should be carefully checked to prevent leakage.

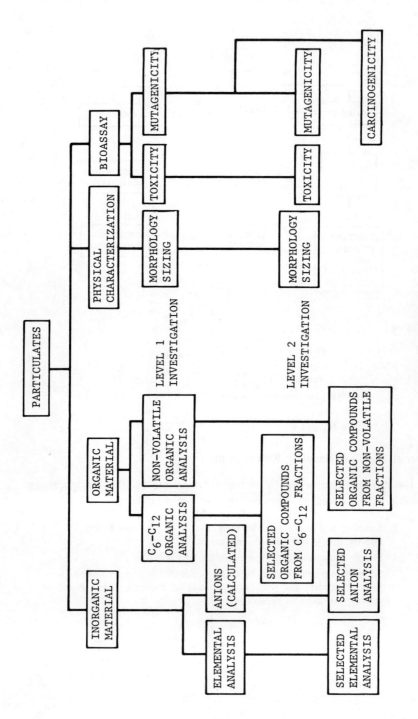

Figure 1-5. Particulate sample analysis program.

2

FLOW MEASUREMENT TECHNIQUES

Flow measurement techniques have been outlined in this section. The devices noted are portable, not for permanent installation.

For gas flow measurement undergoing a Level 1 investigation, in general, the "S"-type pitot tube is the desired sampling device. For the more refined Level 2 investigation, the ellipsoidal-nosed standard pitot tube is desired. The latter device is known for its high accuracy and low sensitivity to interferences.

Methods to estimate or calculate flow rates are also discussed in this section.

_ _

METHOD: STANDARD PITOT TUBE

Description:

1. A pressure probe method. Point velocity determination.
2. Differential pressure at a point in the duct or stack is sensed by a pitot tube connected to a pressure transducer and then related to the velocity at that point.
3. A transverse is performed with points selected according to EPA Method 1. Flow rate is determined from resulting velocity profile.
4. Effect of compressibility is usually handled through use of a correction factor.
5. Method is manual and is not ordinarily used for continuous monitoring.

Advantages:

1. 0.5% accuracy from 1.1-60.9 m/sec.

2. Less sensitive to probe orientation than "S"-type pitot tube.
3. Not easily clogged in high-particulate, high-moisture gas streams for short time periods.
4. Stainless steel construction.
5. Can be water-cooled to withstand temperatures up to 2000°F.
6. Relatively inexpensive.
7. May be used for liquids.

Disadvantages:

1. Lower differential pressure than "S"-type.
2. When used in conjunction with a long sampling probe, it is difficult to insert the assembly into the duct.
3. Must obtain alignment of probe axis with flow direction.

Remarks:

1. A port must be cut into the duct or stack. Size of port is primarily determined by probe assembly diameter.
2. This is the most accurate method for point velocity determinations.
3. Ellipsoidal nose type is preferred because it is more accurate than hemispherical type.

— —

METHOD: "S"-TYPE PITOT TUBE

Description:

A pressure probe method (see Standard Pitot Tube discussion). Point velocity determined.

Advantages:

1. Higher differential pressure than standard pitot tube.
2. 0.5% accuracy, from 2.7-30.9 m/sec.
3. Greater resistance to clogging in high-particulate, high-moisture streams than Standard Pitot Tube.
4. Same as 4, 5, 6 and 7 for Standard Pitot Tube.
5. Recommended in EPA Method 2.
6. No insertion problem as with Standard Pitot.
7. Bidirectional; will show effect of reversal in flow.

Disadvantages:

1.   High yaw sensitivity.

Remarks:

Port must be cut into the duct, as with the Standard Pitot Tube. Due to the availability of electronic pressure transducers, the advantage of the "S" pitot is that a higher differential pressure is reduced.

_ _

METHOD: HASTINGS RAYDIST FLARE GAS FLOW PROBE

Description:

1.   A pressure probe method. Point velocity determination.
2.   Probe is variation of pitot tube with two openings at probe tip that are connected by an internal stainless steel tube. A portion of tube is heated and thermoelectric sensors measure temperature gradients along the wall of the tube, external to the flow stream. Purge gas is injected into the tubing in an arrangement that forms a pneumatic bridge. At zero velocities the bridge is balanced. As flow across the tip occurs, the differential pressure developed unbalances the bridge, and the sensors measure the shift in temperature gradients along the heated portion of the tube. The sensor output voltage is related to the gas velocity by a calibration curve.

Advantages:

1.   Corrosion and abrasion effects are negligible.
2.   Clogging is almost impossible.
3.   Can operate in gas streams up to 315°C.

Disadvantages:

1.   Accuracy not better than 4.8% of reading for range of 2.74-30.48 m/sec.
2.   High sensitivity at low velocities accentuates zero drift.
3.   Must be isolated from vibrations.

Remarks:

1.   Port must be cut into duct to accommodate device.

2. Accuracy inadequates make this device unacceptable for use unless other options are unavailable.
3. Sensitivity to vibrations would indicate that single point probe or rake must be used rather than a reciprocating traversal mechanism.

METHOD: ELLISON ANNUBAR

Description:

A pressure probe method. Pressure velocity averaging. The stagnation pressure is sensed by four upstream holes opening into the probe body while the static pressure is sensed by a single hole facing downstream and located in the center of the probe.

Advantages:

1. 7.2% accuracy for flow.
2. Corrosion and abrasion effects are negligible due to stainless steel construction.
3. Inexpensive compared to point sensor array.

Disadvantages:

1. Must be constructed for specific location since length of device and positioning of high-pressure ports are duct-specific.
2. Rear orifice is subject to clogging.

Remarks:

1. Installed in line.
2. Can be used in both circular and rectangular ducts although greater calibration factor variation can be expected for rectangular case.
3. Holes in annular are located according to specific duct dimensions.

METHOD: FLUID DRAG METER

Description:

Uses a bonded strain gauge bridge to translate deflection due to fluid forces on target disc into an electrical signal proportional to the square of the flow rate. Point velocity determination.

Advantages:

1. 1% accuracy from 2.7-30.48 m/sec.
2. Resistant to chemical attack if internal purge is used.
3. Bidirectional.
4. Can be used for continuous monitoring applications.
5. Can operate in gas streams up to 260°C.

Disadvantages:

1. Sensitive to orientation in duct.
2. Two probes are necessary to cover velocity range of 1.5-38 m/sec.

Remarks:

1. A port must be cut into duct to accommodate device.
2. This is the most accurate method for point velocity determinations, readily applicable to continuous monitoring situations.
3. The probe is attached to an automatic reciprocating traversal mechanism to obtain measurements at various points. This is less expensive than employing a multiple sensor rake.

_ _

METHOD: HOT-WIRE ANEMOMETER

Description:

Point velocity determination. A short length of fine platinum wire is mounted at the end of a probe and heated by electricity. The electrical resistance of the wire is a function of its temperature. Flow of a gas around the hot wire cools it, changing its resistance. By holding constant either the voltage or current across or through the wire the change in amperes or voltage, respectively, becomes a function of gas velocity. Probe is connected to a Wheatston bridge.

Traverse across duct to obtain velocity profile from which volume flow rate can be determined.

Advantages:

1. Very quick response to velocity changes (low thermal inertia).
2. Device is very small.
3. Can be used to obtain velocities near walls.

Disadvantages:

1.   Calibration is complicated (done by placement in stream of same gas of known velocity).
2.   Very delicate device.
3.   Cannot be used successfully with liquids.

Remarks:

1.   Port must be cut into duct to accommodate device.
2.   Best practical method for measuring turbulent fluctuations at a point.
3.   Complete commercial setups available.

_ _

METHOD: HOT-FILM PROBE

Description:

Point velocity determination.   Variation of Hot-Wire Anemometer.   A thin, electrically conducting film is coated on the insulator located at the probe tip.   Film acts as resistance.

Advantages:

1.   Can be used for liquids.
2.   More durable than Hot-Wire device (suitable for high-velocity liquid flows).

Disadvantages:

1.   Same as 1 for Hot-Wire Anemometer.

Remarks:

1.   See Hot-Wire Anemometer.

_ _

METHOD: VENTURI METER

Description:

An obstruction device.   It acts as an obstacle placed in the path of a flowing fluid causing localized changes in velocity and, concurrently, pressure changes. Pressure differential between inlet and throat section of device is measured by taps located in the wall and connected to a pressure transducer. Differential pressure is then related to ideal velocity and volume flow rate. A portion of the pressure drop becomes irrecoverable because of friction and turbulence losses. Actual volume flow rate is obtained by multiplying ideal flow rate by an empirical coefficient which depends on flow, Reynolds number and duct geometry as well as on characteristics of device. The effect of compressibility is handled through the use of a theoretical expansion factor.

Advantages:

1. Most accurate of all obstruction devices (1% of total flow).
2. No solids accumulation occurs.
3. Pressure recovery greater than for other flow obstruction devices.
4. Suitable for continuous flow measurements.
5. Can be installed in any pipe orientation as long as it is completely filled with fluid.
6. Can be used for gases.
7. Good resistance to abrasion.

Disadvantages:

1. More expensive than other obstruction devices.
2. Requires more space than other obstruction devices.
3. More difficult to install than other obstruction devices.

Remarks:

1. A major disadvantage of obstruction devices is that pressure drop varies with the square of the flow rate; thus, for use over a wide range of flow rates, pressure-measuring equipment of very wide range is required, usually resulting in poorer accuracy at the lower flow rates.

— —

METHOD: FLOW NOZZLE

Description:

An obstruction device (see Venturi Meter discussion).

Advantages:

1. 3-4% accuracy.
2. Same as 4, 5, 6, and 7 under Venturi Meter.
3. Requires less space than Venturi Meter.
4. Less expensive than Venturi Meter.
5. Can be installed at discharge point.

Disadvantages:

1. Lower pressure recovery.
2. Relatively difficult to install properly.

Remarks:

1. Installed in line.
2. See Venturi Meter discussion.

_ _

METHOD: ORIFICE METER

Description:

An obstruction device (see Venturi Meter discussion). The effect of compressibility is handled through use of an empirical expansion factor. Several possible pressure tap locations can be used.

Advantages:

1. 2-3% accuracy.
2. Same as 4, 5 and 6 under Venturi Meter.
3. Least expensive of all obstruction devices.
4. Least space requirement of all obstruction devices (may often be installed between existing pipe flanges).
5. Simpler installation than other obstruction devices.

Disadvantages:

1. Poorest pressure recovery of all obstruction devices.
2. Especially susceptible to inaccuracies resulting from wear and abrasion.
3. Lower accuracy than other obstruction devices.
4. Lower physical strength than other obstruction devices (may be damaged by pressure transients, solids).

Remarks:

1. See Venturi Meter discussion.
2. Installed in line.
3. Should only be used in streams of low solids content. Same location requirements as Venturi Meter and Flow Nozzle.
4. Various types of orifices (*e.g.*, sharp edge, square edge, etc.) are available.

_ _

METHOD: ROTAMETER

Description:

A flow meter operating on the physical principle of fluid drag. Flow enters the bottom of a tapered vertical tube causing a bob or "float" to nose upward. For a given rate of flow, the flow assumes a position in tube where the drag forces are just balanced by the weight and buoyancy forces. Through careful design, effects of changing viscosity or density may be minimized. The position of the bob in the tube is taken as an indicator of volume flow rate. Meter is installed vertically.

Advantages:

1. 1-2% accuracy.
2. Meter indication is essentially linear with flow rate.
3. Pressure loss fixed at all flow rates.
4. Condition of flow readily visible.
5. Many corrosive fluids can be handled without complication.
6. Capacity can be changed easily by changing bob and/or tube.

Disadvantages:

1. Relatively expensive.
2. Must be installed in vertical position.
3. Bob not visible when opaque fluids are used.
4. Cannot be used for applications carrying large percentages of suspended solids.

Remarks:

1. Installed in line.

_ _

‒ ‒

METHOD: TURBINE METER

Description:

Meter uses the principle that the change in momentum in a flow through a set of curved vanes causes a torque to be exerted on the vanes. As fluid moves through the meter, it causes rotation of a small turbine wheel. A permanent magnet housed in the wheel hub rotates with the wheel, yielding a varying magnetic field, which is detected by a reluctance pickup in the meter casing. The frequency of magnetic pulses indicates the flow rate.

Advantages:

1. Accuracies ± ½% within specific flow range.
2. Good transient response.

Disadvantages:

1. Bearing maintenance required.
2. Reduced accuracy at low rates.
3. Fluid must have very low solids content.

Remarks:

1. Installed in line.
2. Available sizes from 1/8-8 in. (line diameter).

‒ ‒

METHOD: POSITIVE DISPLACEMENT METER

Description:

A displacement method. For the nutating-disk type, the fluid enters the meter and strikes a disk which is eccentrically mounted. The disk wobbles about its vertical axis, and the fluid moves through the meter. The volumetric flow is proportional to the number of times the disk wobbles in a given period of time. For the rotary-vane type, a fixed amount of fluid is trapped in a section of a rotating eccentric drum, the rotations of which are proportional to flow.

Advantages:

1.   1% accuracy.
2.   Direct flow measurement.

Disadvantages:

1.   Usually limited to clean streams.
2.   If used for slurry applications, the solids would damage the seals around the blades.

Remarks:

1.   Installed in line.

— —

METHOD: ESTIMATION USING PUMP CHARACTERISTICS AND OPERATING DATA

Description:

Estimation Method. Horsepower, head, or rpm readouts for pump can be related to flow rate.

Advantages:

1.   Simple; does not involve flow measurement equipment.

Disadvantages:

1.   Inaccurate.

Remarks:

1.   A number of readings should be taken over the time period of interest to maximize accuracy.

— —

3

EQUIPMENT FOR STACK SAMPLING PARTICULATES*

INTRODUCTION

To define air pollution problems, the sources of pollutants must first be quantified; and, therefore, source characterization is extremely important. This may have the added weight of a legal requirement, when obtaining an operating permit on a new facility, or demonstrating compliance with an existing facility.

For particulates, source characterization means locating sample ports in an emitting duct or stack, and extracting *a representative sample* of the particles in the gas stream, both in terms of their quantity and size distribution. This is not easy to accomplish.

Stack sampling equipment is generally bulky, and the actual sampling locations are often abominable. Typically, 100 or 200 lbs of equipment may have to be carried or hoisted up to a sampling platform that might be 50-150 ft off the ground. The platform will usually be too small, the stack uninsulated, and the location exposed to the weather.

The procedure itself involves shoving a long (perhaps 6-10 ft) probe into a sample port, and obtaining gas samples at specified locations along the stack diameter. Attached to the end of the probe is a sample train, and the entire apparatus must hang from a cantilevered support rack.

If the gas stream to be sampled is under a slight vacuum, safety requirements will be minimal. However, if it is under a slight positive pressure, the sample probe may have to be inserted through a valve, and sealed on the outside; and respiratory equipment may be required for sampling personnel.

---

*By: Gordon A. Lewandowski.

Nevertheless, although bulky equipment is employed under adverse conditions, *laboratory precision must be exercised in the conduct of the sampling runs* if the data are to be meaningful. Although a plant will often hire an outside consulting firm to actually perform the sampling and data analysis, plant engineers should nevertheless familiarize themselves with the procedures and techniques, even if only in a supervisory role.

ISOKINETIC SAMPLING

In measuring the properties of a system, the measuring techniques themselves will influence the results obtained. For example, in measuring the voltage difference between two electrodes, the voltmeter itself will draw a small amount of current. That current will affect the surface characteristics of the electrodes and, therefore, the voltage being measured. The object, then, of any program of source characterization is to minimize, or compensate for, the effect of the measuring tool.

In stack sampling for particulates, the "measuring" tool is a sample probe, which is inserted in the stack to withdraw a gas sample. The effect of this probe on the measured particle properties is minimized by resorting to isokinetic sampling. This means that gas samples are drawn through the probe at the same velocity as that in the stack or duct.

If the sample were drawn at a lower velocity than that in the duct (Figure 3-1a), gas would back up in front of the probe tip, and the smaller particles (diameters $< 5\mu$m) would follow the gas streamlines around the probe. However, the larger particles (diameters $> 5\mu$m), by virtue of their greater inertia, would leave their streamlines and enter the probe tip. Therefore, additional particles would enter with the gas sample (the gas sample being represented by the streamlines entering the probe tip in Figure 3-1). In this case, when the solids content of the sample is measured and extrapolated to the entire gas stream, the net result would be to *overestimate* the amount of particulate present in the stack or duct. Furthermore, the sample would show a greater proportion of larger particles than actually exists in the gas stream.

On the other hand, if the sample were drawn too fast, the smaller particles would again follow the gas streamlines, and be sucked into the probe, while the larger particles would leave their streamlines and miss the probe altogether (Figure 3-1b). This means that for the gas sampled (represented by the streamlines entering the probe tip), the solids content would be *underestimated,* and the particle size distribution biased toward the smaller particles.

Since the solids content of the gas being sampled is always considered on a weight basis, even the presence of a relatively small percentage of

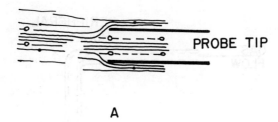

A

Figure 3-1a. Sampling velocity $<$ steam velocity.

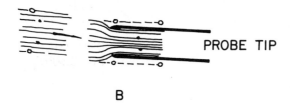

B

Figure 3-1b. Sampling velocity $>$ steam velocity.

larger particles may have a significant impact on the reported values for those cases that are nonisokinetic.

The maintenance of isokinetic conditions is probably the most important consideration in particle sampling. Unless the particles are all small ($< 5\mu m$), the results of a sampling program will be meaningless if the sampling velocity is outside the range of 0.9-1.1 times the gas velocity in the duct at the sample point (*i.e.*, the maximum allowable deviation from isokinetic conditions is ± 10%).[1]

THE PITOT TUBE

For stack velocities less than 10 ft/sec, hot-wire or vane anemometers must be used for velocity measurements. However, in the general case, gas velocities are much higher than this, and the preferred measuring tool is the pitot tube.

Figure 3-2a shows a "standard" pitot tube. The open end measures the so-called "dynamic pressure" (P_d), while the holes around the periphery of the outside tube (perpendicular to the direction of flow) measure the "static pressure" (P). These are related to the gas velocity as follows:

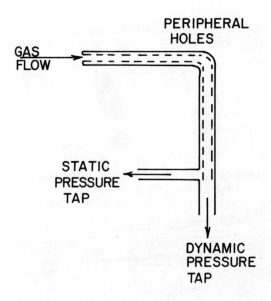

Figure 3-2a. Standard-type pitot tube.

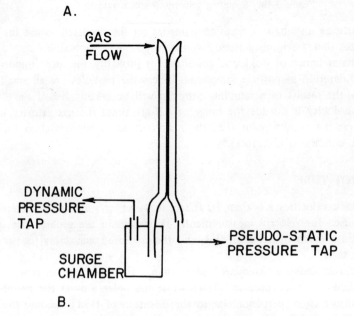

Figure 3-2b. "S-type" pitot tube.

$$V = k\sqrt{2g_c\frac{\triangle P}{\rho}}$$

where		
V	=	gas velocity
g_c	=	Newton's Law conversion factor
ρ	=	gas density
$\triangle P$	=	$P_d - P$
k	=	pitot tube calibration factor

However, the standard-type pitot tube has two disadvantages: (1) the right angle bend requires a large sample port opening; and (2) the small holes for measuring static pressure can plug with particulate.

Therefore, an "S-type" pitot tube (Figure 3-2b) is often preferred for stack sampling. With this type, the opening facing the flow again measures the dynamic pressure (P_d), but the other opening is not oriented perpendicular to the gas flow, and so does not measure the true static pressure. However, this is accounted for in the manufacturer-supplied calibration factor (k). This factor should be determined with the pitot tube attached to the sampling probe (see "pitobe" below), and must be recalibrated if the pitot tube tip has been altered or damaged in any way. Such recalibration can be most easily accomplished by obtaining a simultaneous velocity reading with a pitot tube of known calibration factor.

In addition, the accuracy of the velocity measurement is directly dependent on the accuracy of the $\triangle P$ measurement between the dynamic and static tubes. Therefore, to increase this accuracy, an inclined manometer is used.

Also, instantaneous fluctuations in the manometer reading must be minimized (*i.e.*, the manometer reading cannot be allowed to constantly "jump around"). Otherwise, the eye can only estimate a mean reading. Such fluctuations can either be caused by continual velocity fluctuations in the main duct, or by moisture or particulates becoming trapped in the pitot tube connections. The first of these can be managed by placing a surge, or damping, chamber in the upflow line (Figure 3-2b). The second problem can largely be eliminated if the pitot tube is hooked-up to the manometer before being inserted in the duct. In this way, it's dead-headed —there is no place for the gas to flow (assuming there are no leaks), and little chance for moisture or particles to enter. Additional insurance would be provided by the surge chamber, which would act as a knock-out pot should moisture or particulates manage to enter.

Note that errors in the pitot tube readings will not only influence isokinetic conditions, but will also cause errors in the calculated gas flow for the stack or duct. Since the particulate emission rate is calculated by multiplying the total gas flow times the concentration, errors in the pitot tube readings will be directly reflected in the emission rate.

THE SAMPLING ORIFICE

The velocity at which a gas sample is drawn through the probe is measured by using an orifice in the sample line. This velocity must then equal the duct velocity to have isokinetic conditions.

An orifice is simply a hole drilled in a plate to provide a pressure drop in a flow line. As with the pitot tube, the velocity through the orifice is a function of the pressure drop across it:

$$V_o = k_o \sqrt{2g_c \frac{\Delta P_o}{\rho_o}}$$

where ΔP_o = pressure drop across the orifice
   ρ_o = gas density at the orifice
   g_c = Newton's Law conversion factor
   k_o = orifice calibration factor
   V_o = gas velocity through the orifice

Since the sampling orifice is usually located downstream of a moisture condensation apparatus (Figure 3-3), gas flow through the orifice is related to gas flow through the probe tip, as follows:

$$Q_o = Q_p \left(\frac{T_o}{T_p}\right)\left(\frac{P_p}{P_o}\right) f_{dg}$$

where Q_p = volumetric gas flow through the probe tip
   T_p = gas temperature at probe tip (= T, gas temperature in the stack or duct)
   P_p = gas static pressure at the probe tip (= P, gas pressure in the stack or duct)
   T_o = gas temperature at the orifice
   P_o = average gas pressure at the orifice
   f_{dg} = volume fraction of incoming sample gas which is dry
   Q_o = volumetric gas flow through the orifice

However,

$$Q_p = V_p \left(\frac{\pi D_p^2}{4}\right) = V \left(\frac{\pi D_p^2}{4}\right)$$

and

$$Q_o = V_o \left(\frac{\pi D_o^2}{4}\right)$$

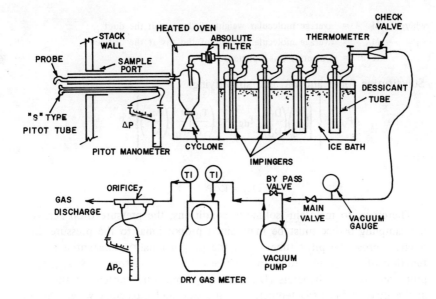

Figure 3-3. EPA Method 5. Sampling train.

where D_o = orifice diameter

D_p = probe tip inside diameter

V_p = velocity in probe tip (= V, velocity in the duct, because of isokinetic condition)

Therefore,

$$k_o\sqrt{2gc\frac{\Delta P_o}{\rho_o}}\left(\frac{\pi D_o^2}{4}\right) = k\sqrt{2gc\frac{\Delta P}{\rho}}\left(\frac{\pi D_p^2}{4}\right)\left(\frac{T_o}{T}\right)\left(\frac{P}{P_o}\right)f_{dg}$$

Or,

$$\Delta P_o = \left[\left(\frac{k}{k_o}\right)\left(\frac{D_p^2}{D_o^2}\right)\left(\frac{T_o}{T}\right)\left(\frac{P}{P_o}\right)f_{dg}\right]^2\left(\frac{\rho_o}{\rho}\right)\Delta P$$

But the gas densities may be approximate by the equations:

$$\rho_o = \frac{P_o M_o}{R T_o}$$

and

$$\rho = \frac{P\,M}{R\,T}$$

where   M   = average molecular weight of the gas in the duct

M_O = average molecular weight of the gas at the orifice

R   = Universal Gas Constant

Substituting for ρ and ρ_O, and simplifying:

$$\Delta P_O = \left\{ \left[\left(\frac{k}{k_O}\right)\left(\frac{D_p^2}{D_O^2}\right) f_{dg} \right]^2 \left(\frac{T_O}{T}\right)\left(\frac{P}{P_O}\right)\left(\frac{M_O}{M}\right) \right\} \Delta P$$

Or,

$$\Delta P_O = K \Delta P$$

Therefore, to maintain isokinetic conditions, the pressure drop across the sampling orifice must be maintained proportional to the pressure difference across the pitot tube. And the proportionality constant is a function of: (1) pitot tube and orifice calibration factors (k & k_O); (2) probe and orifice diameters (D_p & D_O); (3) moisture content of the gas in the duct (1-f_{dg}); (4) temperature, pressure and molecular weight of the gas in the duct (T, P & M); and (5) temperature, pressure and molecular weight of the gas at the orifice (T_O, P_O & M_O).

The calibration factors are supplied by the equipment manufacturers[2-4] and should be checked prior to sampling; however, the other items must be determined by a preliminary test program, as described below.

SAMPLE LOCATION

To characterize a source of pollutants, *a representative sample* is required. However, in a stack or duct there may be considerable nonuniformity of gas flow and, therefore, of particle concentration and size. The first problem that must be tackled in a sampling program is how to locate a sample port and obtain a sufficient number of samples to properly characterize such a heterogeneous source.

For compliance sampling (*i.e.*, for obtaining the total amount of solids), two ports must be located 90° apart at the sample location[1] (Figure 3-4). Where possible, ports should be made of 4-in.-diameter nipples, threaded on the outside, and welded to the stack or duct.

To avoid the nonuniformities in flow caused by tees, elbows, flames, etc., sample ports should be located eight stack diameters upstream, and two downstream, of any such flow disturbance. For a duct of rectangular cross section:

$$\text{equivalent diameter} = 4 \times \frac{\text{cross section}}{\text{perimeter}}$$

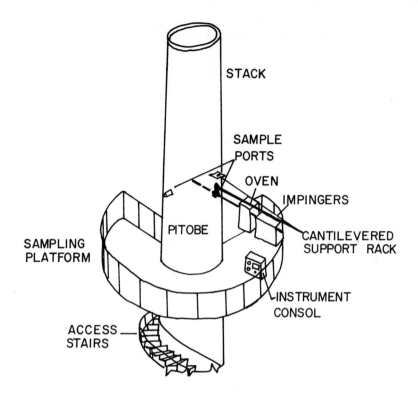

Figure 3-4. Sampling location.

However, it may not always be possible to achieve such an ideal location of the sample port. For example, a 4-ft-diameter duct would require a total straight run of 40 ft (10 duct diameters). In the absence of such conditions, the sample port should be located in the longest straight run of duct available, and a 4:1 ratio maintained between upstream and downstream distances to nearby bends, or flow disturbances.

Figure 3-5 shows the relationship of sample port location to the number of samples required.[1] For the ideal location, that number is 12, and increases as the straight run of duct shortens (for ducts less than 2 ft in diameter, multiply the required number of sample, or "traverse" points by 2/3).

Along with the number of samples to be taken, their location in the duct is also specified by federal regulations.[1] For example, in an "ideal" situation (where the minimum of 12 traverse points is required) Figure 3-6a shows the required sample points along the diameters of a circular duct, and Figure 3-6b is for a duct of rectangular cross section. Table 3-1 gives the general criteria for traverse point location, for those situations

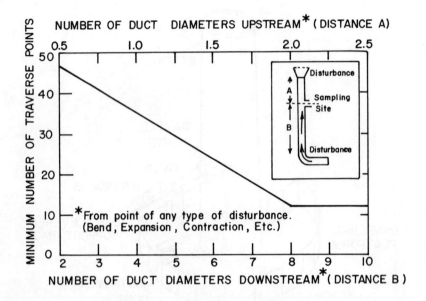

Figure 3-5.   Minimum number of traverse points.

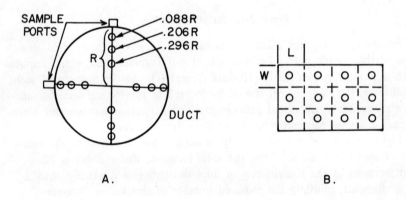

A.

B.

Figure 3-6a.   Placement of 12 traverse points along circular duct diameter.

Figure 3-6b.   Placement of 12 traverse points in rectangular duct.

Location of Traverse Points in Circular Stacks.
(Percent of Stack Diameter From Inside Wall to Traverse Point)

Traverse Point Number	Number of Traverse Points on a Diameter									
	6	8	10	12	14	16	18	20	22	24
1	4.4	3.3	2.5	2.1	1.8	1.6	1.4	1.3	1.1	1.1
2	14.7	10.5	8.2	6.7	5.7	4.9	4.4	3.9	3.5	3.2
3	29.5	19.4	14.6	11.8	9.9	8.5	7.5	6.7	6.0	5.5
4	70.5	32.3	22.6	17.7	14.6	12.5	10.9	9.7	8.7	7.9
5	85.3	67.7	34.2	25.0	20.1	16.9	14.6	12.9	11.6	10.5
6	95.6	80.6	65.8	35.5	26.9	22.0	18.8	16.5	14.6	13.2
7		89.5	77.4	64.5	36.6	28.3	23.6	20.4	18.0	16.1
8		96.7	85.4	75.0	63.4	37.5	29.6	25.0	21.8	19.4
9			91.8	82.3	73.1	62.5	38.2	30.6	26.1	23.0
10			97.5	88.2	79.9	71.7	61.8	38.8	31.5	27.2
11				93.3	85.4	78.0	70.4	61.2	39.3	32.3
12				97.9	90.1	83.1	76.4	69.4	60.7	39.8
13					94.3	87.5	81.2	75.0	68.5	60.2
14					98.2	91.5	85.4	79.6	73.9	67.7
15						95.1	89.1	83.5	78.2	72.8
16						98.4	92.5	87.1	82.0	77.0
17							95.6	90.3	85.4	80.6
18							98.6	93.3	88.4	83.9
19								96.1	91.3	86.3
20								98.7	94.0	89.5
21									96.5	92.1
22									98.9	94.5
23										96.8
24										98.9

where the duct is circular in cross section. For ducts of rectangular cross section, divide the cross-sectional area into n equal segments (where "n" is the number of traverse points), and sample at the center of each segment (Figure 3-6b).

It should be emphasized that when sampling is done to satisfy a regulatory requirement, the sampling procedures are also governed by regulations. The procedures outlined above are the same as those specified by the U.S. Environmental Protection Agency,[1] and are in general use. However, regulations normally require only the total amount of particulate leaving an exhaust stack—they usually do not require inlet duct sampling (to a control device), nor the particle size distribution. The latter are only needed to design and optimize equipment (or make process recommentations) to reduce particulate emissions. Therefore, for engineering or design purposes, some of the sampling requirements may be relaxed. For example, if it is inconvenient to locate two sample ports, a single port may be similarly located, and a traverse made. Or, if the velocity profile is fairly uniform, only two samples may be needed for particle size analysis.

However, the object of an engineering design is usually to being a facility into compliance with state regulations. The definition of compliance is supplied by the sampling. Therefore, where at all possible, design data should be obtained by the same techniques prescribed by the state regulatory agency. Furthermore, since control equipment can be quite costly, reliable engineering data can reap substantial capital savings. Also, it is wise not to cut too many corners during a sampling program.

Finally, sampling involves the use of cumbersome equipment. EPA Method 5 (Figure 3-3), requires a probe, oven and impinger train to be supported from a cantilevered rack which, in turn, is clamped to the stack or duct (Figure 3-4). The probe itself might be 6-10 ft long, with the rack extending out to 10 or 15 ft. To assemble this equipment and shove the probe into the stack to the various traverse points is often an interesting procedure when conducted from a standard 3-ft-wide platform. Therefore, the reader is urged to include adequately sized sampling platforms at the sample port locations (Figure 3-4), with reasonable access to them for sampling personnel and equipment. The platforms must also be supplied with 115 V ac outlets.

PRELIMINARY TESTING

Once the sample port is located, gas properties are required to determine the proportionality constant (K) used to maintain isokinetic conditions (see Isokinetic Sampling, pg. 20). This, in turn, requires coordination between sampling and operating personnel. Tests must be designed for "normal" operating conditions, and perhaps also for anticipated swings in process operation.

After coordination of the sampling program, gas properties must be measured at the sample port location, for each set of process conditions. The gas temperature and pressure may be measured easily, but the humidity and composition require special procedures.

For most combustion sources (*i.e.,* where the average molecular weight of the gas is close to that of air), a psychrometer may be used for a preliminary measurement of gas humidity. For other sources, a condensation procedure must be used, as defined by "EPA Method 4."[1]

The use of a psychrometer in a modified Method 5 sample train is shown in Figure 3-7. The vacuum pump draws gas from the stack into a short, insulated probe, and through a psychrometer, which includes a wet-bulb, and an ordinary (or "dry-bulb") thermometer. The insulation is necessary to prevent moisture condensation upstream of the psychrometer, and the associated impingers and dessicant tube are needed to remove moisture from the gas prior to the vacuum pump.

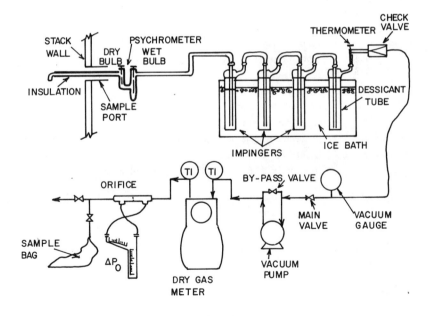

Figure 3-7. Determination of stack humidity with a psychrometer.

A wet-bulb thermometer, as the name implies, has its bulb wrapped in a piece of cloth that is continually wetted. The cloth dips down into a well, and water rises up the cloth and onto the bulb by a wicking action. Since the bulb is wet, as the gas passes over it, water will evaporate and cool the bulb below the actual temperature of the gas. The amount of cooling will depend on the rate of evaporation and, therefore, on the humidity of the gas. For 100% relative humidity, where the gas is already saturated with water vapor, no water will evaporate from the wet bulb, and the two thermometers will read the same temperature. Otherwise, the temperature of the wet-bulb thermometer will always be lower.

Once the vacuum pump is turned on, these two temperatures will rise until they reach equilibrium with the gas, and then remain constant. The gas flow should be adjusted so that the linear velocity passing the wet bulb is at least 10 ft/sec.

The final wet- and dry-bulb temperatures can then be used to calculate the volume percent moisture in the gas. This is preferable to using a psychrometric chart, since at the temperatures normally encountered, the accuracy of a chart reading will be very poor:

$$\frac{f_w}{1 - f_w} = \frac{P_w^o}{P - P_w^o} - \left[\frac{\alpha(T_d - T_w)}{\lambda_w}\right]\left(\frac{M_{dg}}{M_w}\right)$$

where
f_w = volume fraction moisture in the gas (= $1 - f_{dg}$)
P_w^o = vapor pressure of water at T_w
P = gas pressure
T_d = dry-bulb temperature
T_w = wet-bulb temperature
M_{dg} = molecular weight of the dry gas
M_w = molecular weight of water = 18
λ_w = latent heat of water at T_w
α = constant = 0.26 Btu/ $\#^o$F, or equivalent units

P_w^o and λ_w can be found in a standard steam table, while M_{dg} can be determined from the dry gas composition. For combustion sources, this composition is measured by Orsat analysis of a sample collected at the vacuum pump discharge (Figure 3-7). Orsat analysis determines the volume percent CO, CO_2, O_2 and (by difference) N_2 of the gas, on a dry basis. Several determinations should be made to ensure accuracy.

For other sources, a gas sample can be sent to a laboratory for analysis, or the composition determined from operating date (if accurate enough).

By measuring the gas pressure, temperature and composition, the average molecular weight can be determined on a wet (M) and dry (M_{dg}, or M_o) basis.

In addition, the barometric pressure must be measured to determine the gas pressure at the orifice (P_o), and a preliminary determination of the gas velocity in the duct should be made with an "S-type" pitot tube. Once this is accomplished, all that remains to be determined before actual sampling commences are the probe tip (D_p) and orifice (D_o) diameters. These are functions of the above gas properties, the average duct velocity and the properties of the sampling system (particularly, the capacity of the vacuum pump). Commercially available sampling systems are generally supplied with a selection of interchangeable tips and orifices, and individual equipment manufacturers will include nomographs, or guides, for appropriate size selection. However, whenever possible, the probe tip inside diameter should be ¼ in. or larger, to minimize wall effects.

Choosing the correct combination of probe tip diameter and sampling time may turn out to be a trial-and-error procedure. For example, with a dilute gas stream and a particular probe tip, it may be necessary to sample for 30 minutes per traverse point to obtain sufficient sample to reduce the significance of weighing errors (see Required Test Accuracy, pg. 38). Thirty minutes per traverse point will mean a minimum of six hours per test (for 12 traverse points). However, a larger probe tip can be substituted for the original, which will enable the same amount of sample to be collected in a shorter time period.

SAMPLING PROCEDURE

There are several possible methods for sampling particulates,[1,5-8] but the one chosen should agree with the sampling requirements of the state in which the facility is located. Each state will require a certain method for compliance testing, or permit validation. In nearly every case, that method is the "EPA Method 5,"[1] or equivalent, and so this section will concentrate exclusively on that method.

"EPA Method 5" is for determining the total amount of particulate in a gas stream—it is not used to determine the particle size distribution (which is a design, rather than a regulatory, requirement). The apparatus is shown in Figure 3-3. The cyclone is optional, and is intended to prevent overloading of the final filter when high particle concentrations are encountered.

As described in the *Federal Register,* much of this equipment is made of glass. However, regulatory agencies have shown some willingness to relax these requirements, and there are commercially available sample trains made of alloys and plastic materials.[2] The acceptability of a particular sample train will depend on the regulatory agency having jurisdiction. If glass equipment is used, some of it is sure to break, and replacement parts (or entire spare assemblies) should be made available at the sampling site.

Once again, the objective is to sample isokinetically, by ratioing the orifice ($\triangle P_O$) to pitot tube ($\triangle P$) pressure differentials. This is accomplished by adjusting the by-pass valve on the vacuum pump, to draw more, or less, sample.

In this regard, the importance of a leak-tight sample train cannot be overemphasized. One must be certain that the gas sample at the orifice is the same as the sample at the probe tip (minus condensed moisture), or jeopardize the validity of the entire test program. Unfortunately, leak-testing can be a tedious procedure, particularly when so many tubing connections are involved. Nevertheless, it must be done patiently and conscientiously before actual sampling begins.

The general procedure for leak-testing is as follows:

1)   Test the system below the main vacuum valve (Figure 3-3) by shutting off the valve, turning on the pump, and watching the dry gas meter. If it registers a flow greater than 0.02 cfm, there is a leak somewhere between the valve and the gas discharge, which will have to be pinpointed and sealed.

2)   Open the main vacuum valve, seal the end of the probe, start the pump, and let the vacuum build up to 15 in. of Hg. Then, close the main vacuum valve (do *not* shut off the pump) and watch the vacuum gauge. If the gauge reading stays constant for at least 15 sec., it indicates a leak-tight system. In running this test, always release the vacuum at the probe tip first; then let the vacuum gauge fall to zero, before opening the main vacuum valve and, finally, shutting off the pump.

Never shut the vacuum pump off while the probe tip is sealed. Despite the check valve, a reverse flow of air might be created through the vacuum pump and impinger train. If violent enough, this reverse flow could carry water from the impingers into the final filter, requiring disassembly and repreparation of the filter and impinger train.

During assembly of the apparatus, care must be taken in weighing the filter paper (to the nearest 0.5 mg), and mounting it in the filter holder. If the paper is torn, or some of the gas allowed to bypass it, results of the test will have to be disregarded. Before weighing, the filter paper must be dried in an oven at 220°F, and allowed to cool in a dessicator.

Also, in preparing the impinger train, the first two contain approximately 100 ml of water each (measured to the nearest ml), the third is empty and the fourth contains approximately 200 g of silica gel (measured to the nearest gram). These impingers are intended to capture all the moisture in the gas sample, both for the purpose of measurement, and for protection of the vacuum pump. For this reason, during sampling, the outlet temperature of the impingers should be kept below 100°F by adding ice to the chest as needed.

Since it is necessary to capture all the moisture in the impingers, and also to protect the filter paper, condensation elsewhere in the system must be prevented. For this reason, the probe is wrapped in a heating element, with the temperature maintained at 250°F before and during sampling. Also, the cyclone and filter assembly are enclosed in a heated oven (Figure 3-3), and brought up to 250°F before sampling is initiated. Heating coil and thermocouple leads from these units are connected to a sampling consol, where the temperatures are read and the heat adjusted accordingly.

To commence sampling, the entire assembly (probe, pitot tube, oven and impinger train) are hung beneath a cantilevered support rack. Then, the probe and pitot tube ("pitobe") must be shoved together through the sample port, to the first sample, or "traverse" point. To make this easier, the pitobe is housed in a single sheath, and the distances to each traverse point marked on the sheath with adhesive tape. Sampling should begin at the point closest to the far wall of the duct, and the pitobe assembly itself must lie along the duct diameter, perpendicular to flow (Figure3-8). Otherwise, gas flow will not enter directly into the probe and pitot tube openings, and errors will result in measuring the total gas flow, and in setting isokinetic conditions. Adhesive tape guides may be used for aligning the pitobe.

To maintain isokinetic sampling, the consol operator watches the pitot tube and orifice manometer readings, and adjusts the by-pass valve accordingly on the vacuum pump. The calculations for this adjustment can be made by using a nomograph (supplied by the equipment manufacturer), or by

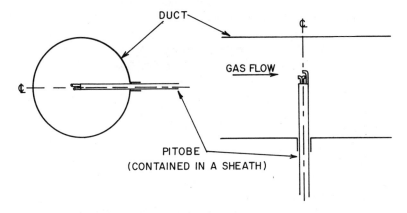

Figure 3-8. Pitobe alignment.

using a hand-held programmable calculator.[9]   The latter is the preferred method, both for ease of calculation and accuracy.

Temperature data, total flow volume (from the dry gas meter), sampling time, etc., must be recorded at least every five minutes, and the records carefully maintained as a part of the regulatory requirements.[1]

When the sampling program is completed, the pump is shut off first, then the pitobe assembly is removed from the duct. The end of the probe is sealed to prevent loss of particulates, and the apparatus is disassembled as follows:

1)   As soon as possible, the probe, cyclone and connecting pieces should be washed out with acetone, and the washings collected in a tared bottle.

2)   The filter holder must be carefully removed from the apparatus, and the openings sealed, until lab analysis can proceed.

3)   The water in the impingers must be measured volumetrically to the nearest ml, and the increase in volume due to moisture condensation determined.

4)   The silica gel must be weighed to the nearest gram, and the amount of water absorbed on it determined by subtracting the original weight.

5)   In the laboratory, the acetone must be evaporated, and the weight of deposited particulate determined to the nearest .0005 g.

6)   The filter holder must be carefully disassembled (avoiding any damage to the filter paper, or loss of particulate), and (after equilibration in a dessicator) the weight of collected particulate determined to the nearest .0005 g.

CALCULATION OF PARTICLE EMISSION RATE, ETC.

The final value of the average moisture content of the gas is calculated from the impinger catch and silica gel pickup:

$$f_w = 1 - f_{dg} = \frac{Q_1}{Q_1 + \left(\frac{P_m}{T_m}\right)\left(\frac{T_{std}}{P_{std}}\right)Q_m}$$

where f_w =   volume fraction of moisture in the stack gas

f_{dg} =   volume fraction of dry gas

Q_1 =   volume of water collected in the impingers, plus the equivalent volume of the silica-gel catch

P_m =   pressure at the dry gas meter

T_m =   temperature at the dry gas meter in degrees absolute

P_{std} =   standard atmospheric pressure

T_{std} =   freezing point of water in degrees absolute ($492°R$ or $273°K$)

Q_m =   dry gas meter reading

The average stack gas velocity is calculated from the pitot tube readings:

$$\overline{V} = \left[k\sqrt{\frac{2g_c \ R \ T}{P \ M}} \right] \left[\frac{\Sigma\sqrt{\triangle P}}{N} \right]$$

where $\overline{V}$ = average gas velocity in the stack, or duct
   T = *average* duct temperature in degrees absolute
   $\triangle$P = pitot tube manometer reading
   Σ = summation of all traverse point readings
   N = number of traverse points

As with every equation, all units must be consistent.

The total particulate emission rate is determined by dividing the total particulate catch by the total sampling volume, and then multiplying by the average duct velocity:

$$W = \left[\frac{W_s \ f_{dg}}{Q_m \left(\frac{P_m}{T_m}\right)\left(\frac{T}{P}\right)} \right] \ \overline{V}\left(\frac{\pi D^2}{4}\right)$$

where   W = particulate emission rate
   W_s = total particulate catch in the sample train
   D = inside stack diameter

And finally, isokinetic conditions must be checked by the following equation:

$$I = \frac{4 \ Q_m \left(\frac{P_m}{T_m}\right)\left(\frac{T}{P}\right)}{f_{dg} \ \overline{V} \theta \ \pi \ D_p^2}$$

where   I = ratio of sample velocity to stack, or duct, velocity
   θ = total sampling time

For example, let:

   T   = $400°F = 860°R$
   P   = $2''H_2O$ draft = 756 mm Hg
   Q_m = 24 ft^3
   P_m = 765 mm Hg
   T_m = $70°F = 530°R$
   Q_1 = 50 ml collected in impingers, plus 10 g absorbed by silica gel
   P_{std} = 760 mm Hg
   T_{std} = $492°R$

$$\theta \quad = \quad 30 \text{ minutes} = 1800 \text{ seconds}$$
$$D_p \quad = \quad \frac{1}{4}'' = 1/48 \text{ ft}$$
$$V \quad = \quad 100 \text{ ft/sec}$$

Q_1 must also be expressed in ft^3:

$$Q_1 = \left[50 \text{ ml } \frac{1 \text{ g}}{\text{ml}} + 10 \text{ g} \right] \left(\frac{1 \text{ g mole}}{18 \text{ g}} \right) \left(\frac{1 \text{ \# mole}}{454 \text{ g mole}} \right) \left(\frac{359 \text{ ft}^3}{\text{\# mole}} \right)$$

$$= 2.64 \text{ ft}^3$$

Therefore,

$$f_w = 1 - f_{dg} = \frac{2.64 \text{ ft}^3}{2.64 \text{ ft}^3 + \left(\frac{765 \text{ mm Hg}}{530^\circ \text{ R}} \right) \left(\frac{492^\circ \text{ R}}{760 \text{ mm Hg}} \right) 24 \text{ ft}^3}$$

$$= 0.105$$

Therefore,

$$f_{dg} = 0.895$$

and,

$$I = \frac{4(24 \text{ ft}^3) \left(\frac{765 \text{ mm Hg}}{530^\circ \text{ R}} \right) \left(\frac{860^\circ \text{ R}}{756 \text{ mm Hg}} \right)}{0.895(100 \text{ ft/sec})(1800 \text{ sec})\pi \, (\frac{1}{48} \text{ ft})^2}$$

$$= 0.72$$

Therefore, this particular sampling run was not successful, since I should fall between 0.9 and 1.1 (a maximum deviation of ± 10% from isokinetic conditions).

REQUIRED TEST ACCURACY

The sampling procedures reemphasize the need to maintain laboratory precision in the conduct of the sampling program. For example, suppose the following set of conditions existed:

1) Actual particulate "loading" (*i.e.,* concentration) in the duct = 0.1 gr/ft^3
2) Duct velocity = 100 ft/sec = sampling velocity at isokinetic conditions
3) Probe diameter = ¼ in.

4)   Sampling time per traverse point = 5 minutes

5)   Number of traverse points = 12

Therefore, total particulate collected in the sample train equals:

$$(12 \text{ samples}) (5 \text{ min/sample}) (0.1 \text{ gr/ft}^3) (1 \text{ \#}/7000 \text{ gr})$$

$$\text{x}(454 \text{ g/ \#}) \frac{\pi}{4} \left(\frac{1/4 \text{ in.}}{12 \text{ in./ft}} \right)^2 (100 \text{ ft/sec}) (60 \text{ sec/min}) = 0.7959 \text{ g}$$

Thus, approximately 0.8 g must be recovered from the sampling apparatus. Although about half the solids will be caught on the final filter, the other half will deposit in the probe, cyclone and connecting pieces. And for every 10 mg left in the apparatus, the error in calculating the particle loading will be about 1.3%.

The particulate not collected on the filter (*i.e.*, that left in the probe and cyclone) must be washed out of the apparatus with acetone, and into a tared jar. The acetone can then be evaporated, and the particulate determined by difference. In the above example, if the jar initially weighs 30 g, the difference in weight must be determined between a 30-g tare and the 30.4-g final weight, with every 10-mg error again causing a 1.3% error in the calculated loading.

SAFETY CONSIDERATIONS

Stack sampling presents a number of hazards to personnel involved:

1) The sampling platform is usually elevated about 50-150 ft off the ground, which presents a danger to the sampling personnel above, and also to those below, from falling equipment or tools. For especially precarious locations, sampling personnel should be attached to safety harnesses, and in all cases, the work area below the sampling platform should be roped off.

2) The sampling platform may become enveloped by the plume from an upwind stack, containing toxic gases such as CO and SO_2. Or, a stack under positive pressure may emit such gases. The best safeguard is to have at least two people on the platform and provide a ready means for evacuation if that becomes necessary. Monitors and respiratory equipment may also be provided on the sampling platform. Note also that the vacuum pump discharge contains dry stack gas, which also may be toxic.

3) The gas being sampled may be explosive, requiring explosion-proof electrical equipment.

4) The potential for condensation, combined with electrical connections, presents an electric shock hazard to personnel. This hazard can be

eliminated by placing a ground fault interrupter between the 115 V outlet and the sampling equipment.

5) An uninsulated stack, stack gases and sample probe will all be hot
enough to cause burns, unless long sleeves, trousers and asbestos gloves
are worn by sampling personnel.

6) During the summer, radiant heat from an uninsulated stack will add
to the loss of body fluids, and may cause nausea, or even fainting. This
can be prevented by keeping on the sampling platform a cooler filled with
soft drinks. The cooler can also serve the dual purpose of storing ice for
the impinger train.

DETERMINATION OF PARTICLE SIZE DISTRIBUTION

To optimize the design of particulate control equipment, knowledge of
the particle size distribution is essential. Without it, the equipment must
be overdesigned to provide a margin of safety, and even then, risks the
possibility of failure.

As noted above, particles will deposit along the length of a sampling
probe and, therefore, techniques to measure the particle size distribution
"out-of-stack" (*i.e.,* by externally analyzing a gas sample), are subject to
deposition errors. These errors will tend to bias the measured size distribution toward the smaller particles.

For this reason, "in-stack" devices are generally preferred. These devices segregate a particulate sample *inside the stack* and, therefore, eliminate
the deposition problems associated with out-of-stack probes. By far the
most important of these devices is the "cascade impactor," and the remainder of this section will concentrate on its use.

Cascade impactors contain several perforated plates, arranged so that
the gas sample must travel in a zig-zag pattern in passing through (Figure
3-9). The perforations decrease in size in the flow direction, with the
largest diameter holes in the plate first encountered by the gas sample.
In passing through these perforations, the gas will form a number of
small jets, whose velocity will increase as the perforations get smaller. In
addition, each plate is covered by a substrate, or filter medium, with holes
that match those in the plate.

As the gas passes through a set of perforations, its velocity increases,
approaching the next plate. It is then forced to make a series of turns
to pass through the next set of perforations. The larger particles in the
gas are unable to make these flow adjustments, and (by virtue of their
inertia) they leave their streamlines, collide with the plate and are collected
on the covering substrate.

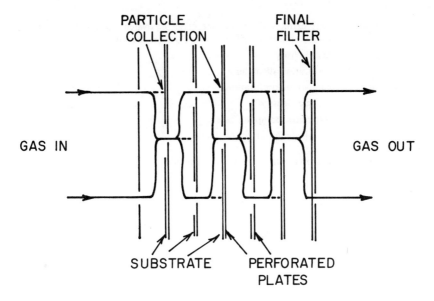

Figure 3-9. Principle of cascade impaction.

With each successive plate, the perforations get smaller, the velocity (and, therefore, the particle inertia) increases, and smaller and smaller particles are forced to leave their streamlines and collect on the following substrate. The combination of plate and following substrate is referred to as an impactor "stage."

An unperforated final filter is located after the last plate, which collects all the remaining particles in the gas. This filter may be an integral part of the impactor, or mounted behind it.

Since particles are being separated *according to their inertia,* cascade impactors do not actually measure size directly. Rather, particle size must be calculated from inertial properties. Inertia is a product of size and density. Small, dense (*i.e.,* "heavy") particles might collect on an upper plate, while larger but "fluffier" particles are being captured on a lower plate. However, assuming a uniform particle density, the cascade impactor will segregate them according to size—the largest on the upper plate closest to flow.

To relate the size of particles collected on a given impactor stage (in this case, the "j^{th}" stage) to their inertial properties, the following formula is used:[10]

$$\overline{D_{pj}} = \left[\frac{2.05 \ \mu \ d_j^3 \ N_j}{C_j \ \rho_p \ Q} \right]^{\frac{1}{2}}$$

where $\overline{D_{pj}}$ = "cut diameter," or the size of an equivalent spherical particle, whose collection efficiency is 50% on the given stage

μ = gas viscosity

d_j = perforation diameter

N_j = number of perforations

ρ_p = particle density

Q = volumetric sampling rate

C_j = Cunningham correction factor

As with all equations, units must be consistent:

$$C_j = 1 + \frac{2\lambda}{\overline{D_{pj}}} \left[1.257 + 0.40\epsilon^{-(1.1\ \overline{D_{pj}}/2\lambda)} \right]$$

where λ = mean free path of the gas molecules

$$\lambda = \frac{\mu}{P\sqrt{\dfrac{2g_c}{\pi}\dfrac{M}{R\ T}}}$$

where P = gas pressure

T = gas temperature in degrees absolute

M = average gas molecular weight

R = universal gas constant

g_c = Newton's Law conversion factor

Since C_j also contains $\overline{D_{pj}}$, the cut diameter must be solved by trial-and-error. As a first approximation, let C_j = 1.0, and solve for $\overline{D_{pj}}$. Then, recalculate C_j, and resolve for $\overline{D_{pj}}$. And so on, until the calculated value of $\overline{D_{pj}}$ remains essentially constant from one iteration to the next.

Choosing Impactor Equipment

The biggest problems encountered in impactor use are overloading of a given stage, and excessive jet velocities. Both will have the result of removing particles that properly belong on one stage, and sending them instead to a downstream stage. When overloaded (*i.e.,* more than about 10 mg accumulated on any single stage), particles will be reentrained into the gas stream and carried to the next stage. And even when not overloaded, excessive jet velocities can erode particles off their proper stage, and deposit them downstream. Both these problems can largely be avoided by appropriate choice of impactor nozzle size and sampling time.

As the nozzle size decreases, the volumetric sampling rate required for isokinetic conditions also decreases, and this, in turn, reduces the jet velocities through the plate perforations. Reduced jet velocities mean reduced plate erosion.

However, a decrease in sample volume also means an increase in the cut diameter of each stage. This may have the effect of depositing most of the particulate on the final filter, where it is unclassified. Therefore, the choice of appropriate nozzle size can be a trial-and-error procedure. In any case, the nozzle diameter should be as large as possible, preferably ¼ in. or more.

An additional variable that must be set is the sampling time. This has a minimum value of 5 minutes, and a maximum value set by the stage capacity ($\sim$10 mg). For example, if the particle concentration in the duct is 0.5 gr/ft^3, and a six-stage impactor is used, the maximum amount of particulate that can be collected is 60 mg. Assuming a 5-minute sampling time, this implies a maximum sampling rate (Q) of:

$$Q = \frac{(60 \text{ mg}) (7000 \text{ gr}/\#)}{(5 \text{ min}) (0.5 \text{ gr/ft}^3) (454 \text{ g}/\#) (1000 \text{ mg/g})} = 0.37 \text{ ft}^3/\text{min}$$

Therefore, a sampling rate of 0.37 cfm must fall within the range of operability of the impactor chosen.

Other considerations in impactor choice are the range of cut diameters, and the purpose of the sampling program. The most difficult control problems and, therefore, the ones requiring the most accurate particle size analysis, are those involving particles below about 2m in size. Therefore, in those cases, the impactor chosen must be able to classify particles in that size range. Also, where fine particle measurements are desired, a precut cyclone may be placed ahead of the impactor to remove particles above about 10 μm in size, allowing a larger sample volume to be taken without overloading the impactor stages.

Finally, for collecting liquid aerosol particles, some impactor manufacturers supply an optional dish design for the collection surfaces. Further information on commercially available impactor types has been documented.[11-13]

Choice of Substrate, or Filter Medium

This is a very important consideration in preparing an impactor for sampling, as the substrate, or filter medium, must have a high capture efficiency, combined with a stable tare weight.

High-vacuum silicone grease coatings have been preferred by some investigators[14] for having a high capture efficiency (i.e., particles that are supposed to remain on a given stage, do so—they are not lost to a lower stage). However, grease coatings are subject to a number of handling errors, which may negate any possible collection advantage. Grease may inadvertantly cover some of the plate perforations, and also may exhibit unpredictable weight losses during sampling, which has the effect of reducing the tare weight.

Another type of substrate often used is the cut and perforated aluminum foil disc; however, these have a poor capture efficiency, so that instead of being collected on the upper stages of the impactor, particles are reentrained and collect on the final filter.[14,15] This will bias the apparent size distribution toward the smaller particles.

For these reasons, the most popular substrate has been glass-fiber filters, placed over the impactor plates, with corresponding perforations. This, too, has disadvantages in that the glass fibers are subject to tears, and may encourage sulfate formation. Tears in the filter medium can be prevented by carefully mounting and disassembling the stages. But the prevention of sulfate formation may require proper selection of the type of glass-fiber and, perhaps, even preconditioning of the filter medium (see Sulfate Formation, pg. 47).

Particle Size Sampling Procedure

As in sampling for total particulates, isokinetic conditions must be maintained. The equipment is shown in Figure 3-10. The diameter of the impactor itself is usually less than 3 in., and may be inserted through a sample port directly into the stack.

The rest of the sample train is similar to that used in the EPA Method 5, with the exception of the cyclone and absolute filter, which are not included. The impingers are used here simply to prevent moisture from damaging the vacuum pump.

Before sampling actually begins, the impactor should be left in the stack for about 30 minutes, to bring its temperature up to that of the stack gas. Otherwise, condensation may occur inside the impactor, and ruin the sampling run. During preheat, the impactor nozzle should point away from the gas flow. Also, if the stack gas is already saturated with moisture, the impactor must be wrapped in heating tape, and heated to 20 or 30°F above the gas temperature, before insertion into the stack.

To initiate sampling, the impactor nozzle should be positioned at a point of average duct velocity, and turned into the gas flow. Sampling must proceed isokinetically, and at least two sampling runs should be made to check for consistency and accuracy of results.

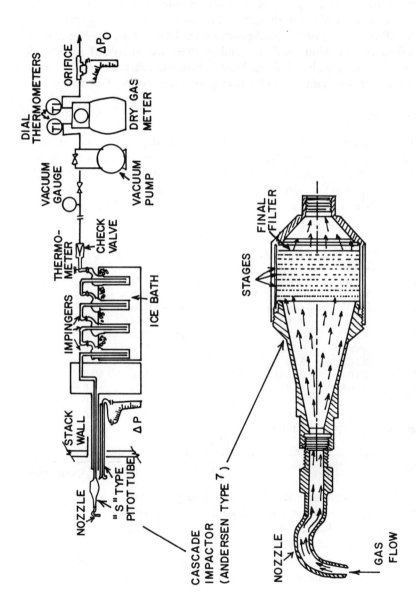

Figure 3-10. Cascade impactor sampling train.

After sampling is completed, great care must be taken to ensure that particles are not displaced from their proper stage as a result of handling. To aid in disassembly, the threads of the impactor head should be wrapped with Teflon®* tape, or antiseizing compound. Tweezers are used to remove glass-fiber or aluminum substrates, and particles are dusted off gaskets and impactor surfaces with a fine hairbrush. Particulate deposits on the substrates should be examined for streaking, or other evidence of erosion.

Data Analysis

The substrates are weighed to the nearest 0.0001g (0.1 mg) to obtain the weight of collected particulate on each stage. Table 3-2 shows an example of the data workup, from which a plot is made on log-probability paper (Figure 3-11). Even though the data will usually show some curvature at each end of the plot, the best straight line should be drawn through the intermediate points.

Table 3-2. Example of Impactor Data Workup

Stage	Tare Wt. (g)	Final Wt. (g)	Net Wt. (mg)	Wt. %	% Cum. Wt.	Cum. Wt. % Less Than	D_{pj} (μm)
1	0.1010	0.1080	7.0	14.0	14.0	86.0	12.9
2	0.1015	0.1045	3.0	6.0	20.0	80.0	8.2
3	0.0990	0.1005	1.5	3.0	23.0	77.0	5.4
4	0.1005	0.1035	3.0	6.0	29.0	71.0	3.7
5	0.1000	0.1055	5.5	11.0	40.0	60.0	2.4
6	0.1000	0.1075	7.5	15.0	55.0	45.0	1.2
7	0.1005	0.1095	9.0	18.0	73.0	27.0	0.75
8	0.0995	0.1090	9.5	19.0	92.0	8.0	0.50
Final Filter	0.1020	0.1060	4.0	8.0	100.0	0.0	<0.50
			50.0				

see Figure 3-11

Such a plot may be characterized by two parameters: the mass median diameter (D_{p50}), and the standard geometric deviation (σg). The mass median diameter signifies that half the particles by weight are larger than D_{p50}, and half smaller, while the standard geometric deviation gives an indication of the spread of the data (σg = 1.0 means all particles have the same size; and the size range increases as σg increases):

---

*Teflon is the registered trademark of E. I. duPont de Nemours and Company, Inc., Wilmington, Delaware.

$$\sigma g = \frac{D_{p84}}{D_{p50}} = \frac{D_{p50}}{D_{p16}}$$

(84% of the particles by weight are smaller than D_{p84}, and 16% are smaller than D_{p16}). Plotted in this fashion, the data can be conveniently used to design control equipment.

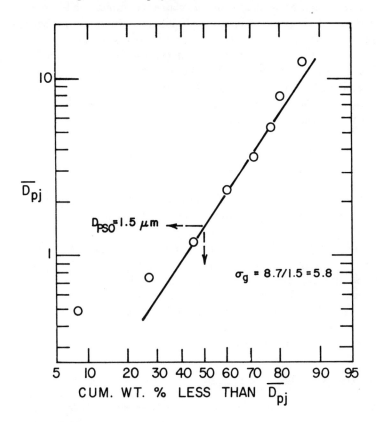

Figure 3-11. Particle size distribution (cumulative weight percent $<\overline{D_{pj}}$).

Sulfate Formation

The presence of moisture, or metallic oxides, in a filter medium, combined with sulfur oxides (SO_2 and SO_3) in the flue gas, can result in the formation of particulate sulfate on the filter. This will then be erroneously measured as particulate in the flue gas.

To avoid this problem, moisture condensation must be prevented upstream of the filter, and both the EPA Method 5, and in-stack cascade impactors, are designed to do so. In addition, a proper choice of filter medium must be made, with perhaps even preconditioning of the filter.

With EPA Method 5, membrane filters (which are relatively inert) should be used, rather than glass-fiber filters.[16] However, glass-fiber may be preferred as a cascade impactor substrate (see above), because of its high capture efficiency. In that case, some investigators[15] claim a reduction in sulfate formation by using low-pH glass-fiber filters, while others[17] claim that low pH alone is not enough. If sulfate formation does become a problem with glass-fiber filters, one can resort to preconditioning.

Preconditioning involves the placement of an absolute prefilter before the cascade impactor, and sampling for about an hour. Since the prefilter will remove all particulates, the glass-fiber substrates are only exposed to the flue gas, and will equilibrate to a new tare weight. The substrates are then reweighed, and sampling proceeds normally.

The extent of sulfate formation may also be determined quantitatively by soaking the filter medium in a dilute hydrochloric acid solution, and then analyzing the solution for sulfate. This can also provide a test of the extent of a sulfate problem, by adding a few drops of barium chloride to the acidified solution. A white precipitate indicates the presence of sulfate, and means that countermeasures must be taken to avoid weighing errors.

Sulfate formation can also occur in the water filled impingers. Originally, Method 5 was to require that this water be evaporated after volumetric measurements were taken, and any solid residue added to the particulate weight. Although this procedure was eventually dropped by the U.S. EPA, it is still included in some state regulations. The so-called "condensible" catch that results can double the apparent (or, calculated) particle loading in the effluent gas.

CONTINUOUS IN-STACK MONITORS

In the last few years, equipment has been developed to continuously monitor particulate concentrations in a stack or duct. This equipment would be useful in monitoring process losses, in optimizing the operation of control equipment, or may even be required by future regulations.

One such instrument is the PILIS V, manufactured by Environmental Systems Corp.,[18] which operates on particle scattering of a collimated beam of light. This instrument is self-contained, and may be inserted into a duct through a 4-in. sample port.

As with any such optical device, the presence of condensed moisture will be recorded as particulate by the instrument. Also, it must be periodically calibrated by using EPA Method 5. Nevertheless, the PILIS V can provide a continuous, real-time response to particle concentrations in a stack, or duct.

AMBIENT AIR SAMPLING

Sampling for air-borne particulates in the environment is hampered to some extent by the inability to proceed isokinetically. The air is essentially still, while high-volume sampling equipment generally operates at a sample rate of 40-60 cfm.

As with stack sampling, criteria have been established by the U.S. EPA[19] for ambient air monitors. Figure 3-12a shows a typical equipment setup for monitoring total air-borne particulate. This equipment is intended to operate 24 hr/day for perhaps several weeks, or months, with daily changes of the filter medium.

The EPA has also modified this equipment to include a cascade impactor of the Anderson type.[14,15,20] This is shown in Figure 3-12b. The impact plates are quite large—approximately 12 in. in diameter—and the filter media are subject to the same errors discussed above.

In addition to a lack of isokinetic sampling, there can be considerable difficulty in maintaining a constant volumetric sampling rate. In practice, the flow rate is simply measured with a rotameter at the beginning and end of the sampling run (which usually spans a 24-hr period), and the average taken. However, a high particle concentration, or condensed moisture, will plug the filter medium, and reduce the volumetric sampling rate. If this were a straight-line reduction, the arithmetic average of the initial and final rates would give a good indication of the actual time-averaged value. However, the sampling rate can exhibit significant fluctuations, for which a simple arithmetic average will not suffice. This will cause a corresponding error in the calculated ambient concentration. One possible alternative is to measure the pressure drop across the sampler with a pressure transducer, and record the output continuously on a circular chart.[19]

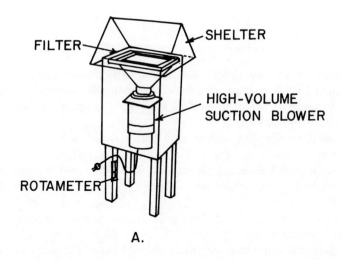

FILTER

SHELTER

HIGH-VOLUME
SUCTION BLOWER

ROTAMETER

A.

Figure 3-12a.   EPA ambient air sampler.

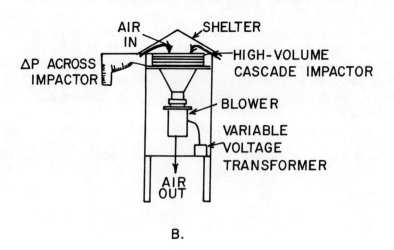

AIR
IN

SHELTER

ΔP ACROSS
IMPACTOR

HIGH-VOLUME
CASCADE IMPACTOR

BLOWER

VARIABLE
VOLTAGE
TRANSFORMER

AIR
OUT

B.

Figure 3-12b.   Ambient air sampler with cascade impactor.

REFERENCES

1. *Federal Register* 36(247):24882-24890 (December 23, 1971).
2. Aerotherm Acurex Corp., 485 Clyde Ave., Mountain View, CA 94042.
3. Scientific Glass and Instruments, Inc., P.O. Box 6, Houston, Texas 77001.
4. Research Appliance Company, Route 8, Gibsonia, PA 15044.
5. *Determining Dust Concentration in a Gas Stream,* Power Test Code 27-1957, American Society of Mechanical Engineers, New York (1957).
6. "Sampling Stacks for Particulate Matter," Standard D2928-71, Part 23 (Water; Atmospheric Analysis) of the *1971 Annual Book of ASTM Standards,* American Society for Testing and Materials, Philadelphia, (1971), pp. 832-858.
7. "Detecting and Measuring Invisible Contaminants," in *Air Pollution Control Field Operations Manual,* M. I. Weisburd, Ed., Los Angeles County Air Pollution Control District, U.S. Dept. of Health, Education and Welfare, Public Health Service Publication No. 937 (1962), Chapter 11, pp. 167-173.
8. *Methods for Determination of Velocity, Volume, Dust, and Mist Content of Gases,* Bulletin WP-50, H. H. Haaland, Ed., Western Precipitation Div., Joy Manufacturing Co., Los Angeles, CA (1968).
9. Byers, R. L. *J. Air Poll. Control Assoc.* 26:143-145 (1976).
10. Pilat, M. J., D. S. Ensor and J. C. Bosch. *Atmos. Environ.* 4:671-679 (1970).
11. Andersen 2000, Inc., P.O. Box 20769, Atlanta, GA 30320.
12. Pollution Control Systems, Inc., Renton, WA 98055.
13. Monsanto Enviro-Chem Systems, Inc., 800 N. Lindbergh Blvd., St. Louis, MO 63166.
14. Dzubay, T. G., L. E. Hines and R. K. Stevens. *Atmos. Environ.* 10:229-234 (1976).
15. Burton, R. M., *et al. J. Air Poll. Control Assoc.* 23:277-281 (1973).
16. Byers, R. L. and J. W. Davis. *J. Air Poll. Control Assoc.* 20:236-238 (1970).
17. Witz, S. and R. D. MacPhee. *J. Air Poll. Control Assoc.* 27:239-241 (1977).
18. Environmental Systems Corp., P.O. Box 2525, Knoxville, TE 37901.
19. *Federal Register* 36(84):8191-8194 April 30, 1971.
20. Lee, R. E. and S. Goranson. *Environ. Sci. Technol.* 6:1019-1024 (1972).

4

PARTICULATE SAMPLING & ANALYSIS

INTRODUCTION

This section outlines the methods for particulate sampling. The complexity and cost of sampling for particulates in stacks or ducts are widely appreciated. The desired sampling methods for particulates for Level 1 and Level 2 investigations is the Source Assessment Sampling System (SASS). Other sampling methods are also discussed in this section.

SAMPLING LOCATION

The sampling location for the Level 1 investigation is not set by rigid requirements, that is, except for the avoidance of gas flow obstructions and disturbances, the sampling site can be of a convenient nature, possibly an existing port opening, and one that requires a minimum of site preparations and funding.

In the Level 2 investigation, which demands higher accuracy, full EPA Method 1 procedures must be followed. Port positions and traverse point positions are defined by this method. In the event of flow obstructions and disturbances, sampling sites should be positioned over eight equivalent diameters downstream. Further, the complete traverse of the stack or vent for a Level 2 investigation must be effected at isokinetic operating conditions. In the Level 1 investigation, the average stack or vent gas velocity is calculated and sampling is effected at the position of the cross section that approximately coincides with this velocity. At this time the proper nozzle is selected and isokinetic sampling is accomplished.

In Tables 4-1 and 4-2, the sample handling and transfer procedures are identified for the SASS train nozzle, the probe, the cyclone and the filter.

Table 4-1. Sample Handling and Transfer—SASS Nozzle, Probe, Cyclones and Filter

Train Component	Probe and Nozzle
Transfer Action	$CH_2Cl_2:CH_3OH$ rinse into amber glass container. Combine with $10\,\mu$ cyclone rinse.
Train Component	$10\,\mu$ cyclone.
Transfer Action	Procedure 1: Tap and brush contents from walls and vane into lower cup receptacle. Remove receptacle and transfer contents into a tared nalgene container.
	Procedure 2: Reconnect lower cup receptacle and rinse walls and vane into cup ($CH_2Cl_2:CH_3OH$). Transfer receptacle rinse to probe and nozzle rinse container.
Train Component	$3\,\mu$ cyclone.
Transfer Action	Procedure 1: Tap and brush contents from walls into lower cup receptacle. Remove receptacle and transfer contents into a tared nalgene container.
	Procedure 2: Reconnect lower cup receptacle and rinse material with $CH_2Cl_2:CH_3OH$ into cup. Remove lower cup receptacle and transfer contents into an amber glass container.
	Procedure 3: Rinse tubing joining 10μ to 3μ with $CH_2Cl_2: CH_3OH$ into other rinse containers.

SAMPLING FOR TOTAL CONCENTRATION AND MASS EMISSION. LOW SAMPLE RATE.

METHOD: EPA METHOD 5

Advantages:

1. Recognized as standard method for testing.
2. Method uses acceptable sampling point traverse system based on flow disturbance (EPA Methods 1 and 2).
3. Simultaneous measurement of moisture (EPA Method 4) give on-line isokinetic sampling conditions.
4. The use of cyclone as a prefilter allows sampling in high-dust loading and moisture-laden streams.
5. An in-stack impactor can be added to the probe for particle size measurement.

Table 4-2. Sample Handling and Transfer—SASS Nozzle, Probe, Cyclones and Filter

Train Component	1 μ cyclone
Transfer Action	Procedure 1: Tap and brush contents from walls into lower cup receptacle. Remove cup and transfer contents into a tared nalgene container.
	Procedure 2: Reconnect lower cup receptacle and rinse adhered material with CH_2Cl_2:CH_3OH into cup. Remove cup and transfer contents into an amber glass container.
	Procedure 3: Rinse with CH_2Cl_2:CH_3OH the tubing joining the 3μ to 1μ into nalgene container. Combine with above rinses.
Train Component	Filter housing
Transfer Action	Procedure 1: Remove filter and seal in tared petri dish.
	Procedure 2: Brush material from housing into a tared nalgene container.
	Procedure 3: Rinse adhered material with CH_2Cl_2:CH_3OH into amber glass container.
	Procedure 4: Rinse tubing joining 1μ to housing with CH_2Cl_2:CH_3OH into above amber glass container.

6. Tries to control the temperature of the dust-laden gas sample and uses in-line glass condensers; thus, gas collection is possible without contamination.

Disadvantages:

1. Does not measure particulate mass in the stack selective to ambient Hi-Vol measurements. Other chemical reactions, condensation and evaporation can change the physical state of particulate prior to ambient measurement.
2. Captured particulate is held at an elevated temperature (320°F) for the duration of the test. This allows reevaporation of some volatile constituents.
3. A manual method that requires field handling of fragile equipment precisely positioned and maintained in a hostile environment.
4. Under high SO_2 concentration conditions, filterable aerosols may be forced from gaseous SO_3, thus introducing a potential error in measurement.
5. Requires glass probe up to a specific temperature and a probe length that requires excessive care to prevent breakage.

Remarks:

The following methods enable the measurement of both concentration and mass emission rate using a sampling rate of less than 2.5 cpm.

1. Material collected before or on a glass-fiber filter media at 320°F is particulate matter.
2. Sampling rate is at 1.0 scfm, isokinetically.
3. Method 5 trains are commercially available.
4. Method involves simultaneous determination of stack gas volumetric flow rate and particulate mass concentration, which allows calculating particulate mass emission rate.
5. Applicable for Level 1 and 2 investigations.

References

Battelle Columbus Laboratories. "Chemical Composition of Particulate Air Pollutants from Fossil-Fuel Combustion Sources" (March 19, 1976).
Smith, F., D. W. Wagoner and A. C. Nelson, Jr. Research Triangle Institute, EPA, 650/4-74-005-C (August 1974).
"Standards of Performances for New Stationary Sources," *Federal Register* 36(247) (December 23, 1971).

- -

METHOD: ASTM D2928-71

Advantages:

1. Fewer simultaneous operating systems and less cumbersome than EPA Method 5.
2. Simpler probe and support system.
3. Recognized as a standard method for testing.
4. Allows discretion relative to sample collectors and number of sampling points.

Disadvantages:

1. Errors in isokinetic sampling rate can occur with separate measurement of sample rate and velocity under the assumption of insignificant change in gas flow rate.
2. Allows different collectors depending on sampling conditions, resulting in a varying collection efficiency, especially with the fine particles ($< 3\mu$).
3. Allows discretion relative to number of sample points in larger ducts.

4.   Care required to prevent condensate from flowing back into filter during handling.
5.   Generally allows more discretion to the user than EPA Method 5, which could result in variability in emission measurements depending on user.
6.   See (3) under EPA Method 5, pg. 55.

Remarks:

Particulate defined as material (dry plus aerosol) collected in the stack at temperature and pressure after oven drying overnight at 102°C.

1.   Difference between ASTM and Method 5 is that ASTM train collects the particulate using an in-stack thimble and filter while the EPA train collects the particulate using an out-of-stack cyclone and filter.
2.   Isokinetic sampling.
3.   Sampling rate is at 1-2 cfm.
4.   Trains are commercially available.
5.   Applicable for Level 1 and 2 investigations.

References

ASTM. "Project Threshold," Progress Report, ASTM Standard, D2036, (February 1973).

_ _

METHOD: ASME PTC-27

Advantages

1.   See 1, 3, 4 and 5 in ASTM: D2928-71, pg. 56.
2.   Offers comments and observations on many questions from experience; however, does not specify exact procedures to be followed under different circumstances.

Disadvantages:

1.   See Method 5 (1), pg. 55.
2.   Allows considerable discretion relative to the number of sample points in large ducts.   Number of points depends on range of velocity at that location with up to 20 points needed in large ducts.
3.   Allows different collectors which results in a varying collector efficiency depending on particle size.
4.   Particles smaller than 1μ are not sampled by definition.

Remarks:

1.   Particulates defined as "dust." Gas-borne solid matter larger than 1 μ mean diameter collected in or out of the stack.
2.   ¼-in. minimum tip diameter.
3.   Isokinetic sampling.
4.   Ten-minute sampling at each point and two circuits of the points.
5.   Collection efficiency is stated to be 99% for size of dust met.
6.   Applicable for Level 1 and 2 investigations.

References:

American Society of Mechanical Engineers. "Determining Dust Concentration in a Loss Stream," Power Test Code 17 (1957).

_____

METHODS: WESTERN PRECIPITATION WP-50; LOS ANGELES COUNTY; SAN FRANCISCO BAY AREA

References:

Cooper, H. B. H., Jr., and A. T. Resszno, Jr. "Source Testing for Air Pollution Control," Environmental Science Services (1971).
Industrial Research Institute. "An Information Search and An Evaluation of Factors Affecting Stack Sampling," University of Windsor (September 1970).

_____

SAMPLING FOR SIZE DISTRIBUTION.
LOW SAMPLING RATE, *IN SITU* IMPACTORS.

METHOD: ANDERSON MODELS II AND III

Advantages:

1.   Extensive data collected allows relative comparison of data.
2.   Fits standard particulate sampling equipment.
3.   By changing inlet nozzle diameter, isokinetic sampling is allowed.
4.   Stainless steel construction allows insertion into high-temperature gas streams.
5.   Since EPA has not adopted a size distribution method, this impactor has been adopted by the testing community.
6.   Commercially available.

7. Options include preseparator glass-fiber substrates. A backup filter can be added.
8. A calibrated, modified preseparator is available for the unit.

Disadvantages:

1. Reported particle bounce off and reentrainment. Upper sampling limit of 10 m/sec prevents this.
2. Tends to overload when used in higher particulate loading situation.
3. Electrostatic effects on bare plates may be substantial.
4. Collection plates are massive relative to material collected resulting in slight loss of sensitivity. Substrates are available, but sulfate uptake on substrate may cause error.
5. Cannot be used for isokinetic total mass measurement at different duct cross section velocity points because it must be maintained at constant velocity.
6. Poor calibration under conditions appropriate for field sampling.
7. "Immediate" weighing after 24 hr dessication to prevent loss is required.
8. Due to size, it frequently cannot be located in gas flow.

Remarks:

1. Samples at < 1.0 cfm.
2. Separates 0.3-20 μ with 9 collection plates with jets.
3. Wall losses may explain part of the difference between EPA Method 5 mass emission and impactor measurements.
4. Impactors are generally very sensitive to overloading. The sample duration must be carefully determined depending on mass loading in the gas stream.
5. When mounted outside the stack, impactors may be inaccurate due to probe wall losses.
6. Generally, impactors rely on theory of impaction developed by Rantz and Wong.
7. Applicable for Level 1 and 2 investigations.

References:

McCain, J. D. *et al.* "Field Measurements of Particle Size Distribution with Inertial Size Devices," Southern Research Institute, EPA-650/120-73-035 (October 1973).

Smith, W. B. *et al.* "Particulate Sizing Techniques for Control Device Evaluation," Southern Research Institute, EPA-650/12-74-102a (October 1974).

Smith, W. B. *et al.* "Anderson Filter Substrate Weight Loss," Southern Research Institute, EPA-650/12-75-022 (February 1975).

Smith, W. B. *et al.* "Particulate Sizing Techniques for Control Device Evaluation," Southern Research Institute, EPA-650/12-74-102a (August 1975).

METHOD: UNIVERSITY WASHINGTON MARK III

Disadvantage:

1.   See 1, 2, 3, 4, 5 and 8 for Anderson Models II and III, pg. 59.
2.   Built-in filter holder.

Remarks:

1.   Samples at < 1.2 cfm.
2.   Separates 0.2-20 μ with 8 stages with multijets.
3.   Applicable for Levels 1 and 2 investigations.

References:

Pilot, M. S. "Submicron Particle Sizing with Cascade Impactors," paper presented at the 66th Annual Meeting of the Air Pollution Control Association, June 1973.

METHOD: MONSANTO BRINK

Disadvantages:

1.   Cannot be easily added to a mass sampling train.
2.   Modification needed for in-stack use.
3.   Low sample rate requires very long periods to obtain a sample in low loading situations.
4.   Often cannot sample isokinetically because of low flow rate and minimum sample tip diameter.
5.   See 2, 3, 5, 6 and 8 for Anderson Models II and III, pg. 59.

Remarks:

1.   Samples at 0.1-0.25 cfm.
2.   Separates 0.3-3 μ with 5 stages.

3. Prevention of reentrainment is accomplished by use of silicone grease substrate. However, grease blowoff can occur at speeds greater than 65 m/sec.

4. Applicable for Level 1 and 2 investigations.

References:

Brink, J. A. *et al.* "Particle Size Measurements with Cascade Impactors," paper presented at the 85th Annual Meeting of American Institute of Chemical Engineers, New York, November 1972.

METHOD: ENVIRONMENTAL RESEARCH CORPORATION TAG

Remarks:

Very similar to Anderson Method with appropriate advantages and disadvantages.

References:

McCain, J. D. *et al.* "Field Measurements of Particle Size Distribution with Inertial Size Devices," Southern Research Institute, EPA-650/120-73-035 (October 1973).

METHOD: MRI MODEL 1502 IMPACTOR

Reference:

TRW Systems Group. "Final Report for Process Measurements Development—Particulate Sulfate Emissions" (March 1975).

SAMPLING FOR SIZE DISTRIBUTION.
LOW SAMPLE RATE CYCLONES.

METHOD: SRI CHANG T2A SRI5-1

Advantages:

1. Cyclone system allows more mass collection for analysis by the appropriate size fraction.

2.   Cyclone performance is independent of orientation.
3.   In-stack system that can be used in 6-in. ports.
4.   Add to available test train to get sample.
5.   Laboratory aerosols are used to calibrate cyclones.
6.   SS construction allows high-temperature sampling.

Disadvantages:

1.   Cyclone "cut" point depends on sample flow rate. Constant rate is required to maintain calibration. Isokinetic sampling requires changing train characteristics to maintain constant flow rate.
2.   Extreme care necessary in cleaning cyclones and totally removing the sample.
3.   Not yet commercially available.
4.   It is difficult to obtain total mass emission rate and size distribution with this system.

Remarks:

1.   Operation consists of points of approximately 0.5 μ, 0.9 μ and 2.6 μ and a flow rate of 1 acfm with particle density of 1.35 g/cm^3. Backup filter is also used.
2.   For stack velocities between 10 and 100 ft/sec, 10 different nozzles are used for isokinetic sampling.
3.   It is not a method, but an experimental assembly of particulate sizing components designed to provide data of interest.
4.   EPA has not yet adopted this system.
5.   Applicable for Level 1 and 2 investigations.

References:

Smith, W. B. *et al.* "Particulate Sizing Techniques for Control Device Evaluation," Southern Research Institute, EPA-650/12-74-102a (August 1975).

SAMPLING FOR SIZE DISTRIBUTION.
HIGH SAMPLE RATE CYCLONES.

METHOD: SRI AEROTHERM HIGH-VOLUME STACK SAMPLER (HVSS)

Advantages:

1. A three-cyclone sizing system that is added to available isokinetic "Method 5 Type" particulate sampler.
2. High flow rate sampling system that takes less time per run to obtain sufficient mass of sample for physical, chemical and biological characteristics.

Disadvantages:

1. In its present form, the 1st cyclone ($< 10 \mu$) is rather large and awkward for typical stack testing.
2. This system requires additional calibration and refinement to avoid excessive wall losses.

Remarks:

1. Three cyclone arrangements with cuts to provide $< 10 \mu$, 3-10 μ and 1-3 μ, at 5.0 scfm, with capacity for collection of large quantities.
2. Not a method *per se*.
3. The three cyclones are being developed by IERL-EPA as part of a combined high sample rate particulate system.
4. Applicable for Level 1 and 2 investigations.

References:

Smith, W. B. *et al.* "Particulate Sizing Techniques for Control Device Evaluation," Southern Research Institute, EPA-650/12-74-102a (August 1975).

TRW Systems Group. "Tentative Procedures for Sampling and Analysis of Coal Classification Processes," (March 1975).

TRW Systems Group. "Interim Report for Fabrication and Calibration of Series Cyclone Sampling Train" (April 1975).

SAMPLING FOR PARTICULATE SULFATE CONCENTRATION, MASS EMISSIONS, AND SIZE DISTRIBUTION.

METHOD: EPA METHOD 8

Advantages:

1. This is one of the reference methods for H_2SO_4 mist sampling.
2. Utilizes EPA Methods 1 and 2.
3. Simultaneous measurement of moisture (EPA Method 4) and velocity (EPA Methods 1 and 2) to give data validation of isokinetic sampling.
4. Proved to be the most efficient in the pilot studies in the field.
5. Probably best method for adoption and ultimate application as sulfate sampling system.
6. Will provide concentration and emission data when used to traverse the stack.

Disadvantages:

1. Fails to measure sulfate in the stack that can be directly compared to ambient sulfate measurements due to evaporation/condensation processes.
2. The impinger solution (isopropanol) was found to absorb SO_2, thus rendering sulfate data suspect.
3. Inadequate for ammonium compounds.
4. New filter material needed with low sulfate blank.
5. The ambient temperature filter allows blowthrough of particulate.
6. Applicable for Level 1 and 2 investigations.

Remarks:

1. Method specifically for H_2SO_4 mist sampling. Adaptation is necessary for other purposes such as sulfate sampling.

References:

"Standards of Performance for New Stationary Sources," *Federal Register* 36(247) (December 23, 1971).
TRW Systems Group. "Final Report for Process Measurements Development—Particulate Sulfate Emissions" (March 1975).

_ _

METHOD: EPA AEROTHERM HVSS

Advantages:

1. High-volume sampling system (5 acfm); very similar to EPA Method 5 train.
2. SS construction and Lexan construction with inside of filter housing coated with Teflon.
3. Moisture does not plug the filter since the filter is operated hot.
4. See 2, 3 and 6 for EPA Method 8, pg. 64.

Disadvantages:

1. See 1, 2 and 4 for EPA Method 8.
2. The precooler scrubs SO_2.

Remarks:

1. Train was essentially set up in a particulate collection operating mode, that is, with water in the impingers for ultimate collection of sulfates.
2. Applicable for Levels 1 and 2 investigations.

References:

TRW Systems Group. "Final Report for Process Measurements Development–Particulate Sulfate Emission " (March 1975).

_ _

METHOD: MRI MODEL 1502 IMPACTOR

Advantages:

1. Simultaneously provides particulate size distribution data and sulfate concentration data.
2. This unit provided good results on sulfate particle size distribution.

Disadvantages:

1. Impactors in that they are calibrated for only one flow rate.
2. A traverse cannot be done isokinetically.

Remarks:

1. Meteorology Research Inc. (MRI) device is a cascade impactor designed for particulate size distribution. It is used to determine ability to size particulate and aerosol sulfates.
2. Train consisted of the Aerotherm probe, the MRI impactor in the heated cyclone-filter oven, and then the Aerotherm coolers.
3. MRI unit was maintained at stack temperature.
4. Applicable for Level 1 and 2 investigations.

References:

See EPA Aerotherm HVSS, pg. 65.

SAMPLING FOR TRACE INORGANIC MATERIAL CONCENTRATION AND MASS EMISSIONS.

METHOD: EPA AEROTHERM HVSS

Advantages:

1. Basically relies on parts of a commercially available sampling system.
2. Similar to EPA Method 5. It may be accepted as reference in the future.
3. See Advantages 2, 3, 4 and 5 under EPA Method 5, pg. 54.
4. High sampling rate (to 6 scfm).
5. Impingers are of rugged Lexan construction.
6. Capable of horizontal and vertical sampling.
7. Well-packaged unit for shipment.
8. Offers an effective sampling system relatively free of contaminating components and capable of collecting both vaporous (or fine particulate As, Hg, Sc, Sb, F, Cl), and nonvolatile trace materials (Ba, B, Be, Ca, Cd, Cn, Cr, Cu, Mn, Ni, NO_3, Pb, PO_4, S, SO_4, Sr, V, Zn).
9. It can be adapted to provide samples by size fraction since it relies on a basic commercial system.

Disadvantages:

1. SS connectors are possible source of Ni and Cr sample contamination.
2. Heated Teflon (filter housing coating) may cause particulate sample alteration.

3. Requires too much time to clean gas cooling coils between runs.
4. Bigger and heavier than standard particulate testing equipment.
5. Traversing is difficult due to size and weight.
6. Kapton probe liner film requires extreme care and patience in inserting into the probe.
7. Ammonium persulfate impinger solution must be prepared immediately prior to a test run.

Remarks:

1. Basic train is made up of an Aerotherm high-volume sampler with a polymer film probe liner, ultraclean filter material and a special sequential exidative scrubbing solution for the impingers.

References:

TRW Systems Group. "Procedures for Process Measurements—Trace Inorganics Materials" (July 1975).

_ _

METHOD: EPA METHOD 5

Advantages:

1. Relies on an established method of measuring mass concentration and emissions.
2. Could be adapted to provide samples by size fraction.
3. Extensive data have been collected using this method.
4. Unofficial procedures of sample removal and cleanup have been drafted for in-house and contractor use that require nitric acid cleaning of portions of the sample train and storage bottles.

Disadvantages:

1. Potential sample contamination from being exposed to SS probe tip and probe.
2. High and variable background concentration in filter media (usually fiberglass), depending on inorganic material.
3. Fine particulate and vaporous inorganic material collection efficiency is variable and generally unknown.

Remarks:

1. It is not a sampling method *per se,* but rather an application of best available tools.
2. Other Method 5-type sampling trains can be used. The construction materials govern the type and amount of potential contamination.

References:

du Pont Company. "Model 460SO_2-NO_x Analyzer," sales literature.
Pinta, M. *Detection and Determination of Trace Elements,* 2nd ed.
　(Ann Arbor, Michigan: Ann Arbor Science Publishers, Inc., 1970).
"Standards of Performance for New Stationary Sources," *Federal Register* 36(247) (December 23, 1971).

METHOD: EPA METHOD 104 (BERYLLIUM)

Advantages:

1. Same train as Method 5.
2. Isokinetic sampling.
3. Tested by EPA.
4. Allows flexibility for filter location in train to account for higher temperature source.

Disadvantages:

1. Needs a millipore filter, which is more difficult to handle.
2. Requires extensive acid washing of train components prior to sampling.
3. See Method 5, pg. 67, for other disadvantages.

Remarks:

1. EPA Method 103 and 104 are specific methods for Beryllium sampling and analysis to determine compliance with EPA hazardous emission standards. Method 104 is a full duct traverse method. Method 103 is a screening method.
2. Applicable for Level 1 and 2 investigations.

References:

"Regulation 8823," *Federal Register* 38 (April 6, 1973).

METHOD: EPA METHOD 105 (MERCURY)

Advantages:

1. Essentially same train as Method 5, pg. 67.
2. Isokinetic sampling for accuracy.

Disadvantages:

1. Serious interference in high concentration SO_2 gas streams.

Remarks:

1. Not recommended for Hg sampling and analysis.
2. Applicable for Level 1 and 2 investigations.

References:

"Regulation 8823," *Federal Register* 38 (April 6, 1973).

METHOD: SRI MA #6 (MERCURY)

Advantages:

1. To isokinetically collect particulates and gaseous Hg this method uses Method 5 type train.
2. Accurate for concentrations up to 0.05 $\mu g/ft^3$.
3. Precision at 5% for 100-ft^3 sample.

Disadvantages:

1. Designed for chloralkali plants and nonferrous metal smelters.
2. When sampling high-concentration (7%) SO_2 sources, the sample volume should be limited to 10 ft^3.

Remarks:

1. This is an adapted version of EPA Method 105 consisting of a probe, pyrolysis furnace, solid scrubber and impingers containing ICl.

2.   Applicable to particulate and gaseous Hg.
3.   Applicable for Level 1 and 2 investigations.

METHOD: SRI MA #4 (CADMIUM)

Advantages:

See SRI MA #3 (Lead) Method, pg. 70.

Disadvantages:

1.   Several interferences overcome by adding EDTA.
2.   NaCl interference.
3.   Further field testing required.
4.   Leaching of filter required.

Remarks:

1.   Designed to collect particulate and gaseous cadmium.
2.   Applicable for Level 1 and 2 investigations.

References:

Coulson, D. C. *et al.* "Survey of Manual Methods of Measurements of Asbestos, Beryllium, Lead, Cadmium, Selenium and Mercury in Stationary Source Emission," Stanford Research Institute (September 1973).

METHOD: SRI MA #3 (LEAD)

Advantages:

1.   Well-defined, well-specified procedures.
2.   Associated accuracy and acceptance of Method 5-type train.
3.   Routine sampling period of 2 hr.

Disadvantages:

1.   Organic matter can interfere with Pb analysis. Can be removed with low-temperature activated oxygen ashing.
2.   Analyze blank filters for Pb.
3.   Further field testing required.

Remarks:

1. Modified EPA Method 5 version, designed to collect particulate and gaseous lead.
2. AA analysis.
3. Uses glass-fiber filter for nitric acid extraction.
4. Nitric acid in impingers.
5. Applicable for Level 1 and 2 investigations.

References:

Coulson, D. C. *et al.* "Survey of Manual Methods of Measurements of Asbestos, Beryllium, Lead, Cadmium, Selenium and Mercury in Stationary Source Emissions," Stanford Research Institute (September 1973).

_ _

METHOD: GOLD AMALGAMATION (MERCURY)

Advantages:

1. Accuracy of Method 5 train to measure total pollutant emission isokinetically.
2. Has more or less replaced EPA Method 105 for high-SO_2 concentration streams.

Disadvantages:

1. Method too sensitive. Maximum sampling period of 5-10 minutes.
2. Further refinement needed.
3. Several tests needed to determine the range of Hg concentration and then the actual concentration.
4. Cannot ship samples back to lab for analysis.

Remarks:

1. It is a modified version of EPA Method 5 designed to collect particulate and gaseous Hg by utilizing two impingers containing 10 g of gold chips to collect Hg vapor as an amalgam.
2. Second amalgamator is a backup to the first.
3. Applicable for Level 1 and 2 investigations.

References:

Kalb, G. W., and C. Baldeck. "The Development of the Gold Amalgamation Sampling and Analytical Procedure for Investigation of Mercury in Stack Gases," Tradit, Inc. (June 1972).

METHOD: SRI MA #5 (SELENIUM)

Advantages:

1.   Well-defined, well-specified procedures.
2.   Analytical accuracy at ± 14% up to 2 μg/ft.
3.   Associated accuracy and acceptance of Method 3-type train.

Disadvantages:

1.   Further field testing required.
2.   Hypochloride, Sn(II), and possibly nitrate ion interference.
3.   Unused filter should be saved for blank determination.
4.   Leaching of filter may be required prior to use.

Remarks:

1.   Modified version of EPA Method 5 designed to collect particulate and gaseous selenium (excluding H_2Se).
2.   Applicable for Level 1 and 2 investigations.

References:

Coulson, D. C. et al. "Survey of Manual Methods of Measurements of Asbestos, Beryllium, Lead, Selenium, and Mercury in Stationary Source Emissions," Stanford Research Institute (September 1973).

METHOD: SRI MA #1 (ASBESTOS)

Advantages:

1.   Well-defined, well-specified procedures with backup references.
2.   Can use low-temperature oxidation of some interfering fibers, and then recollect on another nuclepore filter.
3.   Allows preparation of photomicrograph for visual characterization of general particulate sample.

Disadvantages:

1. At least 10% of collected material must be chrysotile asbestos.
2. Identification of specific fibers is difficult and tedious.
3. Major errors caused by inexperienced particulate counter.
4. Very difficult to transfer particles to microscope grid for counting.
5. Trial and error in obtaining optimum amount of sample.

Remarks:

1. Collection on a nuclepore membrane filter at rate of 0.1 cfm.
2. Materials transferred to electron microscope grid for visual identification, sizing and counting. Use standard reference samples of asbestos from International Union Against Cancer.
3. Applicable for Level 1 and 2 investigations.

References:

Coulson, D. C. *et al.* "Survey of Manual Methods of Measurements of Asbestos, Beryllium, Lead, Cadmium, Selenium, and Mercury in Stationary Source Emissions," Stanford Research Institute (September 1973).

SAMPLING FOR TOTAL CONCENTRATION AND MASS EMISSION. HIGH SAMPLE RATE.

METHOD: EPA AEROTHERM HVSS

Advantages:

1. Commercially available.
2. It has been adopted to simultaneously collect particulate mass samples by three cut sizes (cyclones) followed by total absorption of volatile organics, and a collector for trace elements; thus, it is a refined method of collecting vast amounts of data during one isokinetic measurement (traversing limitation when the three cyclones are used).
3. Highly researched and tested unit is well constructed.
4. Wide range of optional equipment available.

Disadvantages:

1. More expensive than EPA Method 5-type equipment.
2. Equipment is heavier and harder to handle.
3. SS construction can cause some test contamination.
4. Uses Lexan impingers instead of glass.
5. Takes too long to clean cooling coils if impingers wash is desired.

Remarks:

1. Meets EPA Method 5 specification with the exception that the probe and filter holder are constructed of SS instead of glass, and flow rate is much higher.

References:

Acurex Aerotherm. "High-Volume Stack Sampler—HVSS," Sales literature, Mountain View, California (March 1976).

Lapson, W. P. *et al.* "The Development and Application of a High Sampling Train for Particulate Measurements of Stationary Sources," *Proc. 67th Annual Meeting of the Air Pollution Control Association* (1974), pp. 74-189.

_ _

METHOD: RADER

Advantages:

1. Samples at very high flow rate (10-20 cfm).
2. Can be hand-held.
3. Uses same filter as in ambient high-volume samplers, which is the reference EPA ambient method.
4. Requires less time to get a sample. This may be important in very low-concentration sources.
5. Adopted as a standard method for cyclone sampling in Oregon.
6. It is the only commercially available automatic isokinetic sampler.

Disadvantages:

1. Can only be used in ambient temperature sources.
2. Filter can be easily plugged by saturated gas streams.
3. Oversized ports are required for in-stack use.
4. Not comparable to EPA Method 5.

Remarks:

Unit is automatic, designed to sample ambient temperature sources since it does not contain impingers for moisture measurement. It is not a method, but a tool to sample gas streams for particulates.

References:

Bonbel, R. W. "A High-Volume Stack Sampler," *J. Air Poll. Control Assoc.* 21(12) (December 1971).
Rader Pneumatics. "Rader Hi-Volume Stack Samples," Sales literature Portland, Oregon (October 1975).
"Emission Monitoring, Standards for New Stationary Sources," *Federal Register* 40(194) (October 6, 1975).

SAMPLING FOR SIMULTANEOUS ORGANIC AND
TRACE INORGANIC MATERIALS.
TOTAL CONCENTRATION, SIZE DISTRIBUTION.

METHOD: SASS

Advantages:

1. In one sampling system, an isokinetic particulate sample is classified into three calibrated size fractions in which each can be analyzed for organics and trace inorganic materials. Noncondensed organics, including POM pollutants, are collected on porous polymer adsorbent; fine particulates and volatile inorganic species are collected in a glass impinger train.
2. Most of the parts are commercially available.
3. Inlet and outlet particulate size from normally encountered control equipment will be handled by three sizing cyclones.
4. High-volume sample rate will allow sample collection in less time.

Disadvantages:

1. Very expensive.
2. System not designed to allow continuous isokinetic traversing of a duct or stack.

3. Due to (2) above, a single point in the stack must be selected for a representative sample.
4. Further refinement is needed to condense moisture without excessive pressure drop.
5. Due to pressure drop, two pumps are needed.
6. More complex temperature and moisture regulation is required due to high flow rate.
7. Has not been extensively field tested.

Remarks:

1. System essentially based on the HVSS system.
2. An isokinetic traverse can be made by changing nozzles at each traverse point.
3. Applicable for Level 1 and 2 investigations.

References:

Acurex Corporation. "Source Assessment Sampler Preliminary Information," Sales literature, Aerotherm Division.

SAMPLING FOR TOTAL CONCENTRATION AND SIZE DISTRIBUTION, HIGH TEMPERATURE, HIGH PRESSURE SYSTEMS (HTHP).

METHOD: HTHP (AEROTHERM)

Advantages:

1. System meets safety standards.
2. Will traverse 50 in. mechanically against high pressure in 2 ½ min, and has an automatic brake.
3. Utilizes a "scalping" cyclone *in situ.*
4. Secondary particulate collection is by a Mark III University of Washington Impactor, which also provides size distribution data.
5. Uses complex temperature-controlling coolant system that appears to be well designed and allows for thermal expansion, yet properly cools the sample.

Disadvantages:

1. Design needs more field testing.
2. Not commercially available.
3. Expensive.

Remarks:

1. High-temperature, high-pressure; involves temperature up to 2000°F and pressures up to 100 atm.

PARTICULATE SIZING AND MORPHOLOGICAL STUDIES

In this section, methods for particulate sizing and morphological studies are outlined. Level 1 and Level 2 investigations both require the two physical characterization techniques be done, but not to the same degree of accuracy.

SAMPLING FOR PARTICULATE SIZING (DL = DETECTION LIMIT).

METHOD: DRY SIEVING: DL TO 74 μ

Disadvantages:

1. Attrition occurs if sieving operation is prolonged.
2. Timing has been standarized by ASTM for coal.
3. Interference by moisture content, relative humidity, electrostatic properties.

Remarks:

1. Care must be taken to minimize agglomeration and attrition. In reducing gross sample for analysis, special concern should be given to the representativeness of the sample to be analyzed.
2. This method can be used as a first step in sizing.
3. Applicable to Level 1 investigations.

References:

ASTM. ASTM Standard D197, D410, D2972, E11.

Stern, A. C. *Air Pollution: V. II–Analysis, Monitoring and Surveying* (New York: Academic Press, 1968).

— —

METHOD: SIEVING: DL 37 μ

Advantages:

1. Agglomerating particles are tested in a more representative manner by wet sieving than dry sieving.

Disadvantages:

1. Attrition takes place if sieving operation is prolonged.
2. Timing standardized ASTM for coal.
3. Moisture content, relative humidity and electrostatic properties can cause interference.

Remarks:

1. See Dry Sieving, pg. 77.
2. Applicable to Level 1 investigation.

References:

Stern, A. C. *Air Pollution: V. II–Analysis, Monitoring and Surveying* (New York: Academic Press, 1968).

— —

METHOD: MICROMESH: DL TO 5 μ

Disadvantages:

1. Agglomeration of fine particles severs due to electrostatic causes.

Remarks:

1. Agglomeration effects can be minimized by preconditioning dust and pans to 50-75% relative humidity.
2. Applicable to Level 2 technique when preceded by dry or wet sieving.

— —

- -

METHOD: MEMBRANE SIEVING: DL TO 37 μ

Disadvantages:

1. Applicable only to liquid suspensions.
2. Multiple backwashings required.

Remarks:

1. See Dry Sieving, pg. 77.
2. Applicable Level 2 technique when preceded by dry or wet sieving.

- -

METHOD: INCREMENTAL SEDIMENTATION: DL TO 2 μ

Disadvantages:

1. Adequate dispersion of the particles in suspension is essential.
2. Convective disturbances and evaporative effects must be minimized.

Remarks:

1. See Dry Sieving, pg. 77.
2. Determines equivalent particle diameter. Examples of this technique are pipet, diver, hydrometic and photoextinction.

- -

METHOD: BULK SEDIMENTATION: DL TO 0.1 μ

Disadvantages:

1. Same as incremental sedimentation method except that, in some cases, air is the settling medium and evaporative effects do not apply.

Remarks:

1. Same as for Dry Sieving, pg. 77.
2. Lower limit given applies only to Beta or piezoelectric techniques.
3. Applicable for Level 2 investigations.

References:

Brenchley, D. L., C. D. Turley and R. F. Yarmac. *Industrial Source Sampling* (Ann Arbor, Michigan: Ann Arbor Science Publishers, Inc., 1974).

- -

METHOD: CENTRIFUGAL SEDIMENTATION: DL TO 0.1 μ

Disadvantages:

1.   Same as incremental sedimentation.

Remarks:

1.   Same as for Dry Sieving, pg. 77.
2.   Since a sedimentation and dispersing liquid are both used in this technique, care is taken so only a small density difference exists. This prevents particle entrapment at the interface between the liquids.
3.   Applicable for Level 2 investigation.

References:

Brenchley, D. L., C. D. Turley and R. F. Yarmac. *Industrial Source Sampling* (Ann Arbor, Michigan: Ann Arbor Science Publishers, Inc., 1974).
Stern, A. C. *Air Pollution: V. II—Analysis, Monitoring, and Surveying* (New York: Academic Press, 1968).

- -

METHOD: AIR ELUTRIATION: DL TO 2 μ

Disadvantages:

1.   Agglomeration effects due to electrostatic attraction is prime interference.
2.   Particle attrition is also common.
3.   Cut fractions contain a large amount of undersized particles.

Remarks:

1.   Same as for Dry Sieving, pg. 77.
2.   Applicable for Level 2 investigations.

References:

Brenchley, D. L., C. D. Turley and R. F. Yarmac. *Industrial Source Sampling* (Ann Arbor, Michigan: Ann Arbor Science Publishers, Inc., 1974).
Stern, A. C. *Air Pollution: V. II–Analysis, Monitoring and Surveying* (New York: Academic Press, 1968).

_ _

METHOD: LIQUID ELUTRIATION: DL TO 2 μ

Advantages:

1.   Electrostatic effects are minimized.

Remarks:

1.   Same as for Dry Sieving, pg. 77.
2.   For more representative results, each fraction can be resuspended and the process repeated several times.
3.   Applicable for Level 2 investigations.

References:

Brenchley, D. L., C. D. Turley and R. F. Yarmac. *Industrial Source Sampling* (Ann Arbor, Michigan: Ann Arbor Science Publishers, Inc., 1974).
Stern, A. C. *Air Pollution: V. II–Analysis, Monitoring, and Surveying* (New York: Academic Press, 1968).

_ _

METHOD: CENTRIFUGAL ELUTRIATION: DL TO 2 μ

Disadvantages:

1.   Attrition is the prime limitation.

Remarks:

1. Same as for Dry Sieving, pg. 77.
2. The most common commercially available centrifugal elutriator is the Bahco microparticle classifier. This is the standard method of particulate sizing specified in ASME Power Test Code. The Bahco must be calibrated against optical methods.
3. Applicable for Level 2 investigations.

References:

ASME. "Power Test Code 18," *Determining the Properties of Fine Particulate Matter,* Book D30 (1965).

Brenchley, D. L., C. D. Turley and R. F. Yarmac. *Industrial Source Sampling* (Ann Arbor, Michigan: Ann Arbor Science Publishers, Inc., 1974).

Stern, A. C. *Air Pollution: V. II—Analysis, Monitoring, and Surveying* (New York: Academic Press, 1968).

_ _

METHOD: CENTRIFUGAL: DL TO 0.025 μ

Disadvantages:

1. Attrition is prime limitation.

Remarks:

1. Same as for Dry Sieving, pg. 77.
2. Various models are available. Calibration against optical methods must be performed.
3. Applicable for Level 1 investigations.

References:

Brenchley, D. L., C. D. Turley and R. F. Yarmac. *Industrial Source Sampling* (Ann Arbor, Michigan: Ann Arbor Science Publishers, Inc., 1974).

Stern, A. C. *Air Pollution: V. II—Analysis, Monitoring and Surveying* (New York: Academic Press, 1968).

_ _

METHOD: COULTER COUNTER: DL TO 0.03 μ

Advantages:

1. Device is automatic, quick.

Disadvantages:

1. An orientation correction factor must be applied when fibrous particles are present. Also, the length:diameter ratio for other particles must be near unity.

Remarks:

1. Same as for Dry Sieving, pg. 77.
2. Calibration is against monodispersed particles of known diameter.
3. Particle diameter should be no more than 30% of the aperture diameter.

References:

Brenchley, D. L., C. D. Turley and R. F. Yarmac. *Industrial Source Sampling* (Ann Arbor, Michigan: Ann Arbor Science Publishers, Inc., 1974).
Stern, A. C. *Air Pollution: V. II—Analysis, Monitoring, and Surveying* (New York: Academic Press, 1968).

METHOD: OPTICAL MICROSCOPE: DL TO 0.4 μ

Advantages:

1. No interference if proper precautions are taken. Accurate size determination is most dependent on the skill of the microscopist.

Remarks:

1. Particles must be dispersed on the slide in a random manner without causing shattering of the particles. Agglomerated particles should be defloculated. A photomicrograph would minimize any handling or storage interference.
2. Applicable for Level 1 and 2 investigations.

SAMPLING FOR MORPHOLOGY.

METHOD: POLARIZING LIGHT MICROSCOPE

Disadvantages:

1. Particle identification depends on the number of characteristics determined and the skill and experience of the microscopist.

Remarks:

1. Same as for Optical Microscope.
2. This instrument is the standard tool for the identification of particles down to 0.5 μ in diameter.
3. Applicable for Level 1 and 2 investigations.

References:

Hamersma, J. W., S. L. Reynolds and R. F. Maddalone. "IERL-RTP Procedures Manual: Level of Environmental Assessment," TRW Systems Group, EPA-600/2-76-160a (June 1976).

METHOD: ELECTRON MICROSCOPY: DL TO 0.001 μ

Disadvantages:

1. Sample exposure to heat and vacuum in the microscope must be taken into account when interpreting the result.
2. Transfer of electrostatic charge to the beam could cause a rupture of the substrate or particle.

Remarks:

1. Methods commonly used for the collection of submicron particles include membrane filters, thermal precipitators and electrostatic precipitators.
2. These techniques are not available in many cases, so interferences from breaking and agglomeration will occur.
3. Applicable for Level 2 investigations.

References:

Brenchley, D. L., C. D. Turley and R. F. Yarmac. *Industrial Source Sampling* (Ann Arbor, Michigan: Ann Arbor Science Publishers, Inc., 1974).

Stern, A. C. *Air Pollution: V. II—Analysis, Monitoring, and Surveying* (New York: Academic Press, 1968).

- -

METHOD: SCANNING ELECTRON MICROPROBE: DL TO 0.001 μ

Disadvantages:

1. Same as for Electron Microscopy, pg. 86.

Remarks:

1. Same as for Electron Microscopy.
2. Allows for greater details in particle morphology analysis.
3. Applicable for Level 2 investigations.

References:

Brenchley, D. L., C. D. Turley and R. F. Yarmac. *Industrial Source Sampling* (Ann Arbor, Michigan: Ann Arbor Science Publishers, Inc., 1974).

Stern, A. C. *Air Pollution: V. II—Analysis, Monitoring, and Surveying* (New York: Academic Press, 1968).

- -

METHOD: ELECTRON MICROPROBE

Advantages:

1. Can be used to identify the chemical nature of individual particles.

Remarks:

1. The mounting of the sample is a critical step in this procedure.
2. If the Polarizing Light and Electron Microscopes fail to provide enough information, this technique can be used.
3. Applicable for Level 2 investigations.

- -

CHAPTER 5

GAS SAMPLING

INTRODUCTION

This chapter outlines methods for gas sampling. In the Level 1 investigation, a single, simple grab sample is employed, using the common displacement gas bomb. The Level 2 investigation may require this and additional sampling in the form of continuous monitoring.

SAMPLING LOCATION

The sampling location for a Level 1 investigation would be at a convenient position, possibly an existing opening, that allows a sample to be taken with relatively little expense and trouble. In the event stratification of the gas stream exists at the site selected, another position should be utilized where this phenomenon is not occurring.

For a Level 2 investigation a single sampling position is all that is required if the gas stream is unstratified. Should stratification exist, a full cross-sectional sampling traverse must be made of the stack or duct. In Table 5-1, Source Assessment Sampling System (SASS) train impinger reagents are given.

Table 5-1. Sample Handling and Transfer—Source Assessment Sampling System (SASS) Impingers

Train Component	Impinger Nos. 1, 2 and 3.
Transfer Action	Add rinses from connecting tubing. Rinse with IPA:H_2O (Distilled) and transfer to nalgene container. Combine with rinses from other impingers. Measure total volume for analysis.
Train Component	Impinger No. 4.
Transfer Action	Discard drierite.

The sample from a Level 1 investigation must be carefully stored, avoiding heat and sunlight, to limit possible reactions. Tables 5-1 and 5-2 outline handling and transfer techniques for the impingers and XAD-2 module on the SASS train for a Level 2 investigation.

Table 5-2. Sample Handling and Transfer—SASS XAD-2 Module

	SASS XAD-2 Module Transfer Action
Procedure 1	Remove XAD-2 cartridge from cartridge holder by releasing clamps. Remove fine mesh screen from cartridge top and discharge resin to glass amber jar. Replace screen and cartridge and clamps. Reservoir condensate valve is opened and aqueous condensate is drained to a 1-liter separatory funnel. CH_2Cl_2 is the extract. The organic phase is combined with the rinses of Procedure 2.
Procedure 2	Close valves and clamps and remove inner well. Rinse inner well surface with wash bottle ($CH_2Cl_2:CH_3OH$). Allow rinse to run through module into condensate collector. Place inner well aside. Rinse entrance tube to module interior allowing solvent to flow down into condensate cup. Release XAD-2 and condensate cup from the condenser. Rinse the empty XAD-2 section into the condensate cup. Remove condensate reservoir containing all rinses and combine with rinses from Procedure 1.

SAMPLE TYPE: SAMPLE ACQUISITION AND CONDITIONING SYSTEMS FOR CONTINUOUS INSTRUMENTAL ANALYSIS.

Acquisition and conditioning of stack gases by removing particulate and moisture and controlling sample temperature is a necessary step prior to sample analysis using most continuous monitoring instruments that require sample extraction and transport to the instrument. Instruments have been developed to monitor selected gases *in situ*, in the stack. Manual sampling methods incorporate conditioning as an integral part of the sample train.

SAMPLE TYPE: ACQUISITION PROBE LOCATION.

METHOD: PERFORMANCE SPECIFICATION #2

Advantages:

1. Written to provide guidance for both extractive and nonextractive sampling.
2. The number of sample points and location is dependent on the upstream distance to an air in-leakage device or measured stratification profile.
3. Represents the first regulatory agency attempt to define sampling techniques for potentially stratified gases.
4. Total sampling and analysis system has been demonstrated in place to be comparable with reference methods.
5. Based on extensive field testing.

Disadvantages:

1. Tends to specify desirable characteristics of results rather than provide clear-cut examples and procedures of obtaining those results.
2. Designed for SO_2 and NO_x from power plant combustion sources with known stratification causing processes.
3. Does not address merging of two or more different gas streams.
4. Specifies sampling at "average" concentrations yet does not fully define them.
5. Requires "multipoint" probe in one instance yet fails to specify number of points.

Remarks:

1. Continuous stack gas monitoring systems have traditionally consisted of a single in-stack filter for sample extraction and transport to an analytical instrument; if the one sample point in the duct cross section is sufficiently downstream from a gas mixing device, such as a fan, the single point will be representative. If gas stratification exists, several sample points in the cross section may be sampled simultaneously, or a different location in the ductwork may be selected.
2. Applicable for Level 2 investigation.

References:

"Emission Monitoring, Standards for New Stationary Sources, *Federal Register* 40(194) (October 6, 1975).
"Stationary Sources, Proposed Emission Monitoring and Performance Testing Requirements," *Federal Register* 39 (177) (September 11, 1974).

- -

METHOD: STRATIFICATION INVESTIGATION POLLUTION RATIO CONCEPT

Advantages:

1. Preferred over multipoint sample extraction.
2. Single probe—simple system.
3. Can be applied to both circular and rectangular ducts.
4. Concentration and flow data can be taken independently and still be used to accurately determine species mass emissions.

Disadvantages:

1. Made up of a simplified field-tested concept that is incomplete as a method which can be applied to other general situations.

Remarks:

1. Some of the measuring techniques include static pressure null probes, thermal null probes, several continuous automatic sample rate adjustment devices, multipoint probe arrays, annubars and constantly traversing probes.
2. Stack monitoring rather than in a duct is preferred for uniform conditioning. Each system must be designed for a specific location due to site condition variability.
3. Applicable for Level 1 investigation.

References:

Brooks, E. F. *et al.* "Continuous Measurements of Total Gas Flow Rate from Stationary Sources," TRW System Group, EPA 650/12-75-020.
Zahek, A. *et al.,* "Procedures for Measurements in Stratified Gases," Vol. I, Walden Research Division, EPA 650/12-74-086a (September 1974).

Zahek, A. *et al.* "Procedures for Measurement in Stratified Gases," Vol. II, Walden Research Division, EPA 650/12-7-086b (September 1974).

METHOD: STRATIFICATION INVESTIGATION ROW AVERAGE CONCEPTS

Advantages:

1. Field testing is promising.
2. Local duct geometry can usually be examined to select a row that will be in the direction of highest stratification.
3. Specific description of the method is given, thus eliminating different interpretations of the techniques and, subsequently, noncomparable results.
4. Highly accurate, simple.
5. Row sample points can easily be incorporated in a single multiport probe.
6. Ellison Annubar can be used.

Disadvantages:

1. Unpredictable.
2. Requires preliminary survey traversing to select optimum system.
3. System hardware is not commercially available.
4. Further field testing needed.
5. Care required in probe port sizing to insure desired flow into probe.

Remarks:

1. See Pollutant Ratio Concept, pg. 90, for general comments on stratification investigations.
2. The row average technique was first developed as a velocity (flow rate) measurement technique. It consists of extracting continuous samples from several selected points on a particular traverse or "row" across the cross section.
3. Primarily for rectangular ducts.
4. Applicable for Level 2 investigations.

References:

Brooks, E. F. *et al.* "Continuous Measurements of Total Gas Flow Rate from Stationary Sources," TRW System Group, EPA 650/12-75-020 (February 1975).

Zahek, A. *et al.* "Procedures for Measurements in Stratified Gases," Vol. I, Walden Research Division, EPA 650/12-74-086a (September 1974).

_ _

METHOD: STRATIFICATION INVESTIGATION "INNER 50% CONCEPT"

Advantages:

1. Based on extensive field testing of a broad representative group of boilers with characteristic duct-stack configurations.
2. Narrows the cross sectional area of concern to the inner 50%.

Disadvantages:

1. Does not provide a clear method of conducting a stratified investigation and selection of a valid number of representative sample points.

Remarks:

1. It has been concluded that sampling confined to traversing the inner 50% of the duct will allow a determination of the flue gas composition which is very close to the actual composition; single-point sampling is inappropriate for obtaining representative gas samples; recommended multipoint sampling within the inner 50% of the duct; each point represents an equal area.
2. Seven different fossil fuel-fired utility boilers were tested. SO_2, NO, CO_2 and O_2 concentration profiles as well as velocity and temperature profiles were obtained.
3. Applicable for Level 2 investigation.

References:

Crawford, A. R. *et al.* "Determination of the Magnitude of SO_2, NO, CO_2 and O_2 Stratification in the Ducting of Fossil Fuel-Fired Power Boilers," *Exxon Res. Eng.* (June 1975).

_ _

- -

METHOD: STRATIFICATION INVESTIGATION " TANGENTIAL" CONCEPT

Advantages:

1.   Follows directly from accepted (Method 5) procedures used for manual sampling.
2.   Flow-proportional sampling is not required. Same sample rate at each selected point.

Disadvantages:

1.   Still requires traversing investigation and then design and selection of a representative number of sample points.

Remarks:

1.   Applicable for Level 2 investigations.

References:

TRW Systems Group. "Technical Manual for Process Stream Volumetric Flow Measurement and Gas Sample Extraction Methodology," (November 1975).

- -

SAMPLE TYPE: PARTICULATE SEPARATION.

METHOD: IN-STACK, ON-LINE FILTERING

Advantages:

1.   Common method of removing particulate using a filter in the stack.
2.   Extensive experience for process monitoring in the larger process industries.
3.   SS or ceramic construction, resistant and reasonably inert relative to sample contamination.
4.   Strongly built to withstand vacuum sampling and high-pressure blow-back at high temperature.

5. Commercially available.
6. Possible installment in a heated chamber outside the stack to provide access.
7. Multiple in-stack units can be designed to provide simulated traverse sampling.

Disadvantages:

1. Sample is exposed to a more concentrated particulate cake on the filter surface that could result in increased losses by adsorption.
2. Requires additional plumbing and controls to provide blowback gases.
3. Prevention of ceramic filter breakage.
4. Requires backup instrument air compressors with associated cleaners for secured continuous operation.

Remarks:

1. Consists of a simple probe with an attached sintered filter and a compressed gas system for timed blowback to periodically clean the filter of particulate buildup. The entire gas sample flow is delivered to the analysis instrument making it an "on-line" filter.
2. Associated with extractive sampling systems.
3. Applicable for Level 1 and 2 investigations.

References:

EPA. "Monitoring Instrument for the Measurement of Sulfur Dioxide in Stationary Source Emissions," (February 1973).
McNulty, K. J. *et al.* "Investigation of Extractive Sampling Interface Parameters," ABCOR, Inc., Walden Research Division, Cambridge, Massachusetts (October 1974).
Milton Roy Company. "Sales Literature on In-Stack Filters," Hays Republic Division.

- -

SAMPLE TYPE: ACQUISITION DIVERTED STREAM.

METHOD: LARGE-VOLUME SUBSEQUENT SPLITTING

Advantages:

1. Convenient to install, check and maintain several connected sample lines.

2. Allows short sample lines to analysis instrument and direct access to the filter for checkout.
3. High-velocity fan needed to deliver the diverted stream provides excellent mixing and a uniform sample in the manifold.
4. Used in the process industry evaluation; thus, various filters, conditioners are commercially available to build the system.

Disadvantages:

1. A high–horsepower blower is required to insure that velocity in pipes and manifold is high enough to prevent particulate buildup.
2. Large transfer pipe, requiring external structural support and additional heating capability.
3. Subject to single-point sampling errors due to stratification in the stack.

Remarks:

1. Consists of a piping system for drawing a high-flow, high-temperature, diverted portion of the stack gas down to a manifold which can, in turn, be sampled in a more convenient manner. High–flow rate prevents particulate settling, and a return line redeposits the gas stream downstream from the diverted sample point.
2. Applicable for Level 2 investigations.

References:

EPA. "Monitoring Instrumentation for the Measurement of Sulfur Dioxide in Stationary Source Emissions," (February 1973).

Postinger, J. V. *et al.* "Instrumentation for Monitoring Specific Particulate Substances in Stationary Source Emissions," Monsanto Research Corp., EPA R2-73-252 (September 1973).

SAMPLE TYPE: CONDITIONING MOISTURE CONTROL.

METHOD: DILUTION

Advantages:

1. Prevents sulfur oxides from being exposed to moisture, which could result in loss of sample.
2. Usually results in simultaneous temperature control of sample gases.

3.   Commercially available probes.

Disadvantages:

1.   Requires large volumes of dry instrument gas which is relatively expensive.
2.   Requires monitoring of the dilution flow rate for concentration correction.
3.   More complicated system than condensation.

Remarks:

1.   Basically consists of adding, under controlled conditions, dry instrument gas that mixes with the sample stream resulting in a lower dew point and typically lower gas temperature.
2.   Applicable for Level 2 investigations.

References:

Mulik, J. D. *et al.* "An Analytical System Designed to Measure Multi-cycle Malodorous Compounds in Kraft Mills," presented at TAPPI Meeting, Boston, Massachusetts, 1971.

Research Appliance Company. "Automated Stack Monitor-Bulletin. 2412," Gibsonics, Pennsylvania.

– –

METHOD: CONDENSATION

Advantages:

1.   Commercially available and is typically part of a total system for stack-gas monitoring.
2.   Knock-out trap condenser is very simple and inexpensive.
3.   Usually results in simultaneous temperature control.

Disadvantages:

1.   Care and maintenance is required to install and operate a condenser to prevent uncontrolled condensation. The condenser must be tuned relative to the specific stack moisture conditions to insure adequate sample line heating and insulation prior to condensation. Wind chill factors should be considered in designing heating requirements.

Remarks:

1. Temperature limitations govern the type of condensation system selected.
2. The trap is located outside but near the stack.
3. Applicable for Level 2 investigations.

References:

Hankison Corporation. "Multi-Stream Refrigerated Gas Dryer-Sales Information," Canonsbury, Pennsylvania.

McNulty, K. J. *et al.* "Investigation of Extractive Sampling Interface Parameters," ABCOR, Inc., Walden Research Division, Cambridge, Massachusetts (October 1974).

_ _

METHOD: PERMEATION DISTILLATION

Advantages:

1. Commercially available.
2. Explosion-proof; no moving parts or electrical connections.
3. When purchased as a coil it can be mounted horizontally or vertically.
4. Applied to sampling systems connected with auto emissions monitoring.
5. Can be operated under high-pressure conditions (to 100 psig sample pressure).
6. Able to handle high-moisture conditions for typical sampling systems (up to 6 scfh).
7. No possibility of sample loss by solution in liquid condensate.

Disadvantages:

1. Requires a continuous purge of dry plant-instrument air or N_2.
2. Particulates and condensable oil vapors must be removed before drying.
3. Performance as a function of temperature has not been fully evaluated.
4. Additional heating of sample gas to prevent condensation.
5. Permeation, or sample loss by other very low concentration gases of interest may occur.
6. Particulate plugging seems to be the biggest problem.

Remarks:

1. Made up of a bundle of tubes that are permeable to water vapor and are enclosed in a shell. Permeated water vapors removed by either recycled dried gas product or other dry instrument air (N_2).
2. Applicable for Level 2 investigations.

References:

McNulty, K.. J. *et al.* "Investigation of Extractive Sampling Interface Parameters," ABCOR, Inc., Walden Research Division, Cambridge, Massachusetts (October 1974).
Perma Pure Dryer. "Sales Information and Instruction and Operation Manual," Oceanport, New Jersey.

_ _

METHOD: DESSICATION

Advantages:

1. Commercially available.
2. "Indicators" available to illustrate saturation, thus necessary disposal.
3. Highly efficient.

Disadvantages:

1. Saturates easily; requires frequent maintenance.
2. SO_2 loss on some dessicant materials has been reported.

Remarks:

1. Essentially a granular material that adsorbs moisture.
2. Applicable for Level 1 and 2 investigations.

References:

Pall Trinity Corp. "Sales Information for Dryers and Filters," Vauxhall, New Jersey.
Sittig, M. *Pollution Detection and Monitoring Handbook* (London: Noyes Data Corp., 1974).

_ _

- -

SAMPLE TYPE: ACID GAS SAMPLING SO$_2$.

METHOD: EPA METHOD 6

Advantages:

1.   Recognized nationally as a standard testing method.
2.   Adaptable to different types of sources with a reasonably wide range of gas concentrations, even with high–particulate concentrations.
3.   Method is reasonably explicit regarding exactly where in the process stream or duct cross section that sample will be collected, the volume required for a sample, the exact train configuration, and how the sample will be recovered and stored.
4.   Commercially available.

Disadvantages:

1.   Manual method.
2.   Single-point approach; could result in errors if gas stratification exists.
3.   Short test time could result in inaccurate measurements for processes that have short term or erratic changes in SO$_2$ concentration.

Remarks:

1.   Minimum sample time is 20 min and a minimum volume is 0.75 standard ft^3. Two samples constitute on run, collected at 1-hr intervals.
2.   Method yields SO$_2$ concentration which must be combined with stack gas flow rate to obtain SO$_2$ mass emission rate.
3.   Applicable for Level 1 and 2 investigations.

References:

"Standards of Performance for New Stationary Sources," *Federal Register* 36 (247) (December 23, 1971).

- -

— —

METHOD: EPA PERFORMANCE SPECIFICATION NO. 2

Advantages:

1.   Measurement system satisfactorily field-tested.
2.   System is compared in the stack with the EPA Reference Method (Method 6).
3.   Performance specifications are provided; innovative instrument procedure can be used instead of specifying any one manufacturer's type of instrument.   Extractive and nonextractive procedures are allowed.
4.   System analyzer could be installed to service more than one stack source.
5.   Nonextractive system can operate at relatively high stack temperatures since the sample does not have to be conditioned for transport and analysis.

Disadvantages:

1.   Relies on vague instructions.
2.   Weakest part is the sampling interface.
3.   Typically a single-point sampling system that could be seriously influenced by stratification and erratic profiles.

Remarks:

1.   Basically requires a sampling system capable of monitoring emission levels within 20% with a confidence level of 95%.
2.   System was developed to allow more or less continuous observations of the compliance of emissions with applicable New Source Performance Standards.
3.   Applicable for Level 2 investigations.

References:

"Emissions Monitoring, Standards for New Stationary Sources," *Federal Register* 40 (194) (October 6, 1975).
"Stationary Sources, Proposed Emission Monitoring and Performance Testing Requirements," *Federal Register* 39(177) (September 11, 1974).

— —

- -

METHOD: EPA METHOD 3 (INTEGRATED GRAB)

Advantages:

1. Simple, convenient, least time-consuming form of sampling is the grab sample.
2. Container is a flexible inert bag that can be transported easily to a field location for SO_2 analysis.
3. Sample period from 30 seconds to several hours, depending on bag size and sample rate.
4. Adoption to obtain a sufficiently large sample of gas for analysis of many species.

Disadvantages:

1. Method 3 pulls the sample through the pump and then discharges into a bag. The train should be evaluated to determine the effect of the condenser.
2. Depending on the gas of interest, sample loss due to reaction condensation can occur in the condenser.

Remarks:

1. Grab sample is primarily designed to obtain a sample for analysis of CO_2, excess air and dry molecular weight. An Orsat analyzer is used for concentration measurement to the nearest 0.1%.
2. Applicable for Level 1 investigations.

References:

"Standards of Performance for New Stationary Sources," *Federal Register* 36(247) (December 23, 1971).

- -

METHOD: DETECTION TUBE

Advantages:

1. Simple, inexpensive and most convenient system for obtaining approximate concentration of many gases.

2.   A direct reading of the gas concentration is obtained.
3.   Instrument is light and easily hand-held.

Disadvantages:

1.   The method is not particularly accurate in representing an average concentration of the gas in the stack because of the quick, direct readings.
2.   Tool is short and hand-held, making it difficult to obtain readings at large ports with negative pressure to get stack gas as opposed to dilution air.
3.   Stack concentrations may exceed the limit of the tubes.
4.   Detection tubes may be subjected to interferences; thus, judgment is needed in selecting tubes and evaluating results.

Remarks:

1.   Instrument made up of hand pump to obtain a sample of about 100 ml which is drawn through a small glass tube fitted with an indicating reaction chemical. The physical length of a discoloration indicates the specific gas concentration.
2.   Shelf life can be extended up to one or two years by storage at or below 25°C.
3.   Applicable for Level 1 investigations.

References:

Leithe, W. *The Analysis of Air Pollutants* (Ann Arbor, Michigan: Ann Arbor Science Publishers, Inc., 1971).
Linch, A. L. *Evaluation of Ambient Air Quality by Personnel Monitoring* (Cleveland: CRC Press, 1974).

— —

METHOD: HAYS-REPUBLIC ORSAT

Advantages:

1.   Contained in a basic Orsat package, thus providing one instrument for multiple gases.
2.   Inexpensive sampling system.

Disadvantages:

1.   Limited accuracy.

Remarks:

1. Consists of a tube (similar to detection tubes) where a sample is drawn through and the length of a stain indicates the gas concentration.
2. Reading to nearest 100 ppm.
3. Applicable for Level 1 investigations.

References:

Milton Roy Company. "Model 621A.38:30 Orsat with SO_2 Option," Sales Literature, Hayes-Republic Division, Michigan City, Indiana.

METHOD: SIMPLE DISPLACEMENT BOMB (ASTM 1605-60)

Advantages:

1. Simple, inexpensive technique for bringing a sample up to 3 liters in size.
2. Little site preparation necessary.
3. Reference method for carbon dioxide, carbon monoxide, methane, oxygen, hydrogen and nitrogen.

Disadvantages:

1. Small sample size.
2. Sample loss on walls of container.
3. Analysis for reactive gases must be performed quickly.
4. Cannot be stored in direct sunlight.

Remarks:

1. This technique includes air displacement and evacuated flask technique.
2. Can be used to sample high-pressure, high-temperature lines by use of a water-cooled heat exchanger.
3. Applicable for Level 1 investigations.

References:

ASTM. ASTM Standards, D1605.
McNulty, K. J. et al. "Investigation of Extractive Sampling Interface Parameters," ABCOR, Inc., Walden Research Division, Cambridge, Massachusetts (October 1974).

_ _

METHOD: EPA METHOD 12

Advantages:

1.   See EPA Performance Specification No. 2, pg. 100.

Disadvantages:

1.   See EPA Performance Specification No. 2, pg. 100.

Remarks:

1.   Method 12 is not exactly a method but rather a specification of measurement and instrument characteristics that are determined by comparing continuous measurements with manual results. The zero and calibration drift is based on percent of the "emission standard" instead of "span" as in No. 2.

References:

"Standards of Performance for New Stationary Sources," *Federal Register* 39(20) (October 16, 1974).

_ _

METHOD: CROSS-STACK *IN SITU* (TYPE I)

Advantages:

1.   Does not require extracting, conditioning and transporting the sample.
2.   Involves a unit placed on opposite sides of stack or duct cross section path that measures the integrated average gas concentration over that "path."
3.   Also measures CO, CO_2, NO.
4.   Solves gas stratification problems connected with single-point sampling systems by measuring over a "path" in the stack or duct cross section.
5.   Response time is short.
6.   Insensitive to alignment.

Disadvantages:

1.   Expensive.
2.   Located entirely on the stack, requiring service and maintenance on the stack.
3.   Stack gas temperature limitation of 600°F.
4.   Ambient temperature limit of 105°F.
5.   Adequate mixing is still necessary upstream of the path to insure a representative sample at the path.
6.   The stack-instrument interface "window" must be kept clean.

Remarks:

1.   Two types of cross-stack *in-situ* systems are available. Type I is a nondispersive, infrared instrument, often called a correlation spectrometer.

References:

Fyock, D. H. *et al.* "Test of the Environmental Research and Technology Stack Gas Analyzer at the Conemaugh Generating Station," paper 75-606, presented at Air Pollution Control Association Annual Meeting, 1975.
Stevens, R. K., and W. F. Herget, Eds. *Analytical Methods Applied to Air Pollution Measurements* (Ann Arbor, Michigan: Ann Arbor Science Publishers, Inc., 1974).

– –

METHOD: CROSS-STACK *IN SITU* (TYPE II)

Advantages:

1.   Same as (1) and (2) for Type I, pg. 104.
2.   Also measures CO, CO_2, NO, hydrocarbons and opacity.
3.   Short response time.
4.   Overcomes gas stratification problems, as in Type I, pg. 104.
5.   98% reliability.
6.   Compared to EPA Method 7, pg. 109.
7.   Excellent discrimination against other stack gases.

Disadvantages:

1.   Expensive.
2.   Located on the stack.
3.   Requires an in-stack alignment tube.
4.   Moisture interference when measuring NO.
5.   Committed to only one source.
6.   More complex calibration equipment required than for extractive analyzers.

Remarks:

1.   Type II is an adsorption spectroscopy system. The Type II absorption spectroscopy technique is implemented by dispersive gas cell correlation and derivative spectroscopy.
2.   Applicable for Level 2 investigations.

References:

Environmental Data Corporation. "Sales Literature," Minrovia, California. "Measuring Stack Gas Pollutants on the Spot," *Business Week* (June 1, 1974).

Stevens, R. K., and W. F. Herget, Eds. *Analytical Methods Applied to Air Pollution Measurements* (Ann Arbor, Michigan: Ann Arbor Science Publishers, Inc., 1974).

- -

METHOD: *IN SITU* **NONEXTRACTIVE**

Advantages:

1.   Commercially available.
2.   Meets EPA specifications for monitoring specifications for SO_2 and NO.
3.   Well-constructed for permanent installation.
4.   High degree of interference rejection, long operational period and high reliability.
5.   Simultaneous temperature measurement for automatic adjustment.
6.   Reported 3-19 months maintenance-free probes.
7.   Can be used as a portable instrument.
8.   Integral data system available.

Disadvantages:

1. Single-point, in-stack sampling; thus, concerned with stratification.
2. On-the-stack mounting requires on-the-stack service and maintenance.
3. A different probe utilizing an air-purge blowback is required for "wet, sticky, particulates" from sources like a scrubber outlet.
4. Response time could be a problem since diffusion through a filter is required and a rapid driving force, compared to the extractive system, is not utilized.

Remarks:

1. Not a method, but a specific, commercially available unit.
2. An *in situ* single-point system for SO_2 and NO using a second derivative spectroscopic technique. A porous filter is exposed to stack gases.
3. Primarily designed to meet EPA specifications to monitor NO/SO_2 only. A different unit will monitor O_2.
4. Applicable for Level 2 investigations.

References:

Lear Siegler, Inc. "Sales Literature—Model SM800," Environmental Technology Division (1976).

- -

SAMPLE TYPE: ACID GAS SAMPLING SO_3.

METHOD: CONTROLLED CONDENSATION (GOKSOYR-ROSS)

Advantages:

1. Well-researched, -applied and -evaluated technique of collecting SO_3.
2. Reasonably accurate with few interferences.
3. Flow rates of up to 20 l/min still result in good collection efficiency.

Disadvantages:

1. Care required in handling glassware and maintaining water bath temperature, otherwise SO_2 is oxidized, leading to high SO_3 values.
2. Not a method; just a writeup of a technique that can be used.

3.   Requires complex plumbing for field work.
4.   Not commercially available.

Remarks:

1.   Method consists of drawing a hot stack gas sample through a glass spiral in a controlled-temperature water bath that selectively passes SO_2 but condenses SO_3 on a glass frit.
2.   An EPA method has not been adopted for SO_3. EPA Model 8 for H_2SO_4 mist measures free mist and converts SO_3 to a mist to be combined with the free mist.
3.   Applicable for Level 2 investigations.

References:

Berger, A. W. *et al.* "Review and Statistical Analysis of Stack Sampling Procedures for the Sulfur and Nitrogen Oxides in Fossil Fuel Combustion," presented at Annual Meeting of the Air Pollution Control Association, 1970.

Driscoll, J. A. *et al. Improved Chemical Methods for Sampling and Analysis of Gaseous Pollutants from the Combustion of Fossil Fuels,* Vol. I, *Sulfur Oxides* Walden Research Crop. (June 1971).

Goksoyr, H., and K. Ross. "The Determination of Sulfur Trioxide in Flue Gases," *J. Inst. Fuel.* 35:177 (1962).

_ _

METHOD: LOS ANGELES COUNTY

Remarks:

1.   Applicable for Level 1 and 2 investigations.

References:

Byers, R. L., and B. B. Crocker. *Stack Samplings and Monitoring* (New York: American Institute of Chemical Engineers, 1972).

_ _

METHOD: SHELL-THORNTON CELL

Advantages:

1.   See Controlled Condensation Method, pg. 107.

Disadvantages:

1. See Controlled Condensation Method, pg. 107.

Remarks:

1. See Controlled Condensation Method, pg. 107.
2. Applicable for Level 1 and 2 investigations.

References:

Morrow, A. L. *et al.* "Sampling and Analyzing Air Pollution Sources," *Chem. Eng.* 99(2):94 (1972).
Shell Development Co. "Determination of Sulfur Dioxide and Sulfur Trioxide in Stack Gases," Emeryville Method, Series 4S16159 (1959).

— —

SAMPLE TYPE: ACID GAS SAMPLING NO_x.

METHOD: EPA METHOD 7

Advantages:

1. Recognized as a standard method.
2. Inexpensive and commercially available.
3. Collection location, exact train configuration and how sample will be recovered and stored, are explicit.

Disadvantages:

1. A manual method requiring field handling of fragile equipment precisely positioned and maintained in a frequently hostile environment.
2. Single-point sampling approach, which could result in errors if gas stratification exists.
3. Batch nature of this method and the short sample period could be influenced heavily by erratic processes or stratification leaks.
4. Precision and reproducibility are constant down to about 100 ppm NO_x. Method begins to get low results below this concentration.

Remarks:

1. Sample collection in an evacuated 2-liter flask containing an absorbing solution.

2.   Sample location same as for SO_2 (Method 6) sampling.
3.   Applicable for Level 1 and 2 investigations.

References:

"Standards of Performance for New Stationary Sources," *Federal Register* 36(247) (December 23, 1971).

———————————————————————————

METHOD: EPA SPECIFICATION NO. 2

Advantages:

1.   See SO_2, pg. 100; same advantages.   The comparable EPA method is Method 7.

Disadvantages:

1.   See SO_2. Same disadvantages (1 - 3).

Remarks:

1.   See SO_2.
2.   Applicable for Level 2 investigations.

References:

duPont Instruments.   "Model 400 Photometric Analyzer Product Information Sheet," Bulletin 5A.
"Emission Monitoring, Standards for New Stationary Sources," *Federal Register* 40(194) (October 6, 1975).
"Stationary Sources, Proposed Emission Monitoring and Performance Testing Requirements," *Federal Register* 39(177) (September 11, 1974).

———————————————————————————

METHOD: DETECTION TUBE

Remarks:

1.   See SO_2, pg. 101.
2.   Applicable for Level 1 investigations.

References:

Leithe, W. *The Analysis of Air Pollutants* (Ann Arbor, Michigan: Ann Arbor Science Publishers, Inc., 1971).

Linch, A. L. *Evaluation of Ambient Air Quality by Personnel Monitoring* (Cleveland: CRC Press, 1974).

METHOD: SIMPLE DISPLACEMENT BOMB

Disadvantages:

1. NO_2 very soluble in water.
2. Wall losses can be considerable (possible total loss of sample).
3. To reduce losses, nylon, EVA and PVC tubing or containers should not be used.

Remarks:

1. Applicable for Level 1 investigations.
2. See discussion under SO_2, pg. 103.

References:

ASTM. ASTM Standard, D1605.

McNulty, K. J. *et al.* "Investigation of Extractive Sampling Interface Parameters," ABCOR, Inc., Walden Research Division, Cambridge, Massachusetts (October 1974).

White, A., and L. M. Beddows. "The Choice of Sampling Tube Material in the Determination of Nitrogen Oxide Concentration in Products of Combustion," *Appl. Chem. Biotech.* 23:759-767 (October, 1973).

METHOD: ASTM D1608-60

Advantages:

1. Widely applicable concentration range (up to several thousand ppm).
2. Similar to EPA Method 7 (evacuated flask).
3. Supported by extensive field testing.
4. Simple sampling train; does not need extensively trained operators.

Disadvantages:

1. Requires handling glassware near the stack port.
2. Requires overnight absorption period prior to analysis.
3. Interferences from inorganic nitrates, nitrites and organic nitrogen.
4. SO_2 may consume part of absorbing solution leaving inadequate amount for NO_X.

Remarks:

1. Applicable for Level 1 and 2 investigations.

References:

ASTM. *Water; Atmospheric Analysis, Part 23*, ASTM Standard D1608-60, Philadelphia, Pennsylvania (1975).

_ _

METHOD: CROSS-STACK *IN SITU*

Disadvantages:

1. Measures NO only.

Remarks:

1. See Cross-Stack *In Situ* discussion under SO_2, pgs. 104 and 105.
2. Applicable for Level 2 investigations.

References:

Environmental Data Corporation. "Sales Literature," Minrovia, California. "Measuring Stack Gas Pollutants on the Spot," *Business Week* (June 1, 1974).

Stevens, R. K., and W. F. Herget, Eds. *Analytical Methods Applied to Air Pollution Measurements* (Ann Arbor, Michigan: Ann Arbor Science Publishers, Inc., 1974).

_ _

METHOD: *IN SITU* NONEXTRACTIVE

Disadvantages:

1. Measures NO only.

Remarks:

1. See *In-Situ*, Nonextractive Method discussion under SO_2, pg. 106.

References:

Tear Siegler, Inc. "Sales Literature—Model SM800," Environmental Technology Division (1976).

- -

SAMPLING FOR: ACID GAS SAMPLING H_2S.

METHOD: EPA METHOD 11

Advantages:

1. Simple sampling train with midget impingers.
2. Can be used on pressurized ducts.
3. Based on techniques extensively developed in refinery industry.

Disadvantages:

1. Caution: thoroughly mix cadmium hydroxide absorbing solution.
2. Standard disadvantages associated with midget bubbler wet chemical method.
3. Handling concentrated acids.

Remarks:

1. Analysis titration should be conducted at sampling location. Titration should never be done in direct sunlight.
2. Applicable for Level 1 and 2 investigations.

References:

"Air Programs: Standards of Performance for New Stationary Sources," *Federal Register* 39(47) (March 8, 1974).

- -

METHOD: DETECTION TUBE

Remarks:

1. See Detection Tubes discussion under SO_2, pg. 101.

2.   Applicable for Level 1 investigations.

References:

Leithe, W. *The Analysis of Air Pollutants* (Ann Arbor, Michigan: Ann
Arbor Science Publishers, Inc., 1971).
Linch, A. L. *Evaluation of Ambient Air Quality by Personnel Monitoring*
(Cleveland: CRC Press, 1974).

-- --

METHOD: EPA METHOD 3

Advantages:

1.   Uses Tedlar or Teflon bag.

Remarks:

1.   See Method 3 discussion under SO_2, pg. 101.
2.   Applicable for Level 1 and 2 investigations.

References:

"Standards of Performance for New Stationary Sources," *Federal Register*
36(247) (December 23, 1971).

-- --

METHOD: ASTM D2725-70

Advantages:

1.   Test Method for H_2S in natural gas.

Disadvantages:

1.   Questionable applicability above 50 ppm or 23 mg/m^3.

Remarks:

1.   Uses neutral $CdSO_4$ solution in place of zinc acetate-absorbing
solution.
2.   Methylene Blue Method.
3.   Applicable for Level 1 and 2 investigations.

References:

McCain, J. D. *et al.* "Field Measurements of Particle Size Distribution with Inertial Size Devices," Southern Research Institute EPA-650/120-73-035 (October 1973).

Tussey, R. S., Jr. *et al.* "Testing Services for the Coke Oven Charging Demonstration Program," Midwest Research Institute (March 1974).

- -

METHOD: SIMPLE DISPLACEMENT BOMB

Remarks:

1.   See Simple Displacement Bomb discussion under SO_2, pg. 103.

References:

ASTM. ASTM Standard, D1605.

McNulty, K. J. *et al.* "Investigation of Extractive Sampling Interface Parameters," ABCOR, Inc., Walden Research Division, Cambridge, Massachusetts (October 1974).

White, A., and L. M. Beddows. "The Choice of Sampling Tube Material In the Determination of Nitrogen Oxide Concentration in Products of Combustion," *J. Appl. Chem. Biotech.* 23:759-767 (October 1973).

- -

SAMPLING FOR: ACID GAS SAMPLING CO_2.

METHOD: EPA METHOD 3

Remarks:

1.   See Method 3 discussion under SO_2, pg. 101.

References:

"Standards of Performance for New Stationary Sources," *Federal Register* 36(247) (December 23, 1971).

- -

METHOD: EPA PERFORMANCE SPECIFICATION NO. 3

Advantages:

1. Reasonable specifications attainable.
2. First and only writeup of a continuous system.
3. Long-term unattended operation period.
4. Addresses sampling location and potential stratification.

Disadvantages:

1. Stratification does not address sampling interface problems.
2. Need for in-stack calibration has not been fully researched.

Remarks:

1. Not a method, but a set of instrument and sampling system performance specifications.
2. Applicable for Level 1 and 2 investigations.

References:

Environmental Protection Agency. "Performance Specifications for Stationary Source Monitoring Systems for Gases and Visible Emissions," EPA-650/2-74-013 (January 1974).
"Emissions Monitoring, Standards for New Stationary Sources," *Federal Register* 40(194) (October 6, 1975).

METHOD: DETECTION TUBE

Remarks:

1. See discussion of Detection Tubes under SO_2, pg. 101.
2. Applicable for Level 1 investigation.

References:

See references for Detection Tubes under SO_2.

— —

METHOD: SIMPLE DISPLACEMENT BOMB

Remarks:

1. See discussion of Simple Displacement Bomb under SO_2, pg. 103.
2. Basic Orsat Analysis.
3. Applicable for Level 1 investigation.

References:

See references for Simple Displacement Bomb under SO_2.

— —

METHOD: CROSS-STACK *IN SITU*

Remarks:

1. See discussion of Cross-Stack *In Situ* under SO_2, pg. 104.
2. Applicable for Level 2 investigations.

References:

See references for Cross-Stack *In Situ* under SO_2.

— —

SAMPLING FOR: ACID GAS SAMPLING HCl.

METHOD: IMPINGER TRAIN

Advantages:

1. Low to high concentration range.
2. Reasonable accuracy (± 10%).
3. Method using distilled H_2O in impingers has few interferences.

Disadvantages:

1. Requires handling glassware and wet chemicals in the field.
2. Silver salts interference.

3.   Possible interference from sulfate or chloride ions in particulates.

Remarks:

1.   Sample drawn through sodium hydroxide.
2.   Halide interferences in method.
3.   Applicable for Level 1 and 2 investigations.

References:

Cooper, H. B. H., Jr., and A. T. Resszno, Jr. "Source Testing for Air Pollution Control," *Environmental Science Services* (1971).
Jacobs, M. B. *The Analytical Toxicology of Industrial Inorganic Poisons* (New York: Interscience Publishers, Inc., 1967).

_ _

METHOD: DETECTION TUBE

Remarks:

1.   See Detection Tube under SO_2, pg. 101.
2.   Applicable for Level 1 investigations.
3.   Range from 2 ppm to 2% volume.

References:

See references under Detection Tube SO_2.

_ _

METHOD: SIMPLE DISPLACEMENT BOMB

Remarks:

1.   See Simple Displacement Bomb discussion under SO_2, pg. 103.
2.   Applicable for Level 1 and 2 investigations.

References:

See references on Simple Displacement Bomb under SO_2.

_ _

METHOD: EPA METHOD 3

Remarks:

1. See Method 3 discussion under SO_2, pg. 101.
2. Applicable for Level 1 and 2 investigations.

METHOD: CROSS-STACK *IN SITU*

Advantages:

1. Only method for continuous measurement of HCl.

Disadvantages:

1. Needs more refinement before it can be commercially available.

Remarks:

1. See Cross-Stack *In Situ* discussion under SO_2.
2. Applicable for Level 2 investigations.

References:

Stevens, R. K., and W. F. Herget, Eds. *Analytical Methods Applied to Air Pollution Measurements* (Ann Arbor, Michigan: Ann Arbor Science Publishers, Inc., 1974).

METHOD: INTERSOCIETY COMMITTEE (REFERENCE METHOD 201)

Advantages:

1. Proper writeup.
2. Specifies exact procedure accuracy and required apparatus.

Disadvantages:

1. Sulfate interferences (correctable).

Remarks:

1. Reference impinger method.
2. Applicable for Level 1 and 2 investigations.

References:

American Public Health Association, Intersociety Committee. "Methods of Air Sampling and Analysis" (1972).

- -

METHOD: ASTM D2036-72

Advantages:

1. Reference method by ASTM.
2. Experience in applying method to various situations.

Disadvantages:

1. Wet chemical method required use of highly toxic reagents.
2. Requires adoption of a "water" method for sampling.

Remarks:

1. Two impingers in series, each with 15 ml $0.5N$ KOH.
2. ASTM method for cyanides in water.
3. Applicable for Level 2 investigations.

References:

ASTM. ASTM Standard, D2009.
Tussey, R. S., Jr. *et al.* "Testing Services for the Coke Oven Charging Demonstration Program," Midwest Research Institute (March 1974).

- -

SAMPLING FOR: ACID GAS SAMPLING HCN.

METHOD: DETECTION TUBE

Remarks:

1. Range from low ppm to 2% by volume.

2. See discussion for Detection Tubes under SO_2, pg. 101.
3. Applicable for Level 1 investigations.

References:

See references for Detection Tubes under SO_2.

- -

METHOD: SIMPLE DISPLACEMENT BOMB

Remarks:

1, See SO_2 discussion for Simple Displacement Bomb, pg. 103.
2. Applicable for Level 1 investigations.

References:

See references for Simple Displacement Bomb under SO_2.

- -

METHOD: EPA METHOD 3

Remarks:

1. See Method 3 discussion under SO_2, pg. 101.
2. Applicable for Level 1 and 2 investigations.

References:

Same as Method 3 under SO_2.

- -

SAMPLING FOR: ACID GAS SAMPLING HF AND F.

METHOD: EPA METHOD 13A

Advantages:

1. Isokinetic sampling for particulate and gaseous total fluoride emissions using established Method 5-type train.
2. 3% relative standard deviation from replicate interlaboratory determination study with range of 39-350 mg/l.

3.   Stable sample.
4.   SS probe.
5.   Method 5-type train of operation.
6.   Automatically obtains necessary data to calculate fluoride emissions.

Disadvantages:

1.   Fluorocarbons such as Freons® not quantitatively collected.
2.   Chloride ion interference.
3.   See Method 5 disadvantages under Particulate Sampling, pg. 67.
4.   Sensitivity not determined.

Remarks:

1.   SPADNS Zirconium Lake Method.
2.   Fluoride determined instead of HF.
3.   Covers range from 0-1.4 g/ml fluoride.
4.   Filter inserted either prior to first impinger or after third.
5.   Whatman-type filter required.
6.   Method designed for testing Primary Aluminum plants.
7.   Applicable for Level 1 and 2 investigations.

References:

Arthur D. Little Co. "Development of Methods for Sampling and Analysis
   of Particulate and Gaseous Fluorides from Stationary Sources," NTIS
   ORDER #PB21 3313 (November 1972).
Dorsey, J. A. et al. "Source Sampling Technique for Particulate and
   Gaseous Fluorides," J. Air Poll. Control Assoc. 18(1):12-14 (1968).
"Standards of Performance for New Stationary Sources, Proposed Rule for
   Primary Aluminum Plants," Federal Register 39(206) (October 23, 1974).

— —

METHOD: EPA METHOD 13B

Advantages:

1.   See (1) and (5) under EPA Method 13A, pgs. 121-122.
2.   Accuracy reported to be 1-5% (electrode measurements).
3.   Routine sample removal and cleanup.

Disadvantages:

1.   See (1) and (4) under EPA Method 13A.

2. During lab analysis, Al in excess of 300 mg/l and silicon dioxide in excess of 300 mg/l will prevent complete recovery of fluoride.
3. Temperature change in sample will affect electrode response.
4. Glass probe required.
5. If probe becomes coated with particulate, gas reaction with glass probe may be incomplete.

Remarks:

1. Specific Ion Electrode Method.
2. Fluorides determined instead of HF.
3. Covers range of 0.02-2000 mg/l.
4. Same sample train as Method 13A.
5. Method designed for testing Primary Aluminum plants.
6. HF and F_2 react with heated glass probe to form gaseous silicon tetrafluoride which hydrolyzes in the water impingers to form fluosilicic acid and insoluble orthosilicic acid.
7. Applicable for Level 1 and 2 investigations.

References:

See references for EPA Method 13A under HF and F, pg. 122.

— —

METHOD: DETECTION TUBE

Remarks:

1. See Detection Tubes discussion under SO_2, pg. 101.
2. 1.5-15 ppm range.
3. Applicable for Level 1 investigations.

References:

See references under Detection Tube Method SO_2.

— —

METHOD: LOS ANGELES COUNTY

Remarks:

1. Caustic solution impinger train.
2. Applicable for Level 1 and 2 investigations.

References:

Los Angeles Air Pollution Control District. "Source Testing Manual" (November 1965).

_ _

METHOD: WET CHEMICAL

Advantages:

1.   Could be adapted for a wet chemical method of sampling.
2.   Applicable above 1 ppm.

Disadvantages:

1.   Requires field updating for current stack sampling in the presence of many other gases that could interfere (Aluminum sulfate and phosphate).

Remarks:

1.   Specific for HF.
2.   Spectrophoto analysis.
3.   Impinger train containing zirconium oxychloride octahydrate-salochrane cyrhine R solution.
4.   Applicable for Level 1 and 2 investigations.

References:

Marshall, B. S., and R. Wood. "A Simple Field Test for the Determination of Hydrogen Fluoride in Air," *Analyst* 93:821 (1968).
Ruch, W. E. *Quantitative Analysis of Gaseous Pollutants* (Ann Arbor, Michigan: Ann Arbor Science Publishers, Inc., 1970).

_ _

SAMPLING FOR: INORGANIC GAS SAMPLING—CO.

METHOD: EPA METHOD 10

Advantages:

1.   Can use silica gel to remove moisture (need to correct sample gas volume).

2. Grab sampling allowed.
3. Either continuous or manual methods can be used.

Disadvantages:

1. H_2O interference.
2. Does not address gas stratification in the continuous mode (weakness of single-point sampling).

Remarks:

1. Uses Ascarite to remove CO_2 prior to NDIR.
2. Uses condenser for moisture removal in continuous sampling.
3. Integrated grab sample technique or continuous sampling.
4. Analysis using NDIR.
5. Applicable for Level 1 and 2 investigations.

References:

Driscoll, J. N. *Flue Gas Monitoring Techniques: Manual Determination of Gaseous Pollutants* (Ann Arbor, Michigan: Ann Arbor Science Publishers, Inc., 1974).
"Air Programs: Standards of Performance for New Stationary Sources," *Federal Register* 39(47) (March 8, 1974).

- -

METHOD: DETECTION TUBE

Disadvantages:

1. Interference gases include CO_2, SO_2 and NO_2.

Remarks:

1. See Detection Tubes discussion under SO_2, pg. 101.
2. Applicable for Level 1 investigations.

References:

See Detection Tubes under SO_2.
Driscoll, J. N. *Flue Gas Monitoring Techniques: Manual Determination of Gaseous Pollutants* (Ann Arbor, Michigan: Ann Arbor Science Publishers, Inc., 1974).

- -

- -

METHOD: SIMPLE DISPLACEMENT BOMB

Remarks:

1.   See Simple Displacement Bomb discussion under SO_2, pg. 103.
2.   Applicable for Level 1 investigations.

References:

See references for Simple Displacement Bomb under SO_2.

- -

METHOD: REDUCTION METHODS

Remarks:

1.   This includes other wet chemical techniques and reduction techniques. CO has a very low solubility in aqueous solutions. Also, too low in organic solvents.
2.   Technique requires grab sampling as described in EPA Method 10.
3.   Applicable for Level 1 and 2 investigations.

References:

Driscoll, J. N. *Flue Gas Monitoring Techniques: Manual Determination of Gaseous Pollutants* (Ann Arbor, Michigan: Ann Arbor Science Publishers, Inc., 1974).

- -

METHOD: OXIDATION METHODS

Remarks:

1.   Method uses catalytic conversion of CO to CO_2.
2.   Applicable for Level 1 and 2 investigations.

References:

Driscoll, J. N. *Flue Gas Monitoring Techniques: Manual Determination of Gaseous Pollutants* (Ann Arbor, Michigan: Ann Arbor Science Publishers, Inc., 1974).

- -

- -

METHOD: COMPLEXATION METHODS

Advantages:

1.　Simple, reasonably quick measurement technique.

Disadvantages:

1.　Typical precision is reading (in %) ± 0.3.

Remarks:

1.　Orsat technique involving selective scrubbing by a liquid reagent.
2.　Applicable for Level 1 and 2 investigations.

References:

Driscoll, J. N. *Flue Gas Monitoring Techniques: Manual Determination of Gaseous Pollutants* (Ann Arbor, Michigan: Ann Arbor Science Publishers, Inc., 1974).

- -

METHOD: CROSS-STACK *IN SITU*

Remarks:

1.　See Cross-Stack *In Situ* discussion under SO_2, pg. 104.
2.　Applicable for Level 1 and 2 investigations.

References:

Fyock, D. H. *et al.* "Test of the Environmental Research and Technology Stack Gas Analyzer at the Conemaugh Generating Station," paper 75-606, presented at Air Pollution Control Association Annual Meeting, 1975.
Smith, F., W. Wagoner and A. C. Nelson, Jr. *Guidelines for Development of Q.A., Vol. III—Determination of Moisture in Stack Gases,* Research Triangle Institute, EPA-650/4-74-005-C (August 1974).

- -

— —

SAMPLING FOR: INORGANIC GAS SAMPLING–O_2.

METHOD: PERFORMANCE SPECIFICATION NO. 3

Advantages:

1.   Polaragraphic or paramagnetic may be used if they meet specifications.

Remarks:

1.   See CO_2 discussion, pg. 116.
2.   Applicable for Level 1 and 2 investigations.

References:

"Emission Monitoring, Standards for New Stationary Sources," *Federal Register* 40(194) (October 6, 1975).
EPA.   "Performance Specifications for Stationary Source Monitoring Systems for Gases and Visible Emissions," EPA-650/2-74-013 (January 1974).

— —

METHOD: DETECTION TUBES

Remarks:

1.   See Detection Tubes discussion under SO_2, pg. 101.
2.   Applicable for Level 1 and 2 investigations.

References:

See references for Detection Tubes under SO_2.

— —

METHOD: SIMPLE DISPLACEMENT BOMB

Remarks:

1.   See Simple Displacement Bomb discussion under SO_2.
2.   Applicable for Level 1 investigations.

References:

See references for Simple Displacement Bomb under SO_2.

- -

METHOD: EPA METHOD 3

Disadvantages:

1. Orsat analysis technique limited to 0.3% precision.

Remarks:

1. See Method 3 discussion under SO_2, pg. 101.
2. Applicable for Level 1 investigations.

References:

See reference for Method 3 under SO_2.

- -

METHOD: PARAMAGNETIC

Advantages:

1. An instrumental technique that has been used on combustion sources for many years.
2. Reasonably selective.

Disadvantages:

1. NO and NO_2 also magnetic but effect is probably on order of 0.2% oxygen by volume.

Remarks:

1. An instrumental technique that could be used to satisfy Performance Specification No. 3.
2. Relies on the magnetic susceptibility of oxygen.
3. Requires appropriate sample point selection and stratification investigation for emission measurements.
4. Applicable for Level 1 and 2 investigations.

References:

Cooper, J. C., H. E. Birdseye and R. Donnelly, Jr. "Cryogenic Separation of Methane from Other Hydrocarbons in Air," *Environ. Sci. and Technol.* 8(7):671-673 (July 1974).

_ _

METHOD: POLAROMETRIC

Advantages:

1. An instrumental technique that has been used on combustion sources for many years.
2. Good for low concentration ranges.
3. Less expensive and more rugged than Paramagnetic.
4. Accuracy of 0.2-0.5 vol % in 0-25% range.

Disadvantages:

1. Less accurate than paramagnetic.

Remarks:

1. Requires appropriate sample point selection and stratification investigation for emission measurements.
2. Applicable for Level 1 and 2 investigations.

References:

Leithe, W. *The Analysis of Air Pollutants* (Ann Arbor, Michigan: Ann Arbor Science Publishers, Inc., 1971).

_ _

SAMPLING FOR: NH_3.

METHOD: KJELDAHL METHOD (NESSLERIZATION)

Advantages:

1. Precision approximately ± 2%.
2. Primary method for sample testing for ammonia.
3. Gives both particulate and gaseous NH_3.

4. Can use commercially available midget impinger sampling systems.
5. Common sampling equipment.

Disadvantages:

1. Interference from certain amines and calcium in excess of 250 mg/l.
2. Problems with handling glassware and reagent chemicals in the field.
3. May need to add potassium permanganate to H_2SO_4 to eliminate H_2S and formaldehyde from reacting with Nessler reagent.

Remarks:

1. Standard impinger collection in $0.1N$ H_2SO_4.
2. The indicator solution should be added to the boric acid solution immediately prior to use, to insure distinct end-point.
3. A continuous ambient type monitor using this technique was developed and used by EPA.
4. Primarily for stack concentration ranges.
5. Applicable for Level 1 and 2 investigations.

References:

Cambi, F. "Sampling, Analysis and Instrumentation in the Field of Air Pollution," World Health Organization Monograph Series #46, Geneva (1961).
Leithe, W. *The Analysis of Air Pollutants* (Ann Arbor, Michigan: Ann Arbor Science Publishers, Inc., 1971).
Stern, A. C. *Air Pollution: V. II–Analysis, Monitoring and Surveying,* (New York: Academic Press, 1968).

— —

METHOD: INDOPHENOL BLUE TECHNIQUE

Advantages:

1. See Kjeldahl Method, pg. 130.
2. Advantage over Nessler reagent when determining low NH_3 concentrations in that H_2S does not interfere.
3. Detection limit and reproducibility correspond to Nessler reaction.

Disadvantages:

1. See Kjeldahl Method.
2. Potential interference from formaldehyde, nitrite, sulfite.

Remarks:

1. Collected the same way as Kjeldahl Method, then reacted with alkaline phenol and sodium hypochlorite and adsorbents measured on a colimeter.
2. Applicable for Level 1 and 2 investigations.

References:

Sittig, M. *Pollution Detection and Monitoring Handbook* (London: Noyes Data Corporation, 1974).

METHOD: DETECTION TUBE

Remarks:

1. From ppm to 10% volume range.
2. See Detection Tubes discussion under SO_2, pg. 101.
3. Applicable for Level 1 investigations.

References:

Leithe, W. *The Analysis of Air Pollutants* (Ann Arbor, Michigan: Ann Arbor Science Publishers, Inc., 1971).
Linch, A. L. *Evaluation of Ambient Air Quality by Personnel Monitoring* (Cleveland: CRC Press, 1974).

METHOD: SIMPLE DISPLACEMENT BOMB

Remarks:

1. See Simple Displacement Bomb discussion under SO_2, pg. 103.
2. Applicable for Level 1 investigations.

References:

See references for Simple Displacement Bomb under SO_2.

SAMPLING FOR: H_2O.

METHOD: EPA METHOD 4

Advantages:

1. Reasonably simple for accuracy obtained.
2. Very simple analysis (volume measurement and silica gel weighing).
3. Applicable to extreme moisture concentrations by changing sample time.
4. Inexpensive to build—or commercially available.
5. Extensively evaluated for quality assurance.

Disadvantages:

1. Requires handling wet chemical-type glassware.
2. Problems if gas stream is saturated and laden with moisture droplets that could penetrate the filter.
3. Note color of condensate. In some gas streams, other gases may also condense out, giving erroneous results.

Remarks:

1. Moisture is condensed and determined volumetrically.
2. Since this train also includes silica gel prior to the pump, this method is a combination of condensation and adsorption.
3. Condensation generally applies to less than 1% moisture.
4. Applicable for Level 1 and 2 investigations.

References:

Brenchley, D. L., C. D. Turley and R. F. Yarmac. *Industrial Source Sampling* (Ann Arbor, Michigan: Ann Arbor Science Publishers, Inc., 1974).

Cooper, H. B. H., Jr., and A. T. Resszno Jr. "Source Testing for Air Pollution Control," *Environmental Science Services* (1971).

"Standards of Performance for New Stationary Sources," *Federal Register* 36(247) (December 23, 1971).

METHOD: ADSORPTION

Advantages:

1. Simple, can test all volumes (large or small) depending on moisture contents.
2. Available and inexpensive-saturation indicators also available.
3. Used in industry for manual and automatic collection.

Disadvantages:

1. Requires care to avoid loss of condensate prior to absorbent if cooling gas required.
2. May also collect some organics.
3. 300°F limitation.

Remarks:

1. Method involves drawing a volume of gas through a tube of adsorbent like silica gel and weigh for collected moisture.
2. Essentially a dessication technique.
3. Removes particulates before absorption.
4. Applicable for Level 1 and 2 investigations.

References:

See references for EPA Method 4 under H_2O, pg. 133.

- -

METHOD: WET AND DRY BULB TEMPERATURE

Advantages:

1. Simple, manual technique.
2. Commercially available.
3. Acceptable technique below 212°F.

Disadvantages:

1. Problems with judging wet bulb temperature with confidence that equilibrium has been reached. (Above 212°F the wick dries out.)
2. After a period of buildup, particulates on wet bulb can cause errors.
3. Limited to 212°F.

Remarks:

1. Consists of two-temperature measurements.
2. Needs psychometric charts.
3. Applicable for Level 1 and 2 investigations.

References:

Cooper, H. B. H., Jr., and A. T. Resszno, Jr. "Source Testing for Air Pollution Control," *Environmental Science Services* (1971).

————————————————————————————

METHOD: DETECTION TUBE

Remarks:

1. See Detection Tubes under SO_2, pg. 101.
2. Applicable for Level 1 and 2 investigations.

References:

See references for Detection Tubes under SO_2.

————————————————————————————

METHOD: NDIR

Advantages:

1. Good technique for continuously monitoring moisture in stack gas.

Disadvantages:

1. Concentration range is limited to 0-5%.
2. Usually applied to ppm range concentrations.

Remarks:

1. Typically, the stack temperature is too high to maintain in a heat-traced sample line-up to an NDIR to get continuous readings prior to any moisture condensing out.
2. Applicable for Level 2 investigations.

References:

Arthur D. Little Co. "Development of Methods for Sampling and Analysis of Particulate and Gaseous Fluorides from Stationary Sources," NTIS Order #PB13313 (November 1972).

_ _

METHOD: EPA METHOD 3

Disadvantages:

1. Difficult determination in presence of H_2S, CO_2, CS_2.

Remarks:

1. COS is odorless and highly toxic. See Method 3 under SO_2, pg. 101.
2. Gas chromatographic analysis.
3. Applicable for Level 1 and 2 investigations.

References:

See Method 3 under SO_2.
Leithe, W. *The Analysis of Air Pollutants* (Ann Arbor, Michigan: Ann Arbor Science Publishers, Inc., 1971).

_ _

METHOD: CRYOGENIC ABSORBING

Advantages:

1. Probable complete collection of COS at cryogenic temperatures.

Disadvantages:

1. Specified technique familiar to most GC operators. The sampler must complete the train design.

2.   Problems in getting and operating a field cryogenic sampling operation at the stack.

Remarks:

1.   Applicable for Level 1 and 2 investigations.

References:

Leithe, W. *The Analysis of Air Pollutants* (Ann Arbor, Michigan: Ann Arbor Science Publishers, Inc., 1971).

METHOD: SIMPLE DISPLACEMENT BOMB

Remarks:

1.   See Simple Displacement Bomb under SO_2, pg. 103.
2.   Applicable for Level 1 investigations.

References:

See references for Simple Displacement Bomb under SO_2.

SAMPLING FOR: CS_2.

METHOD: EPA METHOD 3

Remarks:

1.   See Method 3 under SO_2, pg. 101.
2.   See COS Above.
3.   Applicable for Level 1 and 2 investigations.

METHOD: SIMPLE DISPLACEMENT BOMB

Remarks:

1.   See Simple Displacement Bomb under SO_2, pg. 103.

2.   Applicable to concentrations up to 60 ppm.
3.   Applicable for Level 1 investigations.

References:

See references for Simple Displacement Bomb under SO_2.

_ _

METHOD: DETECTION TUBE

Remarks:

1.   See Discussion for Detection Tube under SO_3.

References:

See references for Detection Tube under SO_3.

_ _

METHOD: WET CHEMICAL

Advantages:

1.   Method can easily be incorporated in a multigas sample train.
2.   Allows determination down to a few ppm. Impinger train pre-scrubs to remove H_2S.
3.   Range of another method is about 5-50 ppm but requires passing gas through lead acetate paper to remove H_2S prior to collection.
4.   Another method allows determination of H_2S, CS_2 and methyl mercaptan in one simple train.

Disadvantages:

1.   Impinger trains are difficult to handle and operate in the field.
2.   Care to select a train to address H_2S interference.
3.   Method may not be sensitive for ambient concentration levels of 2-3 ppm.

Remarks:

1. CS_2 is not very soluble in aqueous solutions but is soluble in many organic solvents.

2. Problem of single-point sampling could be overcome by conducting a gas stratification investigation.
3. Applicable for Level 1 and 2 investigations.

References:

Driscoll, J. N. *Flue Gas Monitoring Techniques: Manual Determination of Gaseous Pollutants* (Ann Arbor, Michigan: Ann Arbor Science Publishers, Inc., 1974).
Jacobs, M. B. *The Analytical Toxicology of Industrial Inorganic Poisons* (New York: Interscience Publishers, Inc., 1967).

-- --

SAMPLING FOR: ORGANIC GASES.

METHOD: SASS

Advantages:

1. Sampling system isokinetically collects particulates and gaseous organics simultaneously.
2. Train designed to collect both volatile and nonvolatile organics simultaneously.
3. System has high sample rate and shorter sampling period.
4. Ability of train to simultaneously collect any pollutants.
5. Describes adsorption trap system for a full range of typical stack temperatures.

Disadvantages:

1. Low-molecular-weight materials may penetrate the absorption trap.
2. System has not been field tested and some parts are not commercially available.
3. A.D. Little quartz filters are recommended but may not be available.

Remarks:

1. Teflon coating for filter holder may not be used.
2. Requires water-cooled probe for high-temperature stacks.

References:

Acurex Corp. "Source Assessment Sample Preliminary Information," Sales Document, Aerotherm Division.

Jones, P. W. *et al.* "Technical Manual for Analysis of Organic Materials in Process Streams," Battelle Columbus Laboratories (January 1976).
Monsanto Research Corp. "Technical Manual for Process Sampling Strategies for Organic Materials," (January 1976).

_ _

METHOD: HOT INTEGRATED GRAB

Advantages:

1.   Confidence in collection of more volatile organics.
2.   Easy to obtain a sample by operating a one-bag system.

Disadvantages:

1.   Cumbersone field instrument, primarily designed for special situation sampling.
2.   Possible organic material loss mechanisms include collection on quartz wool and condensation on bag walls.

Remarks:

1.   System consists of a heated particulate filter (quartz wool) followed by a short heat-traced Teflon line leading to a four-compartment bag.
2.   System maintained at 300°F during transport to the field lab and manifold sample for extraction. Only used for TCH, CO and CH_4 in the field.
3.   Applicable for Level 1 and 2 investigations.

References:

Tussey, R. S., Jr. *et al.* "Testing Services for the Coke Oven Charging Demonstration Program," Midwest Research Institute (March 1974).

_ _

METHOD: INTEGRATED GRAB SAMPLING

Advantages:

1.   Eliminates concern for loss of low-molecular-weight materials.

2. On-site GC can be used for quick and convenient field analysis.

Disadvantages:

1. Reactions may occur on walls or in condenser.
2. Bag leakage.

Remarks:

1. EPA Method 3 sample train.
2. Appropriate method for certain classes of organics, depending on volatility.

References:

"Standards of Performance for New Stationary Sources," *Federal Register* 36(247) (December 23, 1971).

METHOD: SOLVENT SCRUBBING

Advantages:

1. Can be combined with a Method 5 particulate train to simultaneously collect volatile and nonvolatile organics.
2. Organic material is extracted from the absorbing solution with a suitable solvent.

Disadvantages:

1. Solvent selection may be difficult.
2. Subsequent evaporation of solvent results in sample concentration.
3. Difficult to collect low-boiling-point materials.

Remarks:

1. Not a highly refined method and thus not generally recommended.
2. Applicable for Level 1 and 2 investigations.

References:

Monsanto Research Corp. "Technical Manual for Process Sampling Strategies for Organic Materials" (January 1976).

_____

METHOD: CONDENSATION

Advantages:

1.   Simple technique unless trying to collect specific species using complex temperature-programming procedures.
2.   Used for both ambient and source emission monitoring.
3.   Commonly applied technique.
4.   A wide range of organics may be collected.

Disadvantages:

1.   Collection efficiencies vary with substance being collected.
2.   Condensed water may plug trap and react with absorbed organics.
3.   Cryogenic trapping requires special equipment.

Remarks:

1.   Not a recommended technique.
2.   Applicable for Level 1 and 2 investigations.

References:

Monsanto Research Corp.   "Technical Manual for Process Sampling Strategies for Organic Materials" (January 1976).

_____

METHOD: POROUS POLYMER ADSORBENTS

Advantages:

1.   Water vapor is not strongly adsorbed. High humidity streams may be sampled.
2.   A variety of adsorbent materials are available—selective sampling is possible.
3.   Porous polymers can be used up to temperatures of 300°C.

Disadvantages:

1.   Needs more field testing.
2.   Some adsorbents are expensive.

3. Usually low sample rate due to pressure drop through adsorber.
4. Higher-molecular-weight compounds are more readily adsorbed than low-molecular-weight compounds.
5. Solvent extraction is required.

Remarks:

1. Potentially the best method for collection (and concentration) of organics.
2. A porous polymer adsorbent is specified as part of the SASS Train.
3. Applicable for Level 1 and 2 investigations.

References:

Monsanto Research Corp. "Technical Manual for Process Sampling Strategies for Organic Materials" (January 1976).

– –– –– – – – –

METHOD: CHEMICAL SUBSTRATE ADSORPTION

Advantages:

1. Activated charcoal adsorbs a broad range of organics.
2. Technique is used by NIOSH.
3. Reasonably simple technique.

Disadvantages:

1. Sample recovery may be incomplete.
2. Charcoal may act as a catalyst.
3. Desorption (by heating) may affect chemical changes (*e.g.*, thermally allowed rearrangements). (Pyrolysis may occur.)

Remarks:

1. Silica gel and activated carbon are commonly used.
2. Applicable for Level 1 and 2 investigations.

References:

Monsanto Research Corp. "Technical Manual for Process Sampling Strategies for Organic Materials" (January 1976).

– –

METHOD: DETECTION TUBE

Advantages:

1.   Inexpensive and commonly used.

Disadvantages:

1.   Interferences are common—detection tubes lack sensitivity.
2.   Detection tubes are available only for a limited number of specific compounds.

Remarks:

1.   See Detection Tubes under SO_2, pg. 101.
2.   Applicable for Level 1 and 2 investigations.

References:

See references for Detection Tubes under SO_2.

METHOD: BATTELLE ADSORBENT

Advantages:

1.   Method 5 train design is used (isokinetic sampling).
2.   Primarily designed to sample Polycyclic Organic Material (POM).
3.   The adsorbent trap is designed to be an integral part of a Soxhlet extraction apparatus.
4.   Validation studies have been conducted.

Disadvantages:

1.   Requires controlled-temperature water bath for absorber.
2.   Some water vapor is also condensed in cooler.
3.   Needs additional field testing.
4.   Aqueous SO_2 may react with and thus alter the collected material.

Remarks:

1.   Method 5-type train; the absorber is placed after the filter and before the impingers.

2.   Applicable for Level 1 and 2 investigations.

References:

Monsanto Research Corp. "Technical Manual for Process Sampling Strategies for Organic Materials," (January 1976).

— —

METHOD: KAISER TUBES

Advantages:

1.   Efficient collection of low-molecular-weight materials is possible.
2.   Kaiser Tube is reasonably simple.

Disadvantages:

1.   System operates at -160°C; cryogenic equipment is needed.
2.   Needs extensive field testing.
3.   Difficult to collect samples isokinetically (combined system).

Remarks:

1.   Consists of a polymer-packed tube cooled to -160°C; designed to sample low-molecular-weight materials.
2.   Applicable for Level 1 and 2 investigations.

References:

Monsanto Research Corp. "Technical Manual for Process Sampling Strategies for Organic Materials" (January 1976).

— —

CHAPTER 6

AMBIENT SAMPLING

To implement the desired air sampling program, the program initiator must be aware of the available instrumentation and its capabilities. Before a decision regarding area of interest, sampling locations, frequency of sampling, or sampling duration can be made, the pollutant to be measured and the measurement technique must be specified.

Previous sections have outlined the sampling and analysis procedures best suited to the various pollutants and sources. In this section, specific analytical and sampling techniques for particular pollutants for ambient testing are supplied.

Methods for atmospheric analysis for particulates, sulfur dioxide and sulfates, carbon monoxide, and photochemical oxidants and their precursors (*i.e.,* hydrocarbons and nitrogen oxides) are presented.

- -

SAMPLING FOR: PARTICULATES, AIRBORNE (FINE).

METHOD SUMMARY:

Benzene is used to extract collected particulate. Then the benzene is evaporated, leaving a residue. This residue weight and volume are used to determine the concentration of benzene-soluble fractions of airborne particulates.

Analytical Procedure:

Benzene soluble fraction—gravimetric and volumetric.

147

Distinctive Characteristics:

1. Range: lower limits 0.4 $\mu g/m^3$ per 8000 m^3 air sample to 1.5 $\mu g/m^3$ per 2000 m^3 air sample.
2. Sensitivity: at least 3 mg of residue must be generated.
3. Interferences: impurities in the benzene.
4. Precautions: benzene is toxic and flammable and can form explosive mixtures with air.

References:

Clements, J. B. "Extraction of Airborne Particulates with Benzene, Ca-1," informal compilation of analytical methods, EPA, National Environmental Research Center, Research Triangle Park, N.C. (October 5, 1972).

_ _

SAMPLING FOR: PARTICULATES, WINDBLOWN (COARSE).

METHOD SUMMARY:

Particulates carried by wind are collected on adhesive paper. The number of particulates per square millimeter (p/mm^2) is estimated by visual comparison with reference standards. Particulates are examined to determine their origins. Analyses are made for eight major wind directions.

Analytical Procedure:

Sticky paper—visual.

Distinctive Characteristics:

1. Range: 2-100 p/mm^2.
2. Sensitivity: 1-15 p/mm^2.
3. Interferences: colored particulates interfere with visual comparison.
4. Precision: standard deviation of ± 1.5.
5. Stability: after treatment with clear enamel, samples are stable for three months.

References:

Clements, J. B. "Windblown Particulates: Sticky Paper," from an informal compilation of analytical methods, EPA, National Environmental Research Center, Research Triangle Park, N.C. (October 5, 1972).

_ _

SAMPLING FOR: SUSPENDED PARTICULATES.

METHOD SUMMARY:

Blower draws air into a covered housing and through a filter that allows particles of less than 100 μm to pass. Samples are collected on glass filters. The mass concentration of particles is determined by measuring the particles collected and the volume of air sample.

Analytical Procedure:

Gravimetric and volumetric–high volume.

Distinctive Characteristics:

1. Range: average flow rate of 170 m^3/min for 24 hr. Adequate sample will be obtained. Low concentration 1 μg/m^3 6-8 hr adequate for high particulate levels.
2. Sensitivity: weight to nearest milligram, air flow to 0.03 m^3/min, time to nearest 8 minus, mass concentration to nearest mg/m^3.
3. Interferences: oily particulates, fog, humidity and glass filters.
4. Accuracy: ± 50% of average concentration.
5. Precision: standard deviation is 3.0%-3.7%.

References:

Federal Register 36(84) (April 30, 1971).

SAMPLING FOR: PARTICULATE ACIDS.

METHOD SUMMARY:

The collected acid aerosol on a filter is dissolved in distilled water by maceration of the filter. Sodium hydroxide is added, and the solution is back titrated with sulfuric acid to pH 7.

Analytical Procedure:

Titrimetric

Distinctive Characteristics:

1.   Range: 0.5-50 $\mu g/m^3$.
2.   Sensitivity: 0.5 $\mu g/m^3$, expressed as sulfuric acid.
3.   Interferences: bases in particulates cause neutralization of the acids.

References:

Clements, J. B.   "Determination of Particulate Acids," from an informal compilation of analytical methods, EPA, National Environmental Research Center, Research Triangle Park, N.C. (October 5, 1972).

- -

SAMPLING FOR: PARTICLE SIZING.

METHOD SUMMARY:

Dust sample is fed into center of a centrifuge with a throttled air supply directed to oppose the outward movement of the particles. Larger particles are retarded by the air stream and drop into a chamber while the smaller particles proceed to the rim of the centrifuge. The sample layer particles are weighed and recorded as percent of loss of weight. Procedure is repeated with several throttles to provide various size groupings.

Analytical Procedure:

BAHCO microparticle classifier—gravimetric.

Distinctive Characteristics:

1.   Range: eight size groupings obtainable with data expressed as weight %. For particles with specific gravity of 2.75, sizes from 0.95-26.8 μ diameter can be measured.
2.   Interferences: agglomeration during feed operation. Density of material can limit number of cuts obtainable.
3.   Precision: instrument useful only for at-weight-% distribution.

References:

Clements, J. B.   "Particle Sizing BAHCO Microparticle Classifier; D-8B," from an informal compilation of analytical methods, EPA, National Environmental Research Center, Research Triangle Park, N.C. (October 5, 1972).

- -

_ _

METHOD SUMMARY:

Electronic scaling and counting is utilized for particulate distribution. Using a dilute suspension of the particles in an electrolyte, the particles are passed through an orifice. Particle size changes effect changes in electrical resistance across the orifice, which are recorded as voltage differentials.

Analytical Procedure:

Coulter counter; electronic scaling.

Distinctive Characteristics:

1. Range: through the use of several apertures ranging in size from 10 μ to 400 μ, particle size from 0.2 μ-250 μ may be analyzed.
2. Interferences: electrical equipment; coagulation of particles after suspension in conducting fluid.
3. Precision: vague.

References:

Clements, J. B. "Particle Sizing—Coulter Counter, D-8-A," from an informal compilation of analytical methods, EPA, National Environmental Research Center, Research Triangle Park, N.C. (October 5, 1972).

_ _

SAMPLING FOR: HYDROCARBONS.

METHOD SUMMARY:

Measured air is delivered semicontinuously to a hydrogen flame ionization detector to measure it total hydrocarbon content (THC). Methane content is subtracted from THC.

Analytical Procedure:

Gas chromatography.

Distinctive Characteristics:

1. Range: THC range is 0-13.1 mg/m^3 (0-20 ppm) carbon (as CH_4). CH_4 range is 0-6.55 mg/m^3. Lower Ranges: THC, 0-1.31 mg/m^3 carbon as (CH_4) and CH_4 range to 1.31 mg/m.
2. Sensitivity: THC is 0.65 mg/m^3 carbon; methane is 0.033 mg/m^3. Lower sensitivity is 0.016 mg/m^3.
2. Accuracy: 1% in higher concentrations and 2% in lower.
3. Precision: ± 0.5% of scale in higher ranges.
4. Stability: varies, and room temperature changes should be taken into account.

References:

Federal Register 36(84) (April 30, 1971).

- -

SAMPLING FOR: NITROGEN DIOXIDE.

METHOD SUMMARY:

Nitrite diazotizes sulfanilamide quantitatively in water solution. A reddish-purple azo dye is produced by coupling the diazotized sulfanilamide with N-napthyl-ethylene diamine dihydrochloride at pH 2.0-2.5; determined colorimetrically.

Analytical Procedure:

Automated, colorimetric.

Distinctive Characteristics:

1. Range: 0.1-2.0 μg NO_2/ml.
2. Sensitivity: 0.02 μg/ml.
3. Precision: standard deviation is ± 0.04 μg NO_2/ml.
4. Measurement Time: 60 samples analyzed/hr. Holdup time for manifold is 8 min.

References:

Morgan, G. B., E. C. Tabor, C. Golden and H. Clements. "Automated Laboratory Procedures for the Analysis of Air Pollutants," *Automation in Analytical Chemistry,* New York, October 19, 1966.

Morgan, G. B., C. Golden and E. C. Tabor. "New and Improved Procedures for Gas Sampling and Analysis in the National Air Sampling Network," presented at the Technicon symposium entitled *Automation in Analytical Chemistry*, New York, October 19, 1966.

SAMPLING FOR: NITROGEN DIOXIDE, NITRIC OXIDE AND OXIDES OF NITROGEN.

METHOD SUMMARY:

Air samples are drawn directly into the analyzer to establish a nitric oxide response; then the air is switched through the converter where the NO_2 is converted to NO. Detector then measures total oxides of NO_x signal. The NO_2 present can be determined by subtracting the NO signal from the NO_x.

Analytical Procedure:

Automated chemiluminscent reaction.

Distinctive Characteristics:

1. Range: 0-12.3 $\mu g/m^3$ and 0-1226 x 10^3 $\mu g/m^3$ (0-1000 ppm).
2. Interefernces: NH_3 at temperatures greater than 250°C.
3. Accuracy: ± 2%.
4. Precision: ± 1.0% for all measurement ranges.
5. Stability: depends on proper pressure in reaction chamber, and cooling of phototube to proper constant cooling temperatures.
6. Measurement time: less than 1 min.

References:

O'Keefe, A. "Nitrogen Dioxide, Nitric Oxide, and Oxides of Nitrogen," from an informal compilation of analytical methods, EPA, National Environmental Research Center, Research Triangle Park, N.C.

-- -- -- -- -- -- -- -- -- -- -- -- -- -- -- -- -- -- -- --

SAMPLING FOR: PHOTOCHEMICAL OXIDANTS (OZONE).

METHOD SUMMARY:

Ambient air and ethylene are delivered to a mixing zone, where the ozone in the air reacts with the ethylene to emit light which is detected by a photomultiplier tube. The photocurrent is amplified and read directly or displayed on recorder.

Analytical Procedure:

Chemiluminescence; photometry.

Distinctive Characteristics:

1. Range: 9.8 μg O_3/m^3 - 1960 μg O_3/m^3.
2. Sensitivity: 9.8 μg O_3/m^3 (0.005 ppm ozone).
3. Accuracy: ± 7%.
4. Precision: average deviation does not exceed 5% of the mean of measurement.

References:

Federal Register 36(84) (April 30, 1971).

-- -- -- -- -- -- -- -- -- -- -- -- -- -- -- -- -- -- --

SAMPLING FOR: SULFATE.

METHOD SUMMARY:

Air drawn through filter. An aliquot of the residue on the filter is extracted with distilled water. The water-soluble sulfate is reacted with a reagent that contains equivalent amounts of barium chloride and methylthymol blue while pH is maintained at 2.8. pH is raised to 12.4 with potassium hydroxide and the unreacted barium forms a chelate with the dye. The excess dye becomes yellow and is determined colorimetrically.

Analytical Procedure:

Automated, colorimetric.

Distinctive Characteristics:

1. Range of applicability: 1.0-150 μg $SO_4^=$/ml or 0.3-45 μg $SO_4^=$/m³ of air.
2. Sensitivity: ± 1.0 μg SO_4/ml.
3. Interferences: heavy metals, particulates, oxidizing substances.
4. Precautions: avoid oxidation.
5. Precision: standard deviation is 0.3 μg $SO_4^=$/ml for samples containing 5.0 μg $SO_4^=$/ml.

References:

Clements, J. B. "Determination of Sulfate: Automated Method, M-15," from an informal compilation of analytical methods, EPA, National Environmental Research Center, Research Triangle Park, N.C. (October 5, 1972).

Morgan, G. B., E. C. Tabor, C. Golden and H. Clements. "Automated Laboratory Procedures for the Analysis of Air Pollutants," presented at the Technicon symposium entitled, *Automation in Analytical Chemistry*, New York, October 19, 1966.

- -

METHOD SUMMARY:

Sulfate is extracted from glass filter by refluxing with water, followed by dilution or concentration. An aqueous extract of the sample is treated with barium chloride in the presence of Sulfa Ver, a stabilizing agent. Barium sulfate crystals of uniform size are formed. The absorbance of the barium sulfate suspension is measured by a spectrophotometer or a filter photometer.

Analytical Procedure:

Sulfa Ver, spectrophotometric or filter photometric.

Distinctive Characteristics:

1. Range of applicability: 0.1-1.0 μg/m³.
2. Interferences: ionic strength of other ions, changes in pH and temperature, concentrations of the reagent.
3. Accuracy: based on recovery of standards is ± 11.2%.
4. Stability: maintained under following conditions: temperature variation not more than ± 5°C; Sulfer Ver added should be 0.25 ± 0.1 g;

pH variation not more than one unit; specific conductance of the aqueous extract variation not more than ± 300 μohms/cm.

References:

Clements, J. B. "Determination of Sulfate–Sulfa Ver Method, M-18," from informal compilation of analytical methods, EPA, National Environmental Research Center, Research Triangle Park, N.C. (October 5, 1972).

METHOD SUMMARY:

Extracts of sulfate in particulate matter are treated with barium chloride to form barium sulfate. The turbidity of the barium sulfate is a measure of the sulfate content. The absorbance of the sulfate suspension is measured by spectropyotometer or filter photometer, or, in the case of very low concentration, a nephelometer may be used.

Analytical Procedure:

Turbidimetric barium sulfate, photometric, filter-photometric or nephelometric.

Distinctive Characteristics:

1.   Range of applicability: 0.6-18 $MgSO_4^{=}/m^3$ of air.
2.   Interferences: size of sulfate particles, sulfate concentration, pH, ionic strength, or large temperature variation can affect the measurement.
3.   Accuracy and precision: the coefficient of variation is 11%.

References:

Clements, J. B. "Determination of Sulfate–Turbidimetric Barium Sulfate Method," from an informal compilation of analytical method, EPA, National Environmental Research Center, Research Triangle, Park, N.C. (October 5, 1972).

SAMPLING FOR: SULFUR DIOXIDE.

METHOD SUMMARY:

Sulfur dioxide in samples reacts with tetrachloromercurate. Acid-bleached paarosoniline and formaldehyde react with the dichlorosulfite-mercurate formed to give a red-purple complex of intensity proportional to the original concentration of sulfur dioxide. A Technicon Auto Analyzer is used.

Analytical Procedure:

Automated, colorimetric.

Distinctive Characteristics:

1. Range of applicability: concentrations up to 2.6 μg of SO_2/ml.
2. Precision: relative standard deviation is 1.0 ± 0.05 μg of SO_2/ml.
3. Measurement time: retention time in flow system is approximately 12 min. Rate of 60 samples/hr.

References:

Morgan, B. G., E. C. Tabor, C. Golden and H. Clements. "Automated Laboratory Procedures for the Analysis of Air Pollutants," presented at the Technicon symposium entitled, *Automation in Analytical Chemistry,* New York, October 19, 1966.

Morgan, G. B., C. Golden and E. C. Tabor. "New and Improved Procedures for Gas Sampling and Analysis in the Rational Air Sampling Network," presented at the Technicon Symposium entitled, *Automation in Analytical Chemistry,* New York, October 19, 1966.

METHOD SUMMARY:

Ambient air is drawn through a coulombic titration cell which contains a solution of a buffered halide. Sulfur dioxide present in the air sample reacts with the halide causing a decrease in current flow through the cell. The current from the reference electrode needed to return the cell to a steady state is directly related to the concentration of sulfur dioxide present. This is just a research method.

Analytical Procedure:

Automated, coulometric–titrimetric.

Distinctive Characteristics:

1. Range of applicability: 0-5 ppm full scale; lowest full-scale range is 0-0.50 ppm full scale.
2. Sensitivity: 0.020 ppm.
3. Interferences: nitrogen dioxide, hydrogen sulfide, methyl mercaptan, ethyl ozone (removable by scrubbers).
4. Accuracy: 5% of full scale.
5. Precision: ± 4% of full scale using standard gases.

References:

O'Keefe, A. "Sulfur Dioxide by Coulometry," from an informal compilation of analytical methods, EPA, National Environmental Research Center, Research Triangle Park, N.C.

— —

METHOD SUMMARY:

The monitoring instrument is based on the use of the intensity of chemiluminescent light emitted by sulfur compounds in a hydrogen-rich, hydrogen-air flame as a measure of the concentration of sulfur dioxide in the air samples. When light from the luminescing sulfur passes through a narrow-band optical filter, only the 384 nm light is permitted to strike the photomultiplier tube. The magnitude of the current produced is recorded.

Analytical Procedure:

Automated, flame emission, spectrophotometric.

Distinctive Characteristics:

1. Range of applicability: calibrated range is 0.1-1.5 ppm. Minimum detectable concentration is 0.05 ppm at a signal-to-noise ratio of 2.

2. Interferences: H_2S and mercaptans removed with scrubbers.
3. Accuracy: instrument accuracy is at ± 5%.
4. Precision: ± 2½% over range of 0.1-1.5 ppm. Response of the instrument is within a maximum of 2 min.
5. Measurement time: assumed to be rapid.

References:

Clements, J. B. "Determination of Sulfur Dioxide with an Automated Flame Emission Photometric Instrument, C-5," from an informal compilation of analytical methods, EPA, National Environmental Research Center, Research Triangle Park, N.C. (October 5, 1972).

— —

METHOD SUMMARY:

Sulfur dioxide in ambient air is absorbed in hydrogen peroxide at pH 5.0. This results in the formation of stable sulfuric acid, which is subsequently titrated with standard alkali.

Analytical Procedure:

Hydrogen peroxide, titrimetric.

Distinctive Characteristics:

1. Range of applicability: 0.01-10.0 ppm.
2. Interferences: strong acid gases; reactive acid solids; sulfuric acid if relative humidity > 85%; alkaline gases and basic solids; sulfur trioxide; large amounts of solid material.
3. Accuracy and precision: ± 10% in range < 0.1 ppm. Accuracy increases in range 0.1-1.0 ppm.
4. Stability: solutions after sample collection are stable for at least one month and can be titrated long after collection.

References:

Clements, J. B. "Determination of Sulfur Dioxide: Hydrogen Peroxide Method, H-23," from an informal compilation of analytical methods, EPA, National Environmental Research Center, Research Triangle Park, N.C. (October 5, 1972).

— —

- -

METHOD SUMMARY:

Sulfur dioxide is absorbed from air in a solution of potassium tetrachloride amercurate. A dichlorosulfitomercurate complex is formed which resists oxidation by air. The complex is then reacted with pararosaniline and formaldehyde to form intensely colored pararosaniline sulfonic acid. The absorbance of the solution is measured spectrophotometrically.

Analytical Procedure:

Pararosaniline, spectrophotometric.

Distinctive Characteristics:

1. Range: 25-1050 $\mu g/m^3$.
2. Interferences: oxides of nitrogen, ozone, heavy metals (eliminated by pretreatment).
3. Precision: relative standard deviation at 95% confidence level is 4.6% using standard samples.
4. Stability: at 22°C, losses of SO_2 occur at the rate of 1% per day. At 5°C no detectable loss occurs. EDTA enhances stability on SO_2 in solution.

References:

Federal Register 36(84) (April 30, 1971).

- -

SAMPLING FOR: SULFUR DIOXIDE, HYDROGEN SULFIDE, METHYL MERCAPTAN, HIGHER MOLECULAR WEIGHT GASEOUS COMPOUNDS.

METHOD SUMMARY:

Air stream containing sulfur compounds is injected on an analytical column, where the sulfur compounds are separated prior to entering the flame detector. Hydrogen sulfide and carbonyl sulfide are eluted first, followed by sulfur dioxide, methyl mercaptan, ethyl mercaptan, dimethyl sulfide and propyl mercaptan. After each analysis cycle, a stripper column is backflushed to ensure that heavier sulfur compounds do not reach the chromatographic column.

Analytical Procedure:

Automated, gas chromatographics, flame photometric.

Distinctive Characteristics:

1. Range of applicability: normal operating range of instrument is 0-2618 $\mu g/m^3$ (0-1.0 ppm). Minimum detectable sensitivity is: H_2S–6.95 $\mu g/m^3$; SO_2–13.09 $\mu g/m^3$; CH_3SH–9.81 $\mu g/m^3$; $(CH_3)_2S$–38.01 $\mu g/m^3$; and $CH_3(CH_2)_2SH$–46.59 $\mu g/m^3$.
2. Interferences: carbonyl sulfide removed by silver wool and scrubbers.
3. Accuracy: 1-2% full-scale range.
4. Precision: within 1%.

References:

O'Keefe, A. "Sulfur Dioxide, Hydrogen Sulfide, Methyl Mercaptan, and Higher Molecular Weight Gases Sulfur Compounds," EPA, National Environmental Research Center, Research Triangle Park, N.C.

_ _

SAMPLING FOR: CARBON MONOXIDE.

METHOD SUMMARY:

Based on IR radiation absorption by carbon monoxide. IR source energy is split into parallel beams and directed through reference and sample cells. Detectors are balanced electronically. Introduction of CO in the sample will absorb radiation, which reduces the temperature and pressure in the detector cell and displaces a diaphragm.

Analytical Procedure:

Nondispersive IR spectrometry, infrared spectrometry.

Distinctive Characteristics:

1. Range: 0-58 mg/m^3 or 0-50 ppm.

2. Sensitivity: 1% full-scale response per 0.6 mg CO/m^3 as high as 12 mg CO/m^3.
3. Accuracy: ± 1% scale in 0-58 mg/m^3 range.
4. Precision: ± 0.5% full-scale in the 0-58 mg/m^3 range.
5. Stability varies with temperature and pressure changes.

References:

Federal Register 36(84) (April 30, 1971).

AIRBORNE FUGITIVE SAMPLING

This section gives methods for airborne fugitive sampling. There are three types of airborne fugitive emission sources:

1. Diffuse area sources are those emissions due to a single source which effect a diffuse cloud over a given area.
2. Diffuse area-wide sources are those emissions that result from several sources, which effect a diffuse cloud over a given area.
3. Point sources are those emissions attributable to stacks or vents and effect a plume.

Generally, for Level 1 and 2 investigations of ambient particulate concentrations, a high-volume sampler is utilized. Variations may develop with site location. Gas bombs and high-volume samplers used upwind and downwind of the source or sources may be required depending on the level of investigation and availability of background data. Background data, when available, can eliminate the need for upwind sampling.

Airborne fugitive sampling methods for gases are the same as for "Gas Sampling."

- -

SAMPLING FOR: FUGITIVE PARTICULATES: DIFFUSE AREA SOURCES.

METHOD: EMISSION FACTOR METHOD

Advantages:

1. Rough but adequate estimates of emission rate.
2. Sampling not required.

Disadvantages:

1. Emissions factors may not be available for the source under consideration or may not be relatable to certain meterological conditions prevalent at the site.

Remarks:

1. Diffuse area source sampling is concerned primarily with aggregate piles and other fugitive sources whose emission rate is controlled largely by meteorological conditions such as rainfall and windspeed.
2. Techniques can also be applied on a plant-wide basis.
3. Applicable to any source where emission factors exists.
4. Applicable for Level 1 investigations.

References:

Cowherd, C. *et al.* "Development of Emission Factors for Fugitive Dust Sources," Midwest Research Institute, EPA-450/3-74-037 (June 1974).
Kalka, P. W. "Technical Manual for the Measurement of Fugitive Emissions Upwind-Downwind Sampling Method for Industrial Fugitive Emissions," draft, Research Corp. of New England (December 1975).

— —

METHOD: DOWNWIND

Advantages:

1. Simple and inexpensive for sampling specific sources such as piles, which have a diffuse emission cloud.

Disadvantages:

1. Must be placed into emission cloud area for accurate results.
2. Nonisokinetic sampling.

Remarks:

1. Most applicable to sources having high-density fugitive cloud emissions.
2. Not applicable to emissions on a plant-wide basis.

3.   For accurate quantitative results, a more sophisticated method must be used.
4.   Applicable for Level 1 investigations.

References:

Hamersma, J. W., S. L. Reynold and R. F. Maddalone. "IERL-RTP Procedures Manual: Level of Environmental Assessment," TRW Systems Group, EPA-600/2-76-160a (June 1976).

— —

METHOD: UPWIND-DOWNWIND: MIDWEST RESEARCH INSTITUTE METHOD

Advantages:

1.   Wind-activated hi-volume samplers are utilized.
2.   Samplers are mounted on a vertical tower in a grid arrangement to determine emissions from specific loading or unloading operations.
3.   Particle size measurements can be performed by attaching an Anderson impactor to the hi-volume samplers, or by microscopic examination of the filter.

Disadvantages:

1.   Sampling rates are preset to isokinetic conditions for specific wind velocities; inaccurate when the wind speed varies during the course of the test.
2.   Results from specific samplers are integrated over the length and height of the pile in question. This may be an inaccurate method.
3.   Particles are subject to reentrainment from the impactor surface due to their dry nature and may bias the particle size measurement.

Remarks:

1.   Can be used for emissions from both diffuse area and specific point sources.
2.   The source should be sampled during high- and low-wind velocity periods such that an estimation may be made of a high- and low-emission rate. If enough data is available, the development of an emission-factor equation should be attempted.
3.   Applicable for Level 2 investigations.

References:

Cowherd, C. *et al.* "Development of Emission Factors for Fugitive Dust Sources," Midwest Research Institute, EPA-450/3-74-037 (June 1974).

_ _

METHOD: UPWIND-DOWNWIND: THE RESEARCH CORPORATION OF NEW ENGLAND METHODOLOGY

Advantages:

1. Standard diffusion is used in conjunction with high-volume sampling to produce emission rates for the source.
2. Various particulate sampling devices can be used, including filter impactor, piezoelectric techniques, etc.
3. Particle size measurements can be performed with an Anderson impactor or by microscopic examination of the filter.

Disadvantages:

1. Nonisokinetic sampling.
2. Diffusion equation assumes a Gaussian distribution in both vertical and horizontal direction.
3. Particulate resuspension is not assumed.
4. No provision for individual sampling of aggregate loading or unloading is made.
5. Variations of wind speed and direction would interfere with accuracy of the results.

Remarks:

1. Applicable for emissions from both diffuse area and specific point sources.
2. Since the various sampling devices other than filter impaction have not been extensively field tested, they should not be used without on-site calibration with hi-volume samplers.
3. Applicable for Level 1 and 2 investigations.

References:

Kalka, P. W. "Technical Manual for the Measurement of Fugitive Emissions—Upwind-Downwind Sampling Method for Industrial Fugitive Emissions," draft, Research Corp. of New England (December 1975).

_ _

– –

METHOD: BACKGROUND OPERATING PLANT METHOD

Advantages:

1.   Most accurate measure of difference between ambient background and source contribution.

Disadvantages:

1.   Time involved is usually twice that of other techniques.
2.   For accuracy, similar meteorological conditions should prevail for both sampling period.

Remarks:

1.   Ambient data collected during an environmental impact study can be used for the background data or an existing plant operation should be halted and any aggregate piles wetted and covered. Applicable for determination of diffuse or multipoint fugitive sources on a plant-wide basis.
2.   Applicable for Level 2 investigations.

– –

SAMPLING FOR: FUGITIVE PARTICULATES: SPECIFIC POINT SOURCES.

METHOD: QUASI-STACK

Advantages:

1.   Most accurate sampling techniques where point source can be hooded.
2.   Particle size and mass emissions measurements can be made directly.

Disadvantages:

1.   Applicable only where hooding is possible.
2.   Low–grain loading and possible low velocities would present a problem if a conventional Method 5 train were used.

3. Where many point sources are to be tested, hooding and testing each individual one would be expensive and time-consuming.
4. The hood must be carefully designed so that all particulate is collected.

Remarks:

1. Should be used when emissions are from specific point source only.
2. To increase accuracy, as many tests as possible should be performed. For this reason, due to the fact that a low-grain loading is anticipated, an isokinetic hi-volume sampler is recommended, such as one developed by Radar Pneumatic. If however, particle size measurements are to be performed, then the SASS train should be used.
3. Applicable for Level 2 investigations.

References:

Kalka, P. W. "Technical Manual for the Measurement of Fugitive Emissions—Upwind-Downwind Sampling Method for Industrial Fugitive Emissions," draft, Research Corp. of New England (December 1975).

— —

METHOD: ROOF-MONITOR

Advantages:

1. Can be used to sample multiple-point sources located within a single structure where the emissions exit from the structure through roof, door and windows.
2. Same as for (2) under Quasi-Stack, pg. 166.

Disadvantages:

1. Cannot be used to isolate single-point emissions.
2. Same as (2) for Quasi-Stack.
3. For cost-effectiveness, this method should be applied only when all emissions exit through one opening.

Remarks:

1. Same as for Quasi-Stack.
2. Applicable for Level 2 investigations.

References:

Kalka, P. W. "Technical Manual for the Measurement of Fugitive Emissions—Upwind-Downwind Sampling Method for Industrial Fugitive Emissions," draft, Research Corp. of New England (December 1975).

METHOD: PLUME SAMPLING

Advantages:

1. Particle size and mass emission measurements can be made directly.

Disadvantages:

1. Testing many point sources would be inexpensive and time-consuming.

Remarks:

1. This method consists of sampling a fugitive plume with an SASS train by placing the nozzle directly into the plume and sampling isokinetically.
2. Applicable for Level 1 investigations.

References:

Hamersma, J. W., S. L. Reynold and R. Maddelone. "IERL-RTP Procedures Manual: Level of Environmental Assessment," TRW Systems Group, EPA-600/2-76-160a (June 1976).

SAMPLING FOR: FUGITIVE GASES.

The methods used for sampling fugitive gases are essentially the same as those identified for ducted gases (see Gas Sampling Options, Table 5-1, pg. 87).

CHAPTER 7

INORGANIC GAS ANALYSIS

INTRODUCTION

In this section the methods for inorganic gas are given. In many instances the stability of a given sample's constituents are unknown; therefore, generally, for a Level 1 investigation, a gas chromatographic analysis should be performed on-site. The listing given is not totally exclusive and should not be considered absolute for all mixtures and concentrations.

Level 2 investigation methods are specific to the inorganic gas under analysis and may require special sampling trains. In many instances a method is applicable to Level 1 and Level 2 investigations; however, the Level 2 method is the governing method with regards to accuracy and sensitivity.

A Level 1 investigation sample is accomplished through the use of a 3-liter glass bomb.

The analyses of gases are outlined for Level 1 and Level 2 investigations in Figures 7-1 and 7-2.

SAMPLING FOR: AMMONIA

PREFERRED ANALYTICAL PROCEDURE FOR LEVEL 1 INVESTIGATION: GAS CHROMATOGRAPH THERMAL CONDUCTIVITY DETECTOR

Advantages:

1. Relatively low-cost method; quick, simple and accurate.

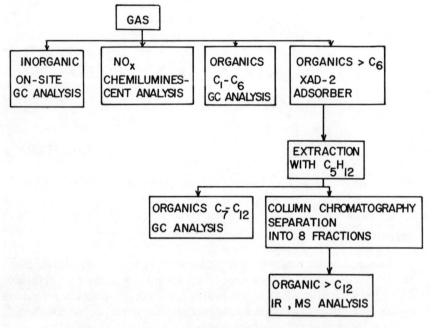

Figure 7-1. Level 1 investigation analysis techniques for gases.

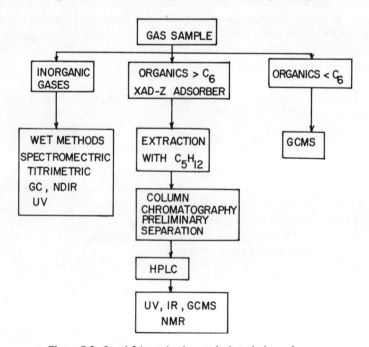

Figure 7-2. Level 2 investigation analysis techniques for gases.

Distinctive Requirements:

1. Sample collected in gas bomb.
2. On-site analysis preferred.
3. Recommended GC column: 6-ft stainless steel Poropak Q; $40°C$ isothermal.

References:

ASTM Standard D1426.

PREFERRED ANALYTICAL PROCEDURE FOR LEVEL 2 INVESTIGATION: FOR GREATER THAN 1 ppm: ABSORPTION TITRATION; FOR 0.05-1.0 ppm: COLORIMETRIC

Advantages:

1. Relatively rapid and quite accurate.

Distinctive Requirements:

1. Sample is collected and buffered at a pH of 9.5.
2. It is then distilled into a boric acid solution.
3. The ammonia in the distillate is determined: a) colorimetrically by nesslerization (MgI_2, KI, NaOH); b) titrimetrically with H_2SO_4 using Methyl Red or Methylene Blue indicator.

Remarks:

1. Interference may result due to specific amines.
2. Final method used is determined by concentration.

OTHER ANALYTICAL PROCEDURES:

1. Colorimetric: Detection Limit (DL) - 25 ppm. Absorption in 1 NH_2SO_4. Interferences Fe^{+2}, Cr^{+2}, Mn^{+2}, Cu^{+1}, Cu^{+2}. Applicable for Level 1 and 2 investigations. References: ASTM Standard D1426.
2. Colorimetric: DL - 25 ppm. Absorption in 1 NH_2SO_4. Interferences Ca^a, Mg^{+2}, Fe^{+2}, Fe^{+3}, S^{-2}, amines, ketones, aldehydes, alcohols. Applicable for Level 1 and 2 investigations.

3.   Specific Ion Electrode (SIE): Dl - 1 ppm. Absorption with 1 NH_2SO_4.
Applicable for Level 1 and 2 investigations. References: Thomas, R. F.,
and R. L. Booth. "Selective Electrode Measurement of Ammonia in
Water and Wastes," *Environ. Sci. Technol.* 7(6):523-525 (June 1973).

SAMPLING FOR: CARBON DIOXIDE

PREFERRED ANALYTICAL PROCEDURE FOR LEVEL 1 INVESTIGATION: GAS CHROMATOGRAPH THERMAL CONDUCTIVITY DETECTOR—DL - 25 ppm

Advantages:

1.   Low cost, accurate, simple and rapid.

Distinctive Requirements:

1.   Sample collected in gas bomb.
2.   On-site analysis preferred.
3.   Recommended GC column—6-ft stainless steel Molecular Sieve 5A;
40°C isothermal.

References:

Driscoll, J. N. *Flue Gas Monitoring Techniques: Manual Determination of
Gaseous Pollutants* (Ann Arbor, Michigan: Ann Arbor Science Publishers,
Inc., 1974).

PREFERRED ANALYTICAL PROCEDURE FOR LEVEL 2 INVESTIGATION: GC NONDISPERSIVE INFRARED—DL - 10 ppm

Advantages:

1.   Rapid, simple, accurate and low in cost.

Distinctive Requirements:

1.   Same as above. Water and particles must be removed.

References:

Same as Level 1, above.

OTHER ANALYTICAL PROCEDURES:

1. Gas Absorption: absorption in 6M NaOH. Interferences: other acid gases. Applicable for Level 1 investigations. References: Driscoll, J. N. *Flue Gas Monitoring Techniques: Manual Determination of Gaseous Pollutants* (Ann Arbor, Michigan: Ann Arbor Science Publishers, Inc., 1974).

- -

SAMPLING FOR: CARBON MONOXIDE

PREFERRED ANALYTICAL PROCEDURE FOR LEVEL 1 INVESTIGATION: GC THERMAL CONDUCTIVITY DETECTOR—DL - 25 ppm

Advantages:

1. Quick, accurate, simple, low in cost.
2. Multicomponent capabilities.

Distinctive Requirements:

1. Sample collected in a gas bomb.
2. On-site analysis preferred.
3. Recommended GC column—6-ft stainless steel Molecular Sieve 5A; 40°C isothermal.

References:

Driscoll, J. N. *Flue Gas Monitoring Techniques: Manual Determination of Gaseous Pollutants* (Ann Arbor, Michigan: Ann Arbor Science Publishers, Inc., 1974).

PREFERRED ANALYTICAL PROCEDURE FOR LEVEL 2 INVESTIGATION: GC NDIR—DL - 10 ppm

Advantages:

1. See Level 1, above.
2. EPA Reference Method.

Distinctive Requirements:

1.   For concentrations below 25 ppm, an HE ionization detector is needed.
2.   Water and particulates must be removed.

Remarks:

1.   NDIR Detector Limit is 10 ppm.

References:

Driscoll, J. N. *Flue Gas Monitoring Techniques: Manual Determination of Gaseous Pollutants* (Ann Arbor, Michigan: Ann Arbor Science Publishers, Inc., 1974).
Federal Register "Air Programs: Standards of Performance for New Stationary Sources," 39(47) (March 8, 1974).

OTHER ANALYTICAL PROCEDURES:

1.   Colorimetric: DL - 1ppm. Absorption train. Applicable for Level 1 and 2 investigations. References: Driscoll, J. N. *Flue Gas Monitoring Techniques: Manual Determination of Gaseous Pollutants* (Ann Arbor, Michigan: Ann Arbor Science Publishers, Inc., 1974). Colket, M. B., D. W. Naegeli, F. L. Dryer and I. Glassman. "Flame Ionization Detection of Carbon Oxides and Hydrocarbon Oxygenates," *Environ. Sci. Technol.* 8(1):43-44 (January 1974).

- -

SAMPLING FOR: HALOGENS AND VOLATILE HALIDES

PREFERRED ANALYTICAL PROCEDURE FOR
LEVEL 1 INVESTIGATION: SSMS

Advantages:

1.   Suitable methods for the rapid routine measurement of halogens and volatile halides.

Distinctive Requirements:

1.   These species may appear in particulates, XAD-2 trap and impingers.

PREFERRED ANALYTICAL PROCEDURE FOR
LEVEL 2 INVESTIGATION (Also see Level 1 Investigation):

Distinctive Requirements:

1. Chlorides cannot be accurately measured in Level 1 Investigation because of sampling problems.

OTHER ANALYTICAL PROCEDURES:

1. Colorimetric: DL - 1 ppm. Train constructed of glass, SS or Teflon. Upper concentration limit of 50 ppm. Halides, CN⁻, NO₂⁻. Mercuric theocyanate method. Applicable for Level 1 and 2 investigations. References: ASTM Standard D1253.

2. Colorimetric: DL - 1 ppm. Train constructed of glass, SS or Teflon. Upper concentration limit of 50 ppm. Chlorides. Colidine method. Applicable for Level 1 and 2 investigations. References: ASTM Standard D1253.

3. Colorimetric: DL - 1 ppm. Train constructed of glass, SS or Teflon. Upper concentration limit of 300 ppm. Chlorides. Applicable for Level 1 and 2 investigations. References: Bethea, R. M. "Improvements in Colorimetric Analysis of Chlorine and Hydrogen Fluoride by Syringe-Sampling Technique," *Environ. Sci. Technol.* 8(6):587-588 (June 1974).

4. GC: DL - 1 ppb. Gas bomb. Electron capture detector. References: Basu, P. K. *et al.* "A Three-Column, Three-Detector Gas Chromatographic Method for the Single-Sample Analysis of a Mixture of $S_1 Cl_4$, Cl_2, AR, N_2, CO and CO_2," *J. Chromatog. Sci.* 10:479-480 (July 1972).

- -

SAMPLING FOR: HYDROGEN CHLORIDE

PREFERRED ANALYTICAL PROCEDURE FOR
LEVEL 1 INVESTIGATION:

See Halogen and Volatile Halides, p. 174.

PREFERRED ANALYTICAL PROCEDURE FOR
LEVEL 2 INVESTIGATION: ABSORPTION/TITRIMETRIC

Advantages:

1. Accurate, rapid.

Distinctive Requirements:

1. Titration with $Hg(NO_3)_2$.
2. Indicator: diphenylcarpazone and bromophenol blue.

Remarks:

1. Bromide and iodide interfere with SIE measurement.

References:

ASTM Standard D512.

OTHER ANALYTICAL PROCEDURES:

1. Titrimetric Potentiometric: DL - 1 ppm. Impinger train. Upper concentration limit of 100 ppm. Chloride is a common contaminant; use deionized water in analysis. References: ASTM Standard D512.
2. Colorimetric: DL - 1 ppm. Impinger train. Upper concentration limit of 10 ppm. Halides, cyanides, nitrites. References: ASTM Standard D2036.
3. Dispersive IR: DL - 0.5 ppm. Remove H_2O. CO_2, H_2O. Suitable for continuous monitoring. For low-concentration detection, long path length required. References: MSA Air Pollution Instrumentation. "Product Information Sheet Bulletin No. 0700-9," MSA Instrument Division.
4. SIE: DL - 1 ppm. Impinger train. Halides. Calibration can reduce interferences. References: Duff, E. J., and J. L. Stuart. *Talanta (Great Britain)* 22:823-826 (1975).

All above are applicable for Level 1 and 2 investigations.

SAMPLING FOR: HYDROGEN CYANIDE (CYANDOGEN)

PREFERRED ANALYTICAL PROCEDURE FOR LEVEL 1 INVESTIGATION: GC (TCD)

Advantages:

1. Rapid, accurate, simple, low in cost.
2. Multicomponent capabilities.

Distinctive Requirements:

1. Collection in 3-liter gas bomb and analysis on-site.
2. Column: 6-ft SS Poropak Q, isothermal at 40°C.

References:

ASTM Standard D2036.

PREFERRED ANALYTICAL PROCEDURE FOR LEVEL 2 INVESTIGATION: ABSORPTION/COLORIMETRIC/TITRIMETRIC

Advantages:

1. Suitable for determining soluble and insoluble cyanides in aqueous solution.

Distinctive Requirements:

1. Absorption in 1 N KOH; sample is acidified and distilled into 1.25 N NaOH.
2. Colorimetric: reaction with chloramine T, color development with pyridine; pyrazolone reagent, measure at 629 nm.
3. Titration: AgNO$_3$ with an appropriate indicator.

References:

ASTM Standard D1253.

OTHER ANALYTICAL PROCEDURES:

1. Potentiometric: DL - 1 ppm. Impinger train. H$_2$S. Applicable for Level 1 and 2 investigations.
2. GC: DL - 1 ppm. Impinger train. SCN⁻. Applied to biological samples and water analysis. Applicable for Level 1 and 2 investigations. References: Natusch, D. F. S., I. R. Sewell, and R. L. Tanner. *Anal. Chem.* 46(3):410-415 (March 1974).
3. GC: DL - 25 ppm. Gas bomb. Poropak Q column.
4. Colorimetric: DL - 1 ppm. Impinger train. Sulfide oxidants. Applicable for Level 1 and 2 investigations. References: ASTM Standard D2036.

SAMPLING FOR: HYDROGEN FLUORIDE (FLUORIDES)

**PREFERRED ANALYTICAL PROCEDURE FOR
LEVEL 1 INVESTIGATION:**

See Halogens and Volatile Halides, p. 174.

**PREFERRED ANALYTICAL PROCEDURE FOR
LEVEL 2 INVESTIGATION: ABSORPTION/
SPECIFIC ION ELECTRODE (SIE)–DL - 1 ppm**

Advantages:

1.   Offers better accuracy and precision and care of analysis than does the
SPADNS Colorimetric Method.

Distinctive Requirements:

1.   Absorption in distilled water.
2.   pH buffered to 5.2-5.5.
3.   Standard addition technique for analysis.

References:

Thomas, J., Jr. and H. J. Gluskoter. *Anal. Chem.* 46(a):1321-1323 (August
1974).

OTHER ANALYTICAL PROCEDURES:

1.   Colorimetric: DL - 1 ppm. Impinger train. Many cations, Cl⁻ temperature control required during analysis. Applicable for Level 1 and 2
investigations. References: MITRE Corp. "National Fluidized-Bed
Combustion Program. Volume III. State-of-the-Art Report," M-75-58
(December 1974).

- -

SAMPLING FOR: HYDROGEN SULFIDE

PREFERRED ANALYTICAL PROCEDURE FOR
LEVEL 1 INVESTIGATION: GC (TCD/FPD)–
DL FOR FPD 10 ppb

Advantages:

1. Rapid, accurate, simple.
2. Multicomponent capabilities.

Distinctive Requirements:

1. Teflon-coated and glass system necessary for analysis.
2. Collection in 3-liter gas bomb.
3. Column: acetone-washed Teflon 18 in. x 1/8 in. OD column packed with 80/100 mesh acetone-washed Poropak QS.
4. Column temperature: 30°C, postinjection delay of 1 min to 210°C at 40°/min.
5. Alternate column: 6-ft glass. 3% OV-1 on chromosorb W, 100/120 mesh; isothermal at 60°.

References:

de Souza, T. L. C., D. C. Lane and S. P. Bhatia. *Anal. Chem.* 47(3):543-545 (March 1975).

OTHER ANALYTICAL PROCEDURES:

1. Paper tape: DL - 10 ppb. Paper must be kept moist; sensitive to light. Upper concentration limit of 50 ppm. RSH, NH_3. $AgNO_3$, $HgCl_2$, $Ag(CN)_2^-$, $Pb(OAc)_2$ tested. AgNO best. References: Bethea, R. M. *J. Air Poll. Control Assoc.* 23(8):710-713 (August 1973). Natusch, D. F. S., J. R. Sewell and R. L. Tanner. *Anal. Chem.* 46(3):410-415 (March 1974).
2. Colorimetric: DL - 1 ppm. Use restricted opening bubbler SO_2 RSH. Rarely used. Lacks sensitivity. SO_2 interference may be removed with $FeCl_3$.

3. Titrimetric: DL - 1 ppm. Impinger train. RSH. Sample must be analyzed immediately. References: ASTM Standard Part 18, Philadelphia (1972).
4. Potentiometric: DL - 1 ppm. Impinger train. SO_2, RSH. References: ASTM Standard Part 18, Philadelphia (1972).
5. Fluorescent: DL - 1 ppb. Collection on tape RSH, HCW. References: Bethea, R. M. *J. Air Poll. Control Assoc.* 23(8):710-713 (August 1973).

All of the above are applicable for Level 1 and 2 investigations.

- -

SAMPLING FOR: MERCURY

PREFERRED ANALYTICAL PROCEDURE FOR LEVEL 1 INVESTIGATION: FLAMELESS ATOMIC ABSORPTION SPECTROSCOPY (FAAS)–DL - 1 ppb

Advantages:

1. GC and SSMS are not suitable techniques for Hg analysis.
2. FAAS provides accurate analysis of Hg.

Distinctive Requirements:

1. Hg is absorbed in SASS train impingers.
2. Hg may also be absorbed by the XAD-2 trap and on particulates.

References:

Kalb, G. W., and C. Baldeck. Tradet, Inc., June 1972.
TRW Systems Group. "Procedures for Process Measurements–Trace Inorganic Materials," (July 1975).

PREFERRED ANALYTICAL PROCEDURE FOR LEVEL 2 INVESTIGATION: FAAS–GOLD AMALGAMATION TECHNIQUE–DL - 1 ppb

Advantages:

1. See Level 1, above.

Distinctive Requirements:

1. Absorption on 10-g gold chips; chips are transferred to furnace and measured by standard FAAS techniques.

Remarks:

1. Method uses EPA isokinetic sampling train.
2. Mercury may also be collected on filter prior to impingers.
3. Normally combined with impingers for analysis.

References:

Kalb, G. W., and C. Baldeck. Tradet, Inc. (June 1972).
TRW Systems Group. "Procedures for Process Measurements—Trace Inorganic Materials" (July 1975).

OTHER ANALYTICAL METHODS:

1. Gas Chromatography: DL - 1 ppm. Organomercury compounds are determined with this method. Can be used for Level 1 and 2 investigations. Interferences are specific. References: Bach, C. A., and D. J. Lisk. *Anal. Chem.* 43(7):950-952 (June 1971).
2. Atomic Fluorescence: DL - 1 ppb. Impinger trains used for adsorption. Interferences are from carbon monoxide and nitrogen disrupt signal making direct determination in air unsatisfactory. Can be used for Level 1 and 2 investigations. References: Chilov, S. *Talanta (Kodak)* 22: 205-232 (January 1975). Novak, J. *et al. Anal. Chem.* 37(6):660-666 (May 1965).
3. FAAS: DL - 1 ppb. At 1350°C the sample is passed through a carbon column. Measurements at 184.9 nm. Can be used for Level 1 and 2 investigations.
4. FAAS: DL - 1 ppb. Iodine, impregnated on charcoal is used as the absorption medium. The charcoal concentrates the mercury removing interference possibilities. Can be used for Level 1 and 2 investigations. References: Hatch, W. R., and W. L. Ott. *Anal. Chem.* 40(14):2085-2087 (December 1968).
5. Cold Vapor: DL - 1 ppb. In this method mercury is converted to Hg^{+2}. Water vapor is an interference that may be removed through the use of drying agents or heating. Can be used for Level 1 and 2 investigations. References: Chilov, S. *Talanta (Kodak)* 22:205-232 (January 1975).
6. Colorimetric: DL - 1 ppm. Several variations of this method are available. Can be used for Level 1 and 2 investigations. References: Chilov, S. *Talanta (Kodak)* 22:205-232 (January 1975).
7. Titrimetric: DL - 1 ppm. Several variations of this method are available. Can be used for Level 1 and 2 investigations. References: Chilov, S. *Talanta (Kodak)* 22:205-232 (January 1975).

8.   Polarographic: DL - 1 ppm. Several variations of this method are available. Can be used for Level 1 and 2 investigations. References: Chilov, S. *Talanta (Kodak)* 22:205-232 (January 1975).

9.   Piezoelectric: DL - 1 ppb. For detection, mercury must be in its elemental form. Particulates, chlorine and water can cause interferences which will be alleviated through use of a filter and trap. This method can be used for continuous monitoring. Can be used for Level 1 and 2 investigations. References: Scheide, E. P., and J. K. Taylor. "Piezoelectric Sensor for Mercury in Air," *Environ. Sci. Technol.* 8(13):1097-1099 (December 1974).

SAMPLING FOR: NITROGEN OXIDES

**PREFERRED ANALYTICAL PROCEDURE FOR
LEVEL 1 INVESTIGATION: CHEMILUMINESCENCE
—DL - 1 ppm**

Advantages:

1.   Reliable and rapid.

Distinctive Requirements:

1.   Collection in 3-liter bomb.
2.   On-site analysis.
3.   Calibration standard is required.

References:

Hamersma, J. W., and S. L. Reynolds. TRW Systems Group, EPA-600/2-76-093b (April 1976).

**PREFERRED ANALYTICAL PROCEDURE FOR
LEVEL 2 INVESTIGATION: PHENOL DI-SULPHONIC
ACID METHOD—DL - 1 ppm**

Advantages:

1.   Accurate measure of nitrogen oxides.
2.   Federal Register Method.

Distinctive Requirements:

1. Absorption in 2-liter flask with 25 ml of 0.1 N H_2SO_4 in 0.1% H_2O_2.
2. Spectrometric measurement at 430 nm.

Remarks:

1. Total oxides of nitrogen are measured.

References:

ASTM Standard. "Water: Atmospheric Analysis," Part 23, D1608-60, Philadelphia (1975).

OTHER ANALYTICAL PROCEDURES:

1. NDIR: DL - 5 ppm dry sample. Interferences aromatic HC, CO_2, H_2O. Suitable for continuous analysis. Applicable for Level 1 and 2 investigations. References: Beckman Instruments, Inc. "Produce Information Bulletin 4129C," Process Instruments Division. MITRE Corp. "Compendium of Analytical Methods" (April 1973). MSA Air Pollution Instrumentation. "Product Information Sheet Bulletin No. 0700-9," MSA Instrument Division.
2. Colorimetric: DL - 1 ppm. Sample dilution required with high levels of NO_2. Applicable for Level 2 investigations.
3. Paramagnetic: DL - 10 ppm. Particulate removal-interferences. Other paramagnetic species, *e.g.*, O_2. NO and NO_2 measured. References: LeRoy, V. M., and A. J. Lincoln. *Anal. Chem.* 46(3):369-373 (March 1974). Sittig, M. *Pollution Detection and Monitoring Handbook* (London: Noyes Data Corp., 1974).
4. SIE: DL - 1 ppm. Interferences HCl, HF, CO, SO_2. Interference may be removed by precipitation. Applicable for Level 1 and 2 investigations. References: Dee, A., H. H. Martens, C. I. Merrill and J. T. Nakamura. *Analyt. Chem.* 45(8):1477-1481 (July 1973). Kneebone, B. M., and H. Freiser. *Anal. Chem.* 45(3):449-452 (March 1973).
5. Electrochemical: DL - 20 ppb. Method is rate-of-flow dependent. H_2S. Applicable for Level 1 and 2 investigations. References: Gunther, F. A., and R. C. Blinn. *Analysis of Insecticides and Acaricides* (New York: Interscience Publishers, Inc., 1955).
6. UV: DL - 1 ppm. Particulate removal.
7. Gas Chromatography: DL - 10 ppb. Gas bomb. Quantitative recovery from column is difficult at low concentration levels. For lower levels, an ionization detector or flame chemiluminescent detector may be used. Applicable for Level 1 and 2 investigations.

8. Fluorescence: DL - 1 ppb. Dilution necessary. HC interverence. References: Bethea, R. M. *J. Air Poll. Control Assoc.* 23(8):710-713 (August 1973).

SAMPLING FOR: OXYGEN

PREFERRED ANALYTICAL PROCEDURE FOR
LEVEL 1 INVESTIGATION: GC (TCD)

Advantages:

1. Rapid, accurate, simple.

Distinctive Requirements:

1. Collection in a 3-liter bomb.
2. On-site analysis.
3. Column: 6-ft SS molecular sieve 5A; isothermal at 40°C.

References:

Driscoll, J. N. *Flue Gas Monitoring Techniques: Manual Determination of Gaseous Pollutants* (Ann Arbor, Michigan: Ann Arbor Science Publishers, Inc., 1974).

PREFERRED ANALYTICAL PROCEDURE FOR
LEVEL 2 INVESTIGATION: GC POLAROGRAPHIC
PARAMAGNETIC

Advantages:

1. See Level 1, above.

Distinctive Requirements:

1. Collection gas bomb.
2. An accurate calibration standard is required.

Remarks:

1. Methods can be used for continuous monitoring.

References:

Driscoll, J. N. *Flue Gas Monitoring Techniques: Manual Determination of Gaseous Pollutants* (Ann Arbor, Michigan: Ann Arbor Science Publishers, Inc., 1974).

Taylor Analytical Instrument Division. "Product Information Sheet," File 18-2A.

Holtzman, J. L. *Anal. Chem.* 48(1):299-230 (January 1976).

OTHER ANALYTICAL PROCEDURES:

1. GC: DL - 25 ppm. Gas bomb. Applicable for Level 1 and 2 investigations. References: Hodges, C. T., and R. F. Matson. *Anal. Chem.* 37(8): 1065-1066 (July 1965). Kusten, W. J. *Anal. Chem.* 48(1):84-87 (January 1976).
2. Combustion: DL - 0.01%. SO_2, acid gas removal. CO, HC interference. Applicable for Level 1 and 2 investigations. References: Bailey Meter Co. "Bailey Receiver-Recorder and Plug-In Modules," Bailey Products Specification E12-5 (1958).

- -

SAMPLING FOR: SULFUR DIOXIDE

PREFERRED ANALYTICAL PROCEDURE FOR
LEVEL 1 INVESTIGATION: GAS CHROMATOGRAPHY
(TCD/FPD)–DL - 1 ppb

Advantages:

1. For SO_2 measurement this is a simple, quick and accurate method.

Distinctive Requirements:

1. See H_2S, p. 179.

PREFERRED ANALYTICAL PROCEDURE FOR
LEVEL 2 INVESTIGATION: GAS CHROMATOGRAPHY
(TCD/FPD) ADSORPTION/TITRIMETRIC

Advantages:

1. All recognized as SO_2 measurement techniques.

Distinctive Requirements:

1. See H_2S.
2. Absorption in 3% hydrogen peroxide.
3. Titration is with barium perchlorate and thorin indicator.

Remarks:

1. Continuous monitoring may be effected with use of NDIR and UV methods.

References:

de Souza, T. L. C., D. C. Lane and S. P. Bhatia. *Anal. Chem.* 47(3):543-545 (March 1975).

OTHER ANALYTICAL METHODS FOR SO_2:

1. Titrimetric: DL - 1 ppm. Requirements include a special sampling train and impingers. Interference is from particulate matter. Can be used for Level 1 and 2 investigations. References: Dewey, C. F., Jr., R. D. Kamm and C. E. Hackett. "Acoustic Amplifier for Detection of Atmospheric Pollutants," *Appl. Physics Lett.* 23(11):633-635 (December 1973).
2. NDIR: DL - 5 ppm. Dry sample is required. Interferences are from aromatics, water and carbon dioxide. Continuous monitoring is obtained from this method. Can be used for Level 1 and 2 investigations. References: Beckman Instruments, Inc. "Process Infrared Analyzer Beckman Model 864," Product Information Bulletin 4129c, Process Instruments Division.
3. Fluorescent: DL - 1 ppb. Dry sample is required. For concentrations greater than 100 ppm, sample dilution is required. Interferences are from aromatics. Continuous monitoring is obtained from this method. Can be used for Level 1 and 2 investigations. References: Thermo Electron Corp. "SO_2 Pulsed Fluorescent Analyzer Model 40," Environmental Instrument Department.
4. UV: DL - 1 ppm. Dry sample is required. Continuous monitoring is obtainable. Can be used for Level 1 and 2 investigations. References: Syty, A. "Determination of Sulfur Dioxide by Ultraviolet Absorption Spectrometry," *Anal. Chem.* 45(9):1744-1746 (August 1973).
5. Electrochemical Transducers: DL - 1 ppm. Method is dependent on flow rate. Interferences are from nitrogen oxide and hydrogen sulfide. Slow response but can be used for continuous monitoring. Can be used

for Level 1 and 2 investigations. References: Hamersma, J. W., and S. L. Reynolds. "Field Test Sampling/Analytical Strategies and Implementation Cost Estimates: Coal Gasification and Flue Gas Desulfurization," TRW Systems Group, EPA-600/2-76-093b (April 1976).

6. Colorimetric: DL - 10 ppb. Several variations of this method are available. Interferences may be from NO_2, H_2S, O_3, CH_3SH, C_2H_4. Interferences may be scrubbed out. Depending on the technique used this method can be used for Level 1 and 2 investigations. References: Robertus, R., and M. J. Schaer. "Portable Continuous Chromatographic Coulometric Sulfur Emission Analyzer," *Environ. Sci. Technol.* 1(9): 849-852 (September 1973). Sittig, M. *Pollution Detection and Monitoring Handbook* (London: Noyes Data Corp., 1974).

7. Piezoelectric: DL - 2 ppm. Interferences are from nitrogen dioxide and particulates. Fatigue is caused by high NO_2 levels. Can be used for Level 1 and 2 investigations. References: Frechette, M. W., J. L. Fasching and D. M. Rosie. "Evaluation of Substrates for Use on a Piezoelectric Detector for Sulfur Dioxide," *Anal. Chem.* 45(9):1765-1766 (August 1973).

8. Conductometric: DL - 1 ppm. Sample conditioning is required. Strong acid gases interfere with analysis. Analysis time is 2 min and can be continuous. Can be used for Level 1 and 2 investigations. References: Hamersma, J. W., and S. L. Reynolds. "Field Test Sampling/Analytical Strategies and Implementation Cost Estimates: Coal Gasification and Flue Gas Desulfurization," TRW Systems Group, EPA-600/2-76-093b (April 1976).

SAMPLING FOR: SULFUR TRIOXIDE

PREFERRED ANALYTICAL PROCEDURE FOR LEVEL 1 INVESTIGATION: SSMS

Advantages:

1. Method capable of quick routine elemental sulfur measurements.

Distinctive Requirements:

1. SO_3 in a sampling train may be found in the XAD-2 trap, impingers, and particulates.

PREFERRED ANALYTICAL PROCEDURE FOR LEVEL 2 INVESTIGATION: CONTROLLED CONDENSATION/TITRIMETRIC

Advantages:

1. The one method providing accurate SO_3 measurement. SO_2 interferes with other methods.

Distinctive Requirements:

1. Flue gas condensed and collected in glass frit at temperature controlled at 60-90°C.
2. Titration with barium perchlorate; thorin indicator.

Remarks:

1. SO_3 absorbed in 80% isopropanol, 20% H_2O.
2. Under these conditions SO_2 may oxidize to SO_3.

References:

Driscoll, J. N., and P. Warneck. *J. Air Poll. Control Assoc.* 23(10):858-863 (October 1973).

OTHER ANALYTICAL METHODS:

1. Spectrophotometric: DL - 1 ppb. Acid-washed filters of glass fiber are used. NH_4 and SO_4 are interferences. This method can be used for Level 1 and 2 investigations. References: West, P. W., and J. J. Chiang. *J. Air Poll. Control Assoc.* 24(7):671-673 (July 1974).
2. Coulometric: DL - less than 1 ppb. Acid-washed filters of glass fiber are used. NH_4 and SO_4 are interferences. This method can be used for Level 1 and 2 investigations. References: Huygen, C. *Atmos. Environ. Great Britain* 9:315-319 (1975).
3. Flame Photometric: DL - less than 1 ppb. Acid-washed filters of glass fiber are used. NH_4 and SO_4 are interferences. This method can be used for Level 1 and 2 investigations. References: Skotnicki, P. A., A. G. Hopkins and C. W. Brown. *Anal. Chem.* 45(13):2291-2292 (November 1973).

SAMPLING FOR: WATER VAPOR

PREFERRED ANALYTICAL PROCEDURE FOR LEVEL 1 INVESTIGATION: GC (TCD)

Advantages:

1. Rapid, simple, water vapor measurement technique.

Distinctive Requirements:

1. GC column is 6 SS, Molecular Sieve 5A.
2. Kept at constant temperature of 45°C.

PREFERRED ANALYTICAL PROCEDURE FOR LEVEL 2 INVESTIGATION: CONDENSATION NDIR

Advantages:

1. See Level 1, above.

Distinctive Requirements:

1. Sample must be kept above the dew point to prevent condensation.
2. Can be used for continuous monitoring.

References:

Smith, F., D. E. Wagoner and A. C. Nelson, Jr. Research Triangle Institute, EPA-650/4-74-005c (August 1974).

CHAPTER 8

ORGANIC ANALYSIS

INTRODUCTION

The basics of an analytical program for gaseous and particulates pollutant identification are summarized in Figures 8-1, 8-2, 8-3, 8-4 and 8-5. More detail on analytical techniques for specific compounds and elements is given in subsequent chapters.

ORGANIC ANALYSIS

This section outlines the methods for organic analysis with comments on the various techniques. In the Level 1 investigation the objective is to determine which organic classes exist in a sample. The Level 2 investigation takes this data and selects those fractions that are environmentally significant. The individual constituents of these fractions are then qualitatively and quantitatively defined.

Depending on the sample type, preparation of the organic sample for analysis will vary. A further variation of sample occurs depending on the level of investigation. For example, the preparation of a sample for a Level 1 investigation involves a separation by gas; while a Level 2 investigation involves an initial class separation.

Pretreatment for C_1 - C_6 hydrocarbons and some organic liquids is not required, yet most samples require solvent extractions. Particulates, solutions, and SASS train components all require pretreatment. Gas chromatography (GC) and infrared (IR) analysis are commonly used.

Gas chromatography is used to separate organic liquids into either fractions; infrared spectroscopy is then used to break down these fractions into classes. Low-resolution mass spectrum (LRMS) is then used for compound determinations. Samples should be handled with care to avoid contamination.

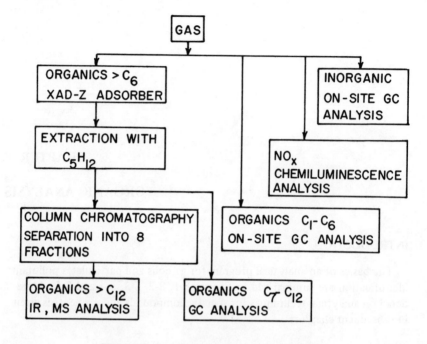

Figure 8-1. Level 1 investigation for gas analysis.

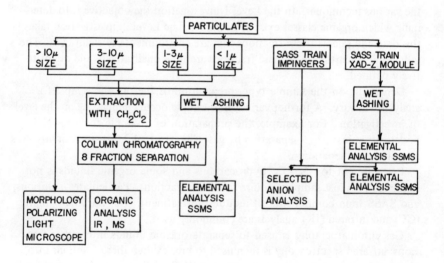

Figure 8-2. Level 1 investigation for particulate analysis.

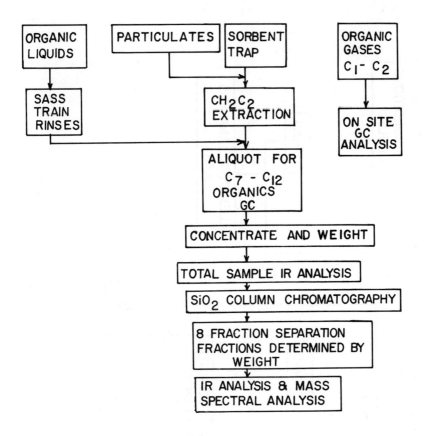

Figure 8-3. Level 1 investigation. Analysis for organic compounds.

The preceding procedure is effective for high-boiling-temperature compounds. C_1-C_6 compounds (b.p. $< 70°C$) must be analyzed with a gas chromatograph on-site. C_7-C_{12} compounds (b.p. 100-200°C) are also analyzed by gas chromatography; however, more elaborate pretreatment techniques are necessary. These may involve special extractions, calibrations and flame ionization detection.

Organic compound qualification and quantification may be effected through the use of the following methods: gas chromatography, gas chromatography-mass spectroscopy, NMR spectroscopy, IR spectroscopy, UV-visible spectroscopy, and high-resolution mass spectroscopy. Figure 8-3 shows a Level 1 investigation.

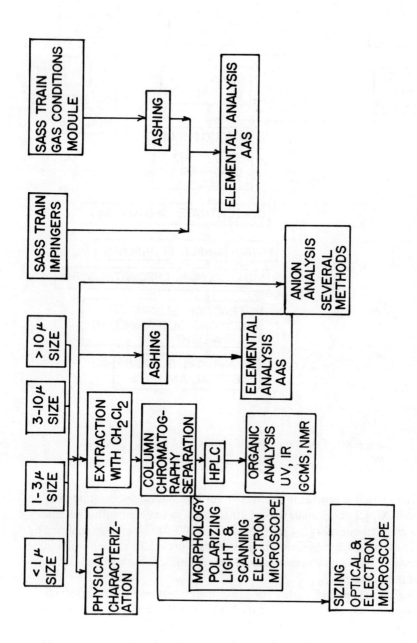

Figure 8-4. Level 2 investigation for particulate analysis.

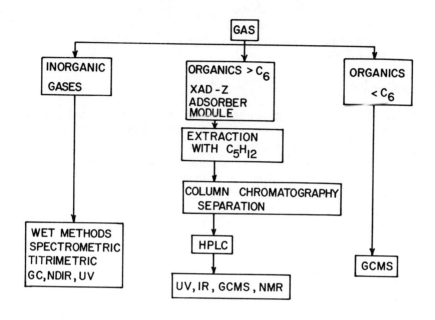

Figure 8-5. Level 2 investigation for gas analysis.

SAMPLING FOR: VOLATILE ORGANIC COMPOUNDS C_1-C_6 RANGE

HC	B.P. ($^\circ$C)	HC	B.P. ($^\circ$C)	HC	B.P. ($^\circ$C)
C_1	-100/-100	C_3	-50/0	C_5	30/60
C_2	-100/- 50	C_4	0/30	C_6	60/90

**PREFERRED ANALYTICAL PROCEDURE FOR
LEVEL 1 INVESTIGATION: GC FLAME IONIZATION
DETECTOR (FID)—DL - 0.1%**

Advantages:

1. Rapid, simple method for organics.

2.   Can be used for on-site analysis.
3.   Designed to separate and analyze organic mixtures based on boiling point range.

Distinctive Requirements:

1.   Collection in 3-liter glass vessel with wool plug to remove particulates.
2.   Column: Poropak Q, 100/120 mesh, 6 ft x 0.125 in. SS FID detector; Temperature 50°C isothermal.
3.   Generally, peaks observed in the chromatogram represent mixtures of compounds within a given b.p. range rather than individual compounds. Peak areas may be calculated by several methods: peak height, triangulation, integration, etc.

References:

Hamersma, J. W., S. L. Reynold and R. F. Maddalone. "EIRL-RTP Procedures Manual: Level of Environmental Assessment," TRW Systems Group, EPA-600/2-76-160a (June 1976).
Jones, P. W. *et al.* "Efficiency Collection of Polycyclic Organic Compounds from Combustion Effluents," Battelle Columbus Laboratories (August 1975).

PREFERRED ANALYTICAL PROCEDURE FOR LEVEL 2 INVESTIGATION: GC (FID)–DL - 1 ppm

Advantages:

1.   Rapid, simple method for organics and reasonably accurate.

Distinctive Requirements:

1.   Collection: same as above.
2.   For C_1 to C_5 HC only.
3.   Column: 8 ft x 1/8 in. o.d. SS, 17% by weight, B, B' oxydiproponitrile on 100-150 mesh activated alumina.
4.   Temperature: isothermal at 0°C.

Remarks:

1.   GC is susceptable to interferences such as identical retention times.

References:

ASTM Standard D1946.

ASTM Standard D2820-69T.
Jones, P. W. *et al.* "Technical Manual for Analysis of Organic Materials in
Process Streams," Battelle Columbus Laboratories (January 1976).

_____

SAMPLING FOR: ORGANIC COMPOUNDS C_7-C_{12} RANGE

HC	B.P. ($^\circ$C)	HC	B.P. ($^\circ$C)	HC	B.P. ($^\circ$C)
C_7	90/100	C_9	140/160	C_{11}	180/200
C_8	110/140	C_{10}	160/180	C_{12}	200/220

**PREFERRED ANALYTICAL PROCEDURE FOR
LEVEL 1 INVESTIGATION: GC (FID)**

Advantages:

1. See Organic Compounds C_1-C_6 Range, p. 195.

Distinctive Requirements:

1. Column: 1.5% OV-101 (or SE-30) on gas chromatograph 100/120 mesh
   0.125 in. x 6 ft SS, FID detector.
2. Temperature: isothermal at 50°C for 5 min, then 10°C/min to 150°C.
   A temperature calibration curve is prepared.
3. Materials are reported within a given b.p. range.

References:

Hamersma, J. W., S. L. Reynolds and R. F. Maddalone. "IERL-RTP Procedures
Manual: Level of Environmental Assessment," TRW Systems Group, EPA-
600/2-76-160a (June 1976).
Jones, P. W. *et al.* "Technical Manual for Analysis of Organic Materials in
Process Streams," Battelle Columbus Laboratories (January 1976).

**PREFERRED ANALYTICAL PROCEDURE FOR
LEVEL 2 INVESTIGATION: GC (FID)**

Advantages:

1. See above.

Distinctive Requirements:

1. Column: the following column to be used for screening purposes. SCOT OV 101 (or OV 17), 100 m, 0.4 mm I.D., FID detector.
2. Temperature 2°C/min ambient to 300°C.
3. No one column can be used for all organic compounds. See other procedures for columns for particular compounds.

Remarks:

1. Additional analysis using infrared or mass spectrometry may be required for compound identification.

References:

ASTM Standard D2682-71.
ASTM Standard D2267, D2427.
Chau, A. S. Y., and R. C. J. Sampson. "Electron Capture Gas Chromatographic Methodology for the Quantitation of Polychlorinated Biphenyls: Survey and Compromise," *Environ. Lett.* 8(2):89-101 (1975).

SAMPLING FOR: NONVOLATILE ORGANIC COMPOUNDS
(b.p. > 220°C)

PREFERRED ANALYTICAL PROCEDURE FOR LEVEL 1 INVESTIGATION: COLUMN CHROMATOGRAPHY EXTRACTION AND FRACTIONATION. INFRARED SPECTROSCOPY AND LOW-RESOLUTION MASS SPECTROMETRY USED FOR CLASS IDENTIFICATION WITHIN EACH FRACTION

Advantages:

1. Column chromatography is a low-resolution technique intended to yield 8 fractions on the basis of class polarity; overlap of class type between many of the fractions is inevitable.
2. IR analysis is used to identify functional groups present in each sample and fraction. LRMS is utilized with IR to assist in class identification.

Distinctive Requirements:

1. Numerous methods of sample preparation.

References:

Hamersma, J. W., S. L. Reynolds and R. F. Maddalone. "IERL-RTP Procedures Manual: Level of Environmental Assessment," TRW Systems Group, EPA-600/2-76-160a (June 1976).

PREFERRED ANALYTICAL PROCEDURE FOR LEVEL 2 INVESTIGATION: HIGH-PERFORMANCE LIQUID CHROMATOGRAPHY (HPLC)

Distinctive Requirements:

1. See above.

References:

Jones, P. W. et al. "Technical Manual for Analysis of Organic Materials in Process Streams," Battelle Columbus Laboratories (January 1976).

OTHER ANALYTICAL PROCEDURES:

Gel Permeation Chromatography

Distinctive Requirements: 1) 20 mg of sample/100 ml of column. 2) Volume molecular weight separation.
References: Kirkland, J. J., Ed. *Modern Practice of Liquid Chromography* (New York: Wiley-Interscience, 1971).

Bonded Phase Chromatography (Reverse)

Distinctive Requirements: 1) Preparative columns can handle up to 1-g samples. Chemical class separation.
References: See Gel Permeation chromatography, above.

Thin-Layer Chromatography (TLC)

Distinctive Requirements: 1) μg quantities of sample required for analysis. 2) TLC support (SiO_2 and Al_2O_3) may need activation by heating at $100°C$ for one hour. 3) Data from this method may be used with HPLC.
References: Sawicki, E., T. W. Stanley, S. McPherson, M. Morgus. "Use of Gas-Liquid and Thin-Layer Chromatography in Characterizing Air Pollutants by Fluorimetry," *Talanta Internat. J. Anal. Chem.* 13:619-629 (1966).

Liquid-Solid Chromatography

Distinctive Requirements: 1) mg quantities of samples are necessary for analysis. 2) Useful for screening studies.
References: Jones, P. W. *et al.* "Efficient Collection of Polycyclic Organic Compounds from Combustion Effluents," Battelle Columbus Laboratories (August 1975).

Ion Exchange Chromatography

Distinctive Requirements: 1) Preparative columns can handle up to 1-g samples. 2) Used for separation of very polar organics.
References: Jones, P. W. *et al.* "Technical Manual for Analysis of Organic Materials in Process Streams," Battelle Columbus Laboratories (January 1976).

Dispersive Infrared Spectroscopy

Distinctive Requirements: 1) Thin film on crystal. 2) For microdetermination, micropressed disk or microfilm technique is used. 3) Used for compound and class identification.
References: Shell Development Co. "Determination of Sulfur Dioxide and Sulfur Trioxide in Stack Gases," Emeryville Method Series 4 S16159 (1959). Gouw, T. H. *Guide to Modern Methods of Instrumental Analysis* (New York: Wiley-Interscience, 1973).

Fourier Transform Infrared Spectroscopy

Distinctive Requirements: 1) Thin film on crystal/KBr Pellet-greater sensitivity obviates need for micro techniques. A very useful screening technique.
References: Gouw, T. H. *Guide to Modern Methods of Instrumental Analysis* (New York: Wiley-Interscience, 1973). Jones, P. W. *et al.* "Technical Manual for Analysis of Organic Materials in Process Streams," Battelle Columbus Laboratories (January 1976).

UV-Visible Spectroscopy

Distinctive Requirements: 1) Sample is dissolved in a suitable solvent (UV/VIS). 2) Inactive in wavelength region of interest. 3) useful in identifying functional groups and certain types of compounds. 4) Used with IR, NMR, and MS.
References: Jones, P. W. *et al.* "Technical Manual for Analysis of Organic Materials in Process Streams," Battelle Columbus Laboratories (January 1976). Shell Development Co. "Determination of Sulfur Dioxide and

Sulfur Trioxide in Stack Gases," Emeryville Method, Series 4S16159 (1959).

Mass Spectroscopy-Chemical Ionization (CI)

Distinctive Requirements: 1) Several injection methods are possible with temperature programming of probe giving additional separation. 2) CH_4 and NH_3 are commonly used as ionizing gases. 3) Limited use as screening technique.
References: See UV-Visible Spectroscopy, above.

- -

Gas chromatography is used extensively for volatile organic compounds. On-site sampling may be had with use of a bag or bomb. Typical sample injection volume for liquids is 1-10 μl; for gas 1-50 ml. As a general guide to GC columns useful in organic analysis, the following list is included:

Sample	Column Type
Acids	
$C_1 - C_9$	Chromosorb 101
$C_1 - C_{18}$	FFAP
Alcohols	
$C_1 - C_5$	Poropak Q, Chromosorb 101
$C_1 - C_{18}$	Silar 5CP, Carbowax 20M, FFAP
Polyalcohols	FFAP
Aldehydes	
$C_1 - C_5$	Poropak N, DC-550, Ethofat
$C_5 - C_{18}$	Carbowax 20M, Silar 5CP
Amines	Poropak Q/PEI, Poropak R
	Chromosorb 103, Pennwalt 223
Amides	Versamid 900, Igepal CO-630
Esters	Poropak Q, Dinonylphthalate
	Chromosorb 101 or 102
Ethers	Carbowax 20M, Silar 5CP
Freons	Poropak Q, Chromosorb 102
Glycols	Chromosorb 107
Halides	OV-210, FFAP
Hydrocarbons	
$C_5 - C_{10}$	OV101, SE-30
Aromatic	Silar 5CP, Carbowax 20M
Olefins C_6 and greater	DC-550, DC-703
POM	Dexsil 300, OV-101, SE-30
Ketones	Poropak Q, Chromosorb 102, FFAP
Pesticides	OV-101, OV-225, OV-1, SE-30
Phenols	OV-17, Silar 5CP, Carbowax 20M

Gas chromatography is widely applied to all types of organic analysis; portable instruments are available for on-site stack gas analysis. It is the most commonly used technique for quantification in organic chemical analysis.

- -

SAMPLING FOR: GENERAL VOLATILE ORGANIC COMPOUNDS

METHOD: GC WITH THERMAL CONDUCTIVITY DETECTOR– DL - 0.1%

Remarks:

1. Sensitivity is limited.
2. Response is linear over a moderate range of concentration.
3. It responds to all compounds and is the most widely used detector.
4. Temperature limit is 450°C.
5. Nondestructive.
6. Used for Level 1 and 2 investigations.

METHOD: GC WITH FLAME IONIZATION DETECTOR (FID)– DL - 1 ppm

Remarks:

1. No response to a number of low-molecular-weight gases, such as air and water.
2. Destructive.
3. Widest linear range of operation gas supply (H_2O and O_2) required.
4. Temperature limit of 400°C.
5. Used for Level 1 and 2 investigations.

References:

Devaux, P., and G. Guiochon. "Determination of the Optimum Operating Conditions of the Electron Capture Detector," *J. Chromatogr. Sci.* 8:502-508 (September 1970).
Bethea, R. M., and M. C. Meador. "Basic Chromatographic Analysis of Reactive Gases in Air," *J. Chromatogr. Sci.* 7:655-664 (November 1969).

- -

SAMPLING FOR: VOLATILE ORGANIC COMPOUNDS CONTAINING PHOSPHORUS, NITROGEN OR SULFUR

METHOD: GC WITH FLAME PHOTOMETRIC DETECTOR– DL - 10 ppb

Remarks:

1. Detector is selective for nitrogen, sulfur and phosphorus-containing species.
2. Destructive.
3. It may also be used in FID mode as a nonselective detector.
4. Temperature limit 300°C.
5. Used in Level 1 and 2 investigations.

SAMPLING FOR: VOLATILE ORGANIC COMPOUNDS LOW-MOLECULAR-WEIGHT GASES

METHOD: GC WITH HELIUM IONIZATION–DL - 10 ppb

Remarks:

1. Nondestructive.
2. Detective limit depends on column bleed.
3. Usually used to measure low-molecular-weight gases.
4. Temperature limit of 100°C.
5. Columns are limited to active solids.
6. Used for Level 1 and 2 investigations.

SAMPLING FOR: VOLATILE ORGANIC COMPOUNDS– SPECIFIC GASES

METHOD: GC WITH COULOMETRIC DETECTOR–DL - 1 ppm

Remarks:

1. No calibration required for direct quantitative results.

2.   Detector may be nonselective at combustion of all effluents.
3.   Destructive.
4.   Selective detector dependent on titrant.
5.   Can be used for Level 1 and 2 investigations.

References:

Schuessler, P. W. H. "Apparatus for Determining Sulfur in Organic Compounds by Means of Gas Chromatography," *J. Chromatogr. Sci.* 763-764 (December 1969).
Struble, D. L. "Quantitative Determination of Elemental Sulfur by GLC With an Electron Capture or a Flame Photometric Detector," *J. Chromatogr. Sci.* 10:57-59 (January 1972).

- -

SAMPLING FOR: VOLATILE ORGANIC COMPOUNDS—
HALOGENS, NITRATES AND CONJUGATED CARBONYLS

METHOD: GC WITH ELECTRON CAPTURE (H^3 OR Ni^{63})

Remarks:

1.   Nondestructive.
2.   Sensitive to water.
3.   Temperature limit for H^3 is 225°C.
4.   Temperature limit for Ni^{63} is 350°C.
5.   Small linear range—selective for electrophiles.
6.   Can be used for Level 1 and 2 investigations.

- -

SAMPLING FOR: VOLATILE AND NONVOLATILE ORGANICS

METHOD: HIGH-PERFORMANCE LIQUID CHROMATOGRAPHY

Remarks:

1.   Detection limit is detector dependent.
2.   Separation of wide variety of compounds especially applicable to high-molecular-weight and thermally sensitive compounds.

3.   Small sample size required.

4.   Can be used for Level 2 investigation.

References:

ASTM Standard D2682-71.

Lao, R. C., R. S. Thomas, II. Oja and L. Dubois. "Application of a Gas Chromatograph-Mass Spectrometer-Data Processor Combination to the Analysis of the Polycyclic Aromatic Hydrocarbon Content of Airborne Pollutants," *Anal. Chem.* 45(6):908-915 (May 1973).

Dong, M., D. C. Locke and E. Ferrand. "High-Pressure Liquid Chromatographic Method for Routine Analysis of Major Parent Polycyclic Aromatic Hydrocarbons in Suspended Particulate Matter," *Anal. Chem.* 48(2):368-372 (February 1976).

SAMPLING FOR: VOLATILE AND NONVOLATILE ORGANIC COMPOUNDS SEPARATION BY MOLECULAR WEIGHT

METHOD: GEL PERMEATION CHROMATOGRAPHY

Remarks:

1.   DL for Refractive Index Detector—1 μg; for UV-visible detector—10^{-9} g.

2.   20 mg of sample/100 ml of column.

3.   Most common column packing is a styrenedivinyl benzene polymer.

4.   Mobile phase must be compatible with support and detection system.

5.   Low refractive index allows more sensitive detection with RI detector.

6.   Used for Level 2 investigations.

SAMPLING FOR: VOLATILE AND NONVOLATILE ORGANIC COMPOUNDS SEPARATION BY CHEMICAL CLASS

METHOD: BONDED-PHASE LIQUID CHROMATOGRAPHY

Remarks:

1.   DL for Refractive Index Detector—1 mg; 10^{-9} g for UV-visible detector.

2.   Preparative columns can handle up to 5-g samples.
3.   Class separation is effected by performing a gradient elution on a reversed phase column.
4.   Detection system and solvents used for gradient elution must be compatible. The preferred packing is microparticle (5-10 μ) reverse phase. With the RI detection system gradient elution is difficult.
5.   Used for Level 2 investigations.

METHOD: THIN-LAYER CHROMATOGRAPHY (TLC)

Remarks:

1.   DL is the μg range.
2.   For analysis, μg sample quantities are required.
3.   Activation requires heating at 110°C for 60 min.
4.   Data obtained from this source may be directly applicable to HPLC.
5.   Class identification is effected through use of color reagents directly on plate.
5.   Can be used for Level 1 and 2 investigations.

METHOD: LIQUID-SOLID CHROMATOGRAPHY

Remarks:

1.   DL for Refractive Index Detector—1 μg; 10^{-9} g for UV-visible detector.
2.   Analysis requires mg sample size.
3.   It is a useful screening study technique, but rather cumbersome.
4.   Difficulty arises in separating homologous series.
5.   Elution order is nonpolar to polar solvents.
6.   It is a preparative method for rough class separation.
7.   Can be used for Level 1 investigations.

References:

Schmidt, J. A. *et al.* "Applications of High-Speed Reversed-Phase Liquid Chromatography," *J. Chromatogr. Sci.* 9:645-650 (November 1971).

Hurtubise, R. J. *et al.* "Instrumentation for Thin-Layer Chromatography," *J. Chromatogr. Sci.* 2:476-491 (September 1973).

Sawicki, E., T. W. Stanley, S. McPherson and M. Morgus. "Use of Gas-Liquid and Thin-Layer Chromatography in Characterizing Air Pollutants by Fluorimetry," *Talanta Internat. J. Anal. Chem.* 13:619-629 (1966).

- -

SAMPLING FOR: VOLATILE AND NONVOLATILE ORGANICS
SEPARATION OF POLAR ORGANIC COMPOUND

METHOD: ION EXCHANGE CHROMATOGRAPHY

Remarks:

1. DL for Refractive Index Detector–1 g; 10^{-9} g for UV-visible detector.
2. Up to 1-g samples can be handled by preparative columns.
3. Used for the separation of very polar organic compounds and as a supplement to reverse-phase liquid chromatography.
4. Used after the ionic fraction has been identified by sequential analysis.
5. It has limited use for complex mixtures.
6. Used for Level 1 and 2 investigations.

- -

SAMPLING FOR: CLASS IDENTIFICATION BY
FUNCTIONAL GROUP COMPOUND IDENTIFICATION

METHOD: DISPERSIVE INFRARED SPECTROSCOPY

Remarks:

1. DL–μg range.
2. Special techniques include thin film on crystal for microdetermination, micropressed disk, or microfilm technique.
3. Dispersive infrared detector systems are widely accessible. They have an average resolution of 4 cm^{-1} over the spectral range (38 cm^{-1} - 600 cm^{-1}).
4. Can be used for Level 1 and 2 investigations.

References:

Silverstein, R. M. *et al. Spectrometric Identification of Organic Compounds* (New York: John Wiley and Sons, Inc., 1967).

METHOD: FOURIER TRANSFORM INFRARED SPECTROSCOPY

Remarks:

1. DL–10-100 mg.

2. Greater sensitivity makes micro techniques necessary.
3. Useful technique for screening.
4. Can be used for Level 1 and 2 investigations.

METHOD: GAS CHROMATOGRAPHY AND INFRARED SPECTROMETRY

Remarks:

1. DL–0.4 g.
2. Special technique involves use of a gas cell.
3. This method is limited by sample volatility.

_ _

SAMPLING FOR: CLASS IDENTIFICATION AND COMPOUND IDENTIFICATION

METHOD: NUCLEAR MAGNETIC RESONANCE (NMR)

Remarks:

1. DL–about 1 mg with computer average transients about 10 μg.
2. After rough separation has been effected it is used for functional group classification.
3. Sensitivity is increased through use of computer average transients.
4. Useful for Level 2 investigations.

References:

Silverstein, R. M., et al. Spectrometric Identification of Organic Compounds (New York: John Wiley and Sons, Inc., 1967).

METHOD: NUCLEAR MAGNETIC RESONANCE WITH FOURIER TRANSFORMS

Remarks:

1. DL–1 mg.
2. This technique has a large chemical shift range (600 ppm) and the ability to identify functional carbons.
3. It is very useful when infrared spectrometry and mass spectroscopy give inadequate information.

4. The solvent for this technique should contain no carbon or only one carbon type.
5. Useful for Level 2 investigations.

_____

SAMPLING FOR: CLASS IDENTIFICATION

METHOD: UV-VISIBLE SPECTROSCOPY (200-1000 nm)

Remarks:

1. DL varies widely between 10^{-2} and 10^{-5}.
2. This technique is useful in functional group identification and also certain compounds.
3. Sensitivities are higher when used with IR, NMR and MS.
4. Can be used for Level 1 and 2 **investigations.**

References:

Silverstein, R. M. *et al. Spectrometric Identification of Organic Compounds* (New York: John Wiley and Sons, Inc., 1967).

_____

SAMPLING FOR: COMPOUND IDENTIFICATION

METHOD: MASS SPECTROSCOPY (MS) AND ELECTRON IMPACT (EI)

Remarks:

1. DL—100-1000 μg depending on the compound.
2. Several injection methods are possible.
3. Additional separation may be obtained by temperature programming of the probe.
4. This technique is limited by sample volatility and is not useful for screening.
5. A large computer data base is imperative.
6. Used for Level 2 investigations.

References:

Silverstein, R. M. *et al. Spectrometric Identification of Organic Compounds* (New York: John Wiley and Sons, Inc., 1967).
Anbar, M., and G. A. St. John. "Field Ionization-Field Desportion Source for Nonfragmenting Mass Spectrometry," *Anal. Chem.* 48(1):198-203 (January 1976).

- -

SAMPLING FOR: COMPOUND IDENTIFICATION/ MOLECULAR WEIGHT

METHOD: MASS SPECTROSCOPY AND CHEMICAL IONIZATION

Remarks:

1. DL is compound-dependent ranging between 10-100 μg.
2. Several techniques for injection are possible, the preferred being probe insertion.
3. Additional separation may be obtained with temperature programming of probe.
4. This method is limited by the sample volatility.
5. Common ionizing gases are methane and ammonia.
6. Used for Level 2 investigation.

- -

SAMPLING FOR: COMPOUND IDENTIFICATION/ MOLECULAR DETERMINATION

METHOD: GAS CHROMATOGRAPHY AND MASS SPECTROSCOPY ELECTRON IMPACT AND CHEMICAL IONIZATION

Remarks:

1. DL is compound dependent ranging from 10-100 μg.
2. Considerations are for typical GC analysis.
3. Method is limited by sample volatility.

4. Data files are extensive for EI analysis but for CI analyses have limited value.
5. Best results are obtained with use of computer.
6. Total ion current or specific ion integration can affect quantification data.
7. Can be used for Level 1 and 2 investigations.

CHAPTER 9

ELEMENTAL ANALYSIS

INTRODUCTION

This section presents, with comments, the methods for elemental analysis. For a Level 1 investigation, the analytical method of spark source mass spectroscopy (SSMS) is used for most elements. For a Level 1 investigation, several methods are available, depending on the element analyzed.

The SSMS technique can be used to screen XAD-2-sorbent trap samples, liquid samples, organic samples and particulate matter samples. Figure 9-1 illustrates the Level 1 investigation for elemental analysis. As with other comparable analytical techniques, pretreatment and care of the sample is an integral part of the analysis. For example, the sample is placed in graphite

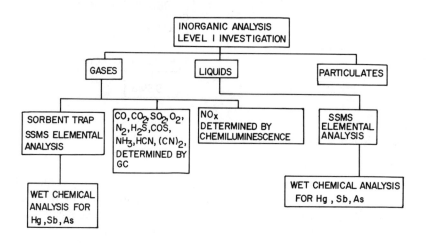

Figure 9-1. Level 1 investigation elemental analysis.

213

(a conducting medium) if it is a nonconductor; otherwise, organic matter contamination can interfere with results. Figure 9-2 shows a typical elemental analysis sample preparation for Level 1 investigation. Photographic plate and electrical detection are the two types of SSMS methods generally employed. SSMS detection because of unreliability or low accuracy and sensitivity is not normally used for arsenic, antimony, mercury, carbon, hydrogen, nitrogen and oxygen identification.

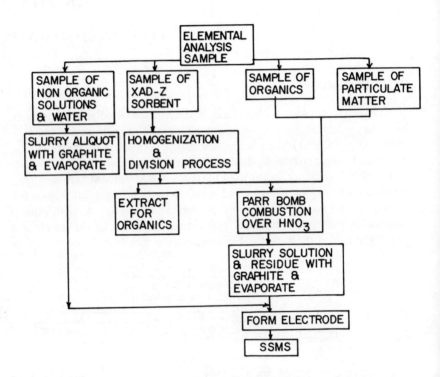

Figure 9-2. Elemental analysis sample preparation for Level 1 investigation.

For a Level 2 investigation, the analytical technique primarily employed is atomic absorption spectroscopy, either conventional (AAS) or flameless (FAAS). Variations of AAS technique can improve detection limits by three orders of magnitude (graphite furnace: FAAS) or allow for increased sensitivity to certain elements (hydride evolution technique, HE). Other modifications to the AAS method such as NAA, X-ray fluorescence (XRF), and specific ion electrode (SIE) can be used for specific areas of interest.

A number of dissolution and ashing pretreatment techniques are used with these methods and mentioned in this and other sections. The references should be consulted for specific procedures.

SAMPLING FOR: ALUMINUM

PREFERRED ANALYTICAL PROCEDURE FOR LEVEL 1 INVESTIGATION: SPARK SOURCE MASS SPECTROMETRY

Advantages:

1.   See Elemental Analysis introduction, p. 213.

Distinctive Requirements:

1.   See Elemental Analysis introduction.

References:

Ahearn, A. J., Ed. *Trace Analysis by Mass Spectrometry* (New York: Academic Press, 1972).
Hamersma, J. W., S. L. Reynolds, and R. F. Maddolone. "IERL-RTP Procedure Manual: Level of Environmental Assessment," TRW Systems Group, EPA-600/2-76-160a (June 1976).

PREFERRED ANALYTICAL PROCEDURE FOR LEVEL 2 INVESTIGATION: ATOMIC ABSORPTION SPECTROSCOPY

Advantages:

1.   See Elemental Analysis introduction, p. 213.

Distinctive Requirements:

1.   Wet   ashing, N_2O-C_2H_2.
2.   Flame, 3093A, Slit 2A.

Remarks:

1.   Detection limit for AAS is 0.1 ppm; for FAAS, 1×10^{-6} ppm.

2.   Interferences may occur due to iron, zinc, copper, calcium, and alkali and alkaline earth metals.

References:

Billings, C. E. *et al. J. Air Poll. Control Assoc.* 23(9):773-777 (September 1973).
Kim, A. G., and I. J. Douglas. *J. Chromatogr. Sci.* 11(12):615-617 (December 1973).

_____

SAMPLING FOR: ANTIMONY

PREFERRED ANALYTICAL PROCEDURE FOR LEVEL 1 INVESTIGATION: RHODAMINE B

Advantages:

1.   For a Level 1 investigation SSMS does not generate reliable data within the desired accuracy limits.

Distinctive Requirements:

1.   Organic samples and XAD-2 sorbent are wet ashed by O_2 combustion in a Parr bomb over HNO_3.
2.   Inorganics, ash and particulates are processed with aqua regia, Na_2CO_3, and then HCl.

References:

Winefordner, J. D., Ed. *Trace Analysis-Spectroscopic Methods for Elements* (New York: John Wiley and Sons, Inc., 1976).

PREFERRED ANALYTICAL PROCEDURE FOR LEVEL 2 INVESTIGATION: ATOMIC ABSORPTION SPECTROSCOPY HYDRIDE EVOLUTION TECHNIQUE (HE)

Advantages:

1.   Compared to the standard AAS method, this method is relatively matrix-independent.

Distinctive Requirements:

1. O_2 bomb combustion; HF/HNO_3 dissolution.
2. Hydride is generated by $TiCl_3$-Mg method.
3. Flame is air-C_2H_2, Z175A, Slit 7A.

Remarks:

1. Rhodamine B spectrometric method can be used in a Level 2 investigation.
2. Detection limit for AAS is 0.03 ppm.
3. Detection limit for AAS/HE is 0.004 ppm.
4. Detection limit for FAAS is 5×10^{-6} ppm.

References:

Babu, S. P. "Trace Elements in Fule," Advances in Chemistry Series, No. 141 (Washington, D. C.: American Chemical Society, 1975).

SAMPLING FOR: ARSENIC

PREFERRED ANALYTICAL PROCEDURE FOR LEVEL 1 INVESTIGATION: SILVER DIETHYL DITHIOCARBAMATE-ARSINE EVOLUTION METHOD

Advantages:

1. Reliable data for a Level 1 investigation cannot be obtained from the SSMS method.

Distinctive Requirements:

1. Organic and XAD-2 trap samples are wet ashed by O_2 combustion in a Parr bomb over HNO_3.
2. Inorganics, ash and particulates are processed with aqua regia, $NaCO_3$, then HCl.

References:

American Public Health Association. *Standard Methods for the Examination of Waste and Wastewater,* 13th ed. (1971).

PREFERRED ANALYTICAL PROCEDURE FOR LEVEL 2 INVESTIGATIONS: ATOMIC ABSORPTION SPECTROSCOPY-HYDRIDE EVOLUTION TECHNIQUE

Advantages:

1.   Compared to the standard AAS method, this method is relatively matrix-independent.

Distinctive Requirements:

1. O_2 bomb combustion; HF/HNO_3 dissolution.
2.   Hydride is generated by $TiCl_3$-Mg method.
3.   Flame is air-C_2H_2, 1937A, Slit 7A.

Remarks:

1.   Silver diethyl dithiocarbamate method can be used in a Level 2 investigation.
2.   Detection limit for AAS is 0.03 ppm.
3.   Detection limit for AAS/HE is 0.004 ppm.
4.   Detection limit for FAAS is 8×10^{-6} ppm.

References:

Babu, S. P. "Trace Elements in Fuel," Advances in Chemistry Series No. 141 (Washington, D.C.: American Chemical Society, 1975).

SAMPLING FOR: BARIUM

PREFERRED ANALYTICAL PROCEDURE FOR LEVEL 1 INVESTIGATION: SPARK SOURCE MASS SPECTROMETRY

Advantages:

1.   See Elemental Analysis introduction, p. 213.

Distinctive Requirements:

1. See Elemental Analysis introduction.

References:

Ahearn, A. J., Ed. *Trace Analysis by Mass Spectrometry* (New York: Academic Press, 1972).

Hamersma, J. W., S. L. Reynolds and R. F. Maddolone. "IERL-RTP Procedures Manual: Level of Environmental Assessment," TRW Systems Group, EPA-600/2-76-160a (June 1976).

PREFERRED ANALYTICAL PROCEDURE FOR LEVEL 2 INVESTIGATION: ATOMIC ABSORPTION SPECTROSCOPY

Advantages:

1. See Elemental Analysis introduction, p. 213.

Distinctive Requirements:

1. Wet ashing.
2. HF/HNO_3 dissolution.
3. Flame is $N_2O-C_2H_2$, 5536A, Slit 40A.

Remarks:

1. Dry ashing is acceptable.
2. Interferences may be from aluminum, silica, alkalai and alkaline earth metals.
3. Detection limit for AAS is 0.02 ppm.
4. Detection limit for FAAS is 6×10^{-6} ppm.

References:

Billings, C. E. *et al. J. Air Poll. Control Assoc.* 23(9):773-777 (September 1973).

Kim, A. G., and I. J. Douglas. *J. Chromatogr. Sci.* 11(12):615-617 (December 1973).

- -

- -

SAMPLING FOR: BERYLLIUM

PREFERRED ANALYTICAL PROCEDURE FOR LEVEL 1 INVESTIGATION: SPARK SOURCE MASS SPECTROMETRY

Advantages:

1. See Elemental Analysis introduction, p. 213.

Distinctive Requirements:

1. See Elemental Analysis introduction.

References:

Ahearn, A. J., Ed. *Trace Analysis by Mass Spectrometry* (New York: Academic Press, 1972).

Hamersma, J. W., S. L. Reynolds and R. F. Maddolone. "IERL-RTP Procedure Manual: Level of Environmental Assessment," TRW Systems Group, EPA-600/2-76-160a (June 1976).

PREFERRED ANALYTICAL PROCEDURE FOR LEVEL 2 INVESTIGATION: ATOMIC ABSORPTION SPECTROSCOPY

Advantages:

1. See Elemental Analysis introduction, p. 213.

Distinctive Requirements:

1. Wet ashing.
2. Flame is N_2O-C_2H_2, 2349A, Slit 20A.

Remarks:

1. Detection limit for AAS is 0.002 ppm.
2. Detection limit for FAAS is 3×10^{-8} ppm.

References:

Billings, C. E., *et al. J. Air Poll. Control Assoc.* 23(9):773-777 (September 1973).
Kim, A. G., and I. J. Douglas. *J. Chromatogr. Sci.* 11(12):615-617 (December 1973).

SAMPLING FOR: BISMUTH

PREFERRED ANALYTICAL PROCEDURE FOR LEVEL 1 INVESTIGATION: SPARK SOURCE MASS SPECTROMETRY

Advantages:

1.   See Elemental Analysis introduction, p. 213.

Distinctive Requirements:

1.   See Elemental Analysis introduction.

References:

Ahearn, A. J., Ed. *Trace Analysis by Mass Spectrometry* (New York: Academic Press, 1972).
Hamersma, J. W., S. L. Reynolds and R. F. Maddolone. "IERL-RTP Procedures Manual: Level of Environmental Assessment," TRW Systems Group, EPA-600/2-76-160a (June 1976).

PREFERRED ANALYTICAL PROCEDURE FOR LEVEL 2 INVESTIGATION: ATOMIC ABSORPTION SPECTROSCOPY

Advantages:

1.   See Elemental Analysis introduction, p. 213.

Distinctive Requirements:

1.   Wet ashing.
2.   Flame is air-C_2H_2, 2231A, Slit 7A.

Remarks:

1. Dry ashing is acceptable.
2. Interferences are from calcium, magnesium, sodium and potassium.
3. Detection limit for AAS is 0.04 ppm.
4. Detection limit for FAAS is 4×10^{-6} ppm.

References:

Billings, C. E. *et al. J. Air Poll. Control Assoc.* 23(9):773-777 (September 1973).
Kim, A. G., and I. J. Douglas. *J. Chromatogr. Sci.* 11(12):615-617 (December 1973).

- -

SAMPLING FOR: BORON

PREFERRED ANALYTICAL PROCEDURE FOR LEVEL 1 INVESTIGATION: SPARK SOURCE MASS SPECTROMETRY

Advantages:

1. See Elemental Analysis introduction, p. 213.

Distinctive Requirements:

1. See Elemental Analysis introduction.

References:

Ahearn, A. J., Ed. *Trace Analysis by Mass Spectrometry* (New York: Academic Press, 1972).
Hamersma, J. W., S. L. Reynolds and R. F. Maddolone. "IERL-RTP Procedures Manual: Level of Environmental Assessment," TRW Systems Group EPA-600/2-76-160a (June 1976).

PREFERRED ANALYTICAL PROCEDURE FOR
LEVEL 2 INVESTIGATION: ATOMIC ABSORPTION SPECTROSCOPY

Advantages:

1.   See Elemental Analysis introduction, p. 213.

Distinctive Requirements:

1.   Wet ashing.
2.   Flame is N_2O-C_2H_2, 2497 A, Slit 2 A.

Remarks:

1.   Detection limit for AAS is 3 ppm.
2.   Detection limit for FAAS is 2×10^{-4} ppm.

References:

Billings, C. E., et al. J. Air Poll. Control Assoc. 23(9):773-777 (September 1973).
Kim, A. G., and I. J. Douglas. J. Chromatogr. Sci. 11(12):615-617 (December 1973).

SAMPLING FOR: BROMINE

PREFERRED ANALYTICAL PROCEDURE FOR
LEVEL 1 INVESTIGATION: SPARK SOURCE MASS
SPECTROMETRY

Distinctive Requirements:

1.   See Elemental Analysis introduction, p. 213.

References:

Ahearn, A. J., Ed. Trace Analysis by Mass Spectrometry (New York: Academic Press, 1972).
Hamersma, J. W., S. L. Reynolds and R. F. Maddolone. "IERL-RTP Procedures Manual: Level of Environmental Assessment," TRW Systems Group EPA-600/2-76-160a (June 1976).

PREFERRED ANALYTICAL PROCEDURE FOR LEVEL 2 INVESTIGATION: SPECTROMETRIC

Advantages:

1.   Accurate, reproducible, rapid technique.

Distinctive Requirements:

1.   Combustion with O_2 bomb, Eschka mixture and $(NH_4)CO_3$.
2.   Spectrometric measurements are at 515 nm.

References:

ASTM Standard D1246.

- -

SAMPLING FOR: CADMIUM

PREFERRED ANALYTICAL PROCEDURE FOR LEVEL 1 INVESTIGATION: SPARK SOURCE MASS SPECTROMETRY

Advantages:

1.   See Elemental Analysis introduction, p. 213.

Distinctive Requirements:

1.   See Elemental Analysis introduction.

References:

Ahearn, A. J., Ed. *Trace Analysis by Mass Spectrometry* (New York: Academic Press, 1972).
Hamersma, J. W., S. L. Reynolds and R. F. Maddolone. "IERL-RTP Procedures Manual: Level of Environmental Assessment," TRW Systems Group EPA-600/2-76-160a (June 1976).

PREFERRED ANALYTICAL PROCEDURE FOR LEVEL 2 INVESTIGATION: ATOMIC ABSORPTION SPECTROSCOPY

Advantages:

1.   See Elemental Analysis introduction, p. 213.

Distinctive Requirements:

1. Wet ashing.
2. Flame is air-C_2H_2, 2288A, Slit 7A.

Remarks:

1. Interference can be from iron.
2. Detection limit for AAS is 0.001 ppm.
3. Detection limit for FAAS is 8×10^{-8} ppm.

References:

Billings, C. E. *et al. J. Air Poll. Control Assoc.* 23(9):773-777 (September 1973).
Kim, A. G., and I. J. Douglas. *J. Chromatogr. Sci.* 11(12):615-617 (December 1973).

- -

SAMPLING FOR: CALCIUM

PREFERRED ANALYTICAL PROCEDURE FOR LEVEL 1 INVESTIGATION: SPARK SOURCE MASS SPECTROMETRY

Advantages:

1. See Elemental Analysis introduction, p. 213.

Distinctive Requirements:

1. See Elemental Analysis introduction.

References:

Ahearn, A. J., Ed. *Trace Analysis by Mass Spectrometry* (New York: Academic Press, 1972).
Hamersma, J. W., S. L. Reynolds and R. F. Maddolone. "IERL-RTP Procedures Manual: Level of Environmental Assessment," TRW Systems Group EPA-600/2-76-160a (June 1976).

PREFERRED ANALYTICAL PROCEDURE FOR LEVEL 2 INVESTIGATION: ATOMIC ABSORPTION SPECTROSCOPY

Advantages:

1. See Elemental Analysis introduction, p. 213.

Distinctive Requirements:

1. Wet ashing.
2. Flame is air-C_2H_2, 4227A, Slit 20A.

Remarks:

1. Dry ashing is acceptable.
2. Interferences from SO_4^{-2}, PO_4^{-3}, Al, Si, Na, and K.
3. A buffer of Sr reduces interferences.

References:

Billings, C. E. *et al. J. Air Poll. Control Assoc.* 23(9):773-777 (September 1973).
Kim, A. G., and I. J. Douglas. *J. Chromatogr. Sci.* 11(12):615-617 (December 1973).

- -

SAMPLING FOR: CARBON

PREFERRED ANALYTICAL PROCEDURE FOR LEVEL 1 INVESTIGATION: COMBUSTION GRAVIMETRIC

Advantages:

1. SSMS is not used for carbon.
2. Relatively cheap, quick and common method.
3. Results are reproducible.

Distinctive Requirements:

1. Na- or K-adsorbent that is impregnated in a 8-20-mesh inert carrier.

References:

ASTM Standard D271.

PREFERRED ANALYTICAL PROCEDURE FOR
LEVEL 2 INVESTIGATION: COMBUSTION GRAVIMETRIC

Advantages:

1.   See Level 1, above.

Distinctive Requirements:

1.   See Level 1, above.

References:

ASTM Standard D271.

— —

SAMPLING FOR: CESIUM

PREFERRED ANALYTICAL PROCEDURE FOR
LEVEL 1 INVESTIGATION: SPARK SOURCE MASS
SPECTROMETRY

Advantages:

1.   See Elemental Analysis introduction, p. 213.

Distinctive Requirements:

1.   See Elemental Analysis introduction.

References:

Ahearn, A. J., Ed. *Trace Analysis by Mass Spectrometry* (New York: Academic Press, 1972).
Hamersma, J. W., S. L. Reynolds and R. F. Maddolone. "IERL-RTP Procedures Manual: Level of Environmental Assessment," TRW Systems Group EPA-600/2-76-160a (June 1976).

PREFERRED ANALYTICAL PROCEDURE FOR LEVEL 2 INVESTIGATION: ATOMIC ABSORPTION SPECTROSCOPY

Advantages:

1. See Elemental Analysis introduction, p. 213.

Distinctive Requirements:

1. Ashing is at high temperatures.
2. Dissolution HF/HNO_3.
3. Flame is air-C_2H_2, 8521A, Slit 40A.

Remarks:

1. Dry ashing is acceptable.
2. Interferences from alkalai metal ionization.
3. Detection limit for AAS is 0.05 ppm.
4. Detection limit for FAAS is 4×10^{-7} ppm.

References:

Billings, C. E., *et al. J. Air Poll. Control Assoc.* 23(9):773-777 (September 1973).
Kim, A. G., and I. J. Douglas. *J. Chromatogr. Sci.* 11(12):615-617 (December 1973).

SAMPLING FOR: CHLORINE

PREFERRED ANALYTICAL PROCEDURE FOR LEVEL 1 INVESTIGATION: SPARK SOURCE MASS SPECTROMETRY

Advantages:

1. See Elemental Analysis introduction, p. 213.

Distinctive Requirements:

1. Dissolution procedure prohibits measurement of chlorine in particulate, ash and inorganic samples.

References:

Ahearn, A. J., Ed. *Trace Analysis by Mass Spectrometry* (New York: Academic Press, 1972).

Hamersma, J. W., S. L. Reynolds and R. F. Maddolone. "IERL-RTP Procedures Manual: Level of Environmental Assessment," TRW Systems Group EPA-600/2-76-160a (June 1976).

PREFERRED ANALYTICAL PROCEDURE FOR LEVEL 2 INVESTIGATION: COMBUSTION TITRIMETRIC

Advantages:

1. Relatively quick and accurate method.
2. Results are reproducible.

Distinctive Requirements:

1. O_2 bomb combustion with $(NH_4)_2 CO_3$ or Eschka mixture.

References:

ASTM Standard D512.

SAMPLING FOR: CHROMIUM

PREFERRED ANALYTICAL PROCEDURE FOR LEVEL 1 INVESTIGATION: SPARK SOURCE MASS SPECTROMETRY

Advantages:

1. See Elemental Analysis introduction, p. 213.

Distinctive Requirements:

1. See Elemental Analysis introduction.

References:

Ahearn, A. J., Ed. *Trace Analysis by Mass Spectrometry* (New York: Academic Press, 1972).

Hamersma, J. W., S. L. Reynolds and R. F. Maddolone. "IERL-RTP Proceddures Manual: Level of Environmental Assessment," TRW Systems Group EPA-600/2-76-160a (June 1976).

PREFERRED ANALYTICAL PROCEDURE FOR LEVEL 2 INVESTIGATION: ATOMIC ABSORPTION SPECTROSCOPY

Advantages:

1.   See Elemental Analysis introduction, p. 213.

Distinctive Requirements:

1.   Wet ashing.
2.   Flame is air-C_2H_2, 3579A, Slit 2A.

Remarks:

1.   Interference from iron. A buffer of NH_4Cl eliminates interferences.
2.   Detection limit for AAS is 0.002 ppm.
3.   Detection limit for FAAS is 2×10^{-6} ppm.

References:

Billings, C. E. *et al. J. Air Poll. Control Assoc.* 23(9):773-777 (September 1973).
Kim, A. G., and I. J. Douglas. *J. Chromatogr. Sci.* 11(12):615-617 (December 1973).

- -

SAMPLING FOR: COBALT

PREFERRED ANALYTICAL PROCEDURE FOR LEVEL 1 INVESTIGATION: SPARK SOURCE MASS SPECTROMETRY

Advantages:

1.   See Elemental Analysis introduction, p. 213.

Distinctive Requirements:

1. See Elemental Analysis introduction, p. 213.

References:

Ahearn, A. J., Ed. *Trace Analysis by Mass Spectrometry* (New York: Academic Press, 1972).

Hamersma, J. W., S. L. Reynolds and R. F. Maddolone. "IERL-RTP Procedures Manual: Level of Environmental Assessment," TRW Systems Group EPA-600/2-76-160a (June 1976).

PREFERRED ANALYTICAL PROCEDURE FOR LEVEL 2 INVESTIGATION: ATOMIC ABSORPTION SPECTROSCOPY

Advantages:

1. See Elemental Analysis introduction, p. 213.

Distinctive Requirements:

1. Wet ashing.
2. Flame is $C_2 H_2$, 2407A, Slit 2A.

Remarks:

1. Dry ashing is acceptable.
2. Interferences are from Ca, Mg, K, Na.
3. Detection limit for AAS is 0.002 ppm.
4. Detection limit for FAAS is 2×10^{-6} ppm.

References:

Billings, C. E. *et al. J. Air Poll. Control Assoc.* 23(9):773-777 (September 1973).

Kim, A. G., and I. J. Douglas. *J. Chromatogr. Sci.* 11(12):615-617 (December 1973).

SAMPLING FOR: COPPER

PREFERRED ANALYTICAL PROCEDURE FOR LEVEL 1 INVESTIGATION: SPARK SOURCE MASS SPECTROMETRY

Advantages:

1. See Elemental Analysis introduction, p. 213.

Distinctive Requirements:

1. See Elemental Analysis introduction.

References:

Ahearn, A. J., Ed. *Trace Analysis by Mass Spectrometry* (New York: Academic Press, 1972).

Hamersma, J. W., S. L. Reynolds and R. F. Maddolone. "IERL-RTP Procedures Manual: Level of Environmental Assessment," TRW Systems Group EPA-600/2-76-160a (June 1976).

PREFERRED ANALYTICAL PROCEDURE FOR LEVEL 2 INVESTIGATION: ATOMIC ABSORPTION SPECTROSCOPY

Advantages:

1. See Elemental Analysis introduction, p. 213.

Distinctive Requirements:

1. Wet ashing.
2. Flame is air-C_2H_2, 3248A, Slit 7A.

Remarks:

1. Interferences are from Ca, Na and K.
2. Detection Limit for AAS is 0.004 ppm.
3. Detection Limit for FAAS is 6×10^{-7} ppm.

References:

Billings, C. E. *et al. J. Air Poll. Control Assoc.* 23(9):773-777 (September 1973).
Kim, A. G., and I. J. Douglas. *J. Chromatogr. Sci.* 11(12):615-617 (December 1973).

_ _

SAMPLING FOR: DYSPROSIUM

PREFERRED ANALYTICAL PROCEDURE FOR LEVEL 1 INVESTIGATION: SPARK SOURCE MASS SPECTROMETRY

Advantages:

1.　See Elemental Analysis introduction, p. 213.

Distinctive Requirements:

1.　See Elemental Analysis introduction.

References:

Ahearn, A. J., Ed. *Trace Analysis by Mass Spectrometry* (New York: Academic Press, 1972).
Hamersma, J. W., S. L. Reynolds and R. F. Maddolone. "IERL-RTP Procedures Manual: Level of Environmental Assessment," TRW Systems Group EPA-600/2-76-160a (June 1976).

PREFERRED ANALYTICAL PROCEDURE FOR LEVEL 2 INVESTIGATION: ATOMIC ABSORPTION SPECTROSCOPY

Advantages:

1.　See Elemental Analysis introduction, p. 213.

Distinctive Requirements:

1.　Wet ashing.
2.　Flame is $N_2O-C_2H_2$, 4212A, Slit 2A.

Remarks:

1.   Detection limit for AAS is 0.4 ppm.
2.   Detection limit for FAAS is 0.007 ppm.

References:

Billings, C. E. *et al. J. Air Poll. Control Assoc.* 23(9):773-777 (September 1973).
Kim, A. G., and I. J. Douglas. *J. Chromatogr. Sci.* 11(12):615-617 (December 1973).

SAMPLING FOR: ERBIUM

PREFERRED ANALYTICAL PROCEDURE FOR LEVEL 1 INVESTIGATION: SPARK SOURCE MASS SPECTROMETRY

Advantages:

1.   See Elemental Analysis introduction, p. 213.

Distinctive Requirements:

1.   See Elemental Analysis introduction.

References:

Ahearn, A. J., Ed. *Trace Analysis by Mass Spectrometry* (New York: Academic Press, 1972).
Hamersma, J. W., S. L. Reynolds and R. F. Maddolone. "IERL-RTP Procedures Manual: Level of Environmental Assessment," TRW Systems Group EPA-600/2-76-160a (June 1976).

PREFERRED ANALYTICAL PROCEDURE FOR LEVEL 2 INVESTIGATION: ATOMIC ABSORPTION SPECTROSCOPY

Advantages:

1.   See Elemental Analysis introduction, p. 213.

Distinctive Requirements:

1. Wet ashing.
2. Flame is N_2O-C_2H_2, 4008A, Slit 2A.

Remarks:

1. Detection limit for AAS is 0.1 ppm.

References:

Billings, C. E. *et al. J. Air Poll. Control Assoc.* 23(9):773-777 (September 1973).
Kim, A. G., and I. J. Douglas. *J. Chromatogr. Sci.* 11(12):615-617 (December 1973).

_ _

SAMPLING FOR: EUROPIUM

PREFERRED ANALYTICAL PROCEDURE FOR LEVEL 1 INVESTIGATION: SPARK SOURCE MASS SPECTROMETRY

Advantages:

1. See Elemental Analysis introduction, p. 213.

Distinctive Requirements:

1. See Elemental Analysis introduction.

References:

Ahearn, A. J., Ed. *Trace Analysis by Mass Spectrometry* (New York: Academic Press, 1972).
Hamersma, J. W., S. L. Reynolds and R. F. Maddolone. "IERL-RTP Procedures Manual: Level of Environmental Assessment," TRW Systems Group EPA-600/2-76-160a (June 1976).

PREFERRED ANALYTICAL PROCEDURE FOR
LEVEL 2 INVESTIGATION: ATOMIC ABSORPTION SPECTROSCOPY

Advantages:

1. See Elemental Analysis introduction.

Distinctive Requirements:

1. Wet ashing.
2. Flame is N_2O-C_2H_2, 4594A, Slit 2A.

Remarks:

1. Detection limit for AAS is 0.2 ppm.
2. Detection limit for FAAS is 0.02 ppm.

References:

Billings, C. E. *et al. J. Air Poll. Control Assoc.* 23(9):773-777 (September 1973).
Kim, A. G., and I. J. Douglas. *J. Chromatogr. Sci.* 11(12):615-617 (December 1973).

- -

SAMPLING FOR: FLUORINE

PREFERRED ANALYTICAL PROCEDURE FOR
LEVEL 1 INVESTIGATION: SPARK SOURCE MASS
SPECTROMETRY

Advantages:

1. See Elemental Analysis, introduction, p. 213.

Distinctive Requirements:

1. See Elemental Analysis, introduction.

References:

Ahearn, A. J., Ed. *Trace Analysis by Mass Spectrometry* (New York: Academic Press, 1972).
Hamersma, J. W., S. L. Reynolds and R. F. Maddolone. "IERL-RTP Procedures Manual: Level of Environmental Assessment," TRW Systems Group EPA-600/2-76-160a (June 1976).

PREFERRED ANALYTICAL PROCEDURE FOR
LEVEL 2 INVESTIGATION: COMBUSTION/SIE

Advantages:

1. Relatively simple, quick, and accurate method for fluorine.
2. Results are reproducible.

Distinctive Requirements:

1. Plasticware is used throughout the analysis.
2. A quartz sample holder is used and absorption is by NaOH.

Remarks:

1. More accurate than the distillation-colorimetric technique.

References:

Babu, S. P. "Trace Elements in Fuel," Advances in Chemistry Series No. 141 (Washington, D.C.: American Chemical Society, 1975).
Thomas, R. F., and R. L. Booth. *Environ. Sci. Technol.* 7(6):523-525 (June 1973).

SAMPLING FOR: GADOLINIUM

PREFERRED ANALYTICAL PROCEDURE FOR
LEVEL 1 INVESTIGATION: SPARK SOURCE MASS
SPECTROMETRY

Advantages:

1. See Elemental Analysis introduction, p. 213.

Distinctive Requirements:

1. See Elemental Analysis introduction.

References:

Ahearn, A. J., Ed. *Trace Analysis by Mass Spectrometry* (New York: Academic Press, 1972).

Hamersma, J. W., S. L. Reynolds and R. F. Maddolone. "IERL-RTP Procedures Manual: Level of Environmental Assessment," TRW Systems Group EPA-600/2-76-160a (June 1976).

PREFERRED ANALYTICAL PROCEDURE FOR
LEVEL 2 INVESTIGATION: ATOMIC ABSORPTION
SPECTROSCOPY

Advantages:

1. See Elemental Analysis introduction, p. 213.

Distinctive Requirements:

1. Wet ashing.
2. Flame is $N_2 O$-$C_2 H_2$, 3684A, Slit 2A.

Remarks:

1. Detection limit for AAS is 4 ppm.

References:

Billings, C. E. *et al. J. Air Poll. Control Assoc.* 23(9):773-777 (September 1973).
Kim, A. G., and I. J. Douglas. *J. Chromatogr. Sci.* 11(12):615-617 (December 1973).

SAMPLING FOR: GALLIUM

PREFERRED ANALYTICAL PROCEDURE FOR
LEVEL 1 INVESTIGATION: SPARK SOURCE MASS
SPECTROMETRY

Advantages:

1. See Elemental Analysis introduction, p. 213.

Distinctive Requirements:

1. See Elemental Analysis introduction.

References:

Ahearn, A. J., Ed. *Trace Analysis by Mass Spectrometry* (New York: Academic Press, 1972).

Hamersma, J. W., S. L. Reynolds and R. F. Maddolone. "IERL-RTP Procedures Manual: Level of Environmental Assessment," TRW Systems Group EPA-600/2-76-160a (June 1976).

PREFERRED ANALYTICAL PROCEDURE FOR LEVEL 2 INVESTIGATION: ATOMIC ABSORPTION SPECTROSCOPY

Advantages:

1. See Elemental Analysis introduction, p. 213.

Distinctive Requirements:

1. Ashing at high temperatures.
2. Dissolution HF/HNO_3.
3. Flame is air-C_2H_2, 2874A, Slit 20A.

Remarks:

1. Interference from aluminum.
2. Detection limit for AAS is 0.05 ppm.
3. Detection limit for FAAS is 4×10^{-4} ppm.

References:

Billings, C. E. *et al. J. Air Poll. Control Assoc.* 23(9):773-777 (September 1973).

Kim, A. G., and I. J. Douglas. *J. Chromatogr. Sci.* 11(12):615-617 (December 1973).

_ _

SAMPLING FOR: GERMANIUM

PREFERRED ANALYTICAL PROCEDURE FOR LEVEL 1 INVESTIGATION: SPARK SOURCE MASS SPECTROMETRY

Advantages:

1. See Elemental Analysis introduction, p. 213.

Distinctive Requirements:

1.   See Elemental Analysis introduction.

References:

Ahearn, A. J., Ed. *Trace Analysis by Mass Spectrometry* (New York: Academic Press, 1972).
Hamersma, J. W., S. L. Reynolds and R. F. Maddolone. "IERL-RTP Procedures Manual: Level of Environmental Assessment," TRW Systems Group EPA-600/2-76-160a (June 1976).

**PREFERRED ANALYTICAL PROCEDURE FOR
LEVEL 2 INVESTIGATION: ATOMIC ABSORPTION
SPECTROSCOPY**

Advantages:

1.   See Elemental Analysis introduction, p. 213.

Distinctive Requirements:

1.   Combustion over HNO_3 in O_2 bomb.
2.   Wet ashing.
3.   Flame is N_2O-C_2H_2, 2652A, Slit 2A.

Remarks:

1.   Losses are increased by excess chloride.
2.   Detection limit for AAS is 0.1 ppm.
3.   Detection limit for FAAS is 3×10^{-6} ppm.

SAMPLING FOR: GOLD

**PREFERRED ANALYTICAL PROCEDURE FOR
LEVEL 1 INVESTIGATION: SPARK SOURCE MASS
SPECTROMETRY**

Advantages:

1.   See Elemental Analysis introduction, p. 213.

Distinctive Requirements:

1. See Elemental Analysis introduction.

References:

Ahearn, A. J., Ed. *Trace Analysis by Mass Spectrometry* (New York: Academic Press, 1972).

Hamersma, J. W., S. L. Reynolds and R. F. Maddolone. "IERL-RTP Procedures Manual: Level of Environmental Assessment," TRW Systems Group EPA-600/2-76-160a (June 1976).

PREFERRED ANALYTICAL PROCEDURE FOR LEVEL 2 INVESTIGATION: ATOMIC ABSORPTION SPECTROSCOPY

Advantages:

1. See Elemental Analysis introduction, p. 213.

Distinctive Requirements:

1. Wet ashing.
2. Flame is $N_2O\text{-}C_2H_2$, 2428A, Slit 20A.

Remarks:

1. Interferences from iron and calcium.
2. Detection limit for AAS is 0.02 ppm.
3. Detection limit for FAAS is 1×10^{-6} ppm.

References:

Billings, C. E., *et al. J. Air Poll. Control Assoc.* 23(9):773-777 (September 1973).

Kim, A. G., and I. J. Douglas. *J. Chromatogr. Sci.* 11(12):615-617 (December 1973).

- -

SAMPLING FOR: HAFNIUM

PREFERRED ANALYTICAL PROCEDURE FOR LEVEL 1 INVESTIGATION: SPARK SOURCE MASS SPECTROMETRY

Advantages:

1.   See Elemental Analysis introduction, p. 213.

Distinctive Requirements:

1.   See Elemental Analysis introduction.

References:

Ahearn, A. J., Ed. *Trace Analysis by Mass Spectrometry* (New York: Academic Press, 1972).
Hamersma, J. W., S. L. Reynolds and R. F. Maddolone. "IERL-RTP Procedures Manual: Level of Environmental Assessment," TRW Systems Group EPA-600/2-76-160a (June 1976).

PREFERRED ANALYTICAL PROCEDURE FOR LEVEL 2 INVESTIGATION: ATOMIC ABSORPTION SPECTROSCOPY

Advantages:

1.   See Elemental Analysis introduction, p. 213.

Distinctive Requirements:

1.   Fusion with $Li_2 B_4 O_7$.
2.   Dissolution HCl-HNO_3.
3.   Flame is $N_2 O$-$C_2 H_2$, 3073A, Slit 2A.

Remarks:

1.   Wet ashing is acceptable.
2.   Detection limit for AAS is 15 ppm.

References:

Billings, C. E. *et al. J. Air Poll. Control Assoc.* 23(9):773-777 (September 1973).
Kim, A. G., and I. J. Douglas. *J. Chromatogr. Sci.* 11(12):615-617 (December 1973).

SAMPLING FOR: HOLMIUM

PREFERRED ANALYTICAL PROCEDURE FOR LEVEL 1 INVESTIGATION: SPARK SOURCE MASS SPECTROMETRY

Advantages:

1.   See Elemental Analysis introduction, p. 213.

Distinctive Requirements:

1.   See Elemental Analysis introduction.

References:

Ahearn, A. J., Ed. *Trace Analysis by Mass Spectrometry* (New York: Academic Press, 1972).
Hamersma, J. W., S. L. Reynolds and R. F. Maddolone. "IERL-RTP Procedures Manual: Level of Environmental Assessment," TRW Systems Group EPA-600/2-76-160a (June 1976).

PREFERRED ANALYTICAL PROCEDURE FOR LEVEL 2 INVESTIGATION: ATOMIC ABSORPTION SPECTROSCOPY

Advantages:

1.   See Elemental Analysis introduction, p. 213.

Distinctive Requirements:

1.   Wet ashing.
2.   Flame is air-C_2H_2, 4104A, Slit 2A.

Remarks:

1. Detection limit for AAS is 0.3 ppm.

References:

Billings, C. E. *et al. J. Air Poll. Control Assoc.* 23(9):773-777 (September 1973).
Kim, A. G., and I. J. Douglas. *J. Chromatogr. Sci.* 11(12):615-617 (December 1973).

. SAMPLING FOR: HYDROGEN

PREFERRED ANALYTICAL PROCEDURE FOR LEVEL 1 INVESTIGATION: COMBUSTION

Advantages:

1. More suitable for hydrogen than SSMS.

Distinctive Requirements:

1. Anhydrous absorbent ($Mg(ClO_4)_2$) of 8-45 mesh size.

References:

ASTM Standard D271.

PREFERRED ANALYTICAL PROCEDURE FOR LEVEL 2 INVESTIGATION: COMBUSTION

Advantages:

1. See Level 1, above.

Distinctive Requirements:

1. See Level 1, above.

Remarks:

1. To remove moisture interference from analysis, sample should be dried at 104-110°C.
2. Water of hydration of inorganic compounds will be included in analysis.

References:

ASTM Standard D271.

- -

SAMPLING FOR: LANTHANUM

PREFERRED ANALYTICAL PROCEDURE FOR LEVEL 1 INVESTIGATION: SPARK SOURCE MASS SPECTROMETRY

Advantages:

1. See Elemental Analysis introduction, p. 213.

Distinctive Requirements:

1. See Elemental Analysis introduction.

References:

Ahearn, A. J., Ed. *Trace Analysis by Mass Spectrometry* (New York: Academic Press, 1972).
Hamersma, J. W., S. L. Reynolds and R. F. Maddolone. "IERL-RTP Procedures Manual: Level of Environmental Assessment," TRW Systems Group EPA-600/2-76-160a (June 1976).

PREFERRED ANALYTICAL PROCEDURE FOR LEVEL 2 INVESTIGATION: ATOMIC ABSORPTION SPECTROSCOPY

Advantages:

1. See Elemental Analysis introduction, p. 213.

Distinctive Requirements:

1. Wet ashing.
2. Flame is $N_2 O-C_2 H_2$, 5501A, Slit 4A.

Remarks:

1. Detection limit for AAS is 2 ppm.

References:

Billings, C. E. *et al. J. Air Poll. Control Assoc.* 23(9):773-777 (September 1973).
Kim, A. G., and I. J. Douglas. *J. Chromatogr. Sci.* 11(12):615-617 (December 1973).

SAMPLING FOR: INDIUM

PREFERRED ANALYTICAL PROCEDURE FOR LEVEL 1 INVESTIGATION: SPARK SOURCE MASS SPECTROMETRY

Advantages:

1. See Elemental Analysis introduction, p. 213.

Distinctive Requirements:

1. See Elemental Analysis introduction.

References:

Ahearn, A. J., Ed. *Trace Analysis by Mass Spectrometry* (New York: Academic Press, 1972).
Hamersma, J. W., S. L. Reynolds and R. F. Maddolone. "IERL-RTP Procedures Manual: Level of Environmental Assessment," TRW Systems Group EPA-600/2-76-160a (June 1976).

PREFERRED ANALYTICAL PROCEDURE FOR
LEVEL 2 INVESTIGATION: ATOMIC ABSORPTION SPECTROSCOPY

Advantages:

1.   See Elemental Analysis introduction, p. 213.

Distinctive Requirements;

1.   Fusion with $Li_2B_4O_7$.
2.   Dissolution $HCl-HNO_3$.
3.   Flame is air-C_2H_2, 3039A, Slit 7A.

Remarks:

1.   Dry ashing is acceptable.
2.   Detection limit for AAS is 0.03 ppm.
3.   Detection limit for FAAS is 4×10^{-7} ppm.

References:

Billings, C. E. *et al. J. Air Poll. Control Assoc.* 23(9):773-777 (September 1973).
Kim, A. G., and I. J. Douglas. *J. Chromatogr. Sci.* 11(12):615-617 (December 1973).

- -

SAMPLING FOR: IODINE

PREFERRED ANALYTICAL PROCEDURE FOR
LEVEL 1 INVESTIGATION: SPARK SOURCE MASS
SPECTROMETRY

Advantages:

1.   See Elemental Analysis introduction, p. 213.

Distinctive Requirements:

1.   See Elemental Analysis introduction.

References:

Ahearn, A. J., Ed. *Trace Analysis by Mass Spectrometry* (New York: Academic Press, 1972).

Hamersma, J. W., S. L. Reynolds and R. F. Maddolone. "IERL-RTP Procedures Manual: Level of Environmental Assessment," TRW Systems Group EPA-600/2-76-160a (June 1976).

PREFERRED ANALYTICAL PROCEDURE FOR
LEVEL 2 INVESTIGATION: COMBUSTION SPECTROMETRIC

Advantages:

1. Common, rapid technique.
2. Results are reproducible.

Distinctive Requirements:

1. Combustion with O_2 bomb.
2. For spectrometric technique measure at 45 nm.

References:

ASTM Standard D1246.

_ _

SAMPLING FOR: IRIDIUM

PREFERRED ANALYTICAL PROCEDURE FOR
LEVEL 1 INVESTIGATION: SPARK SOURCE MASS
SPECTROMETRY

Advantages:

1. See Elemental Analysis introduction, p. 213.

Distinctive Requirements:

1. See Elemental Analysis introduction.

References:

Ahearn, A. J., Ed. *Trace Analysis by Mass Spectrometry* (New York: Academic Press, 1972).
Hamersma, J. W., S. L. Reynolds and R. F. Maddolone. "IERL-RTP Procedures Manual: Level of Environmental Assessment," TRW Sytems Group EPA-600/2-76-160a (June 1976).

PREFERRED ANALYTICAL PROCEDURE FOR
LEVEL 2 INVESTIGATION: ATOMIC ABSORPTION
SPECTROSCOPY

Advantages:

1.   See Elemental Analysis introduction, p. 213.

Distinctive Requirements:

1.   Wet ashing.
2.   Flame is $N_2O-C_2H_2$, 2640A, Slit 2A.

Remarks:

1.   Detection limit for AAS is 1 ppm.

References:

Billings, C. E. *et al. J. Air Poll. Control Assoc.* 23(9):773-777 (September 1973).
Kim, A. G., and I. J. Douglas. *J. Chromatogr. Sci.* 11(12):615-617 (December 1973).

SAMPLING FOR: IRON

PREFERRED ANALYTICAL PROCEDURE FOR
LEVEL 1 INVESTIGATION: SPARK SOURCE MASS
SPECTROMETRY

Advantages:

1.   See Elemental Analysis introduction, p. 213.

Distinctive Requirements:

1.   See Elemental Analysis introduction.

References:

Ahearn, A. J., Ed. *Trace Analysis by Mass Spectrometry* (New York: Academic Press, 1972).

Hamersma, J. W., S. L. Reynolds and R. F. Maddolone. "IERL-RTP Procedures Manual: Level of Environmental Assessment," TRW Systems Group EPA-600/2-76-160a (June 1976).

PREFERRED ANALYTICAL PROCEDURE FOR LEVEL 2 INVESTIGATION: ATOMIC ABSORPTION SPECTROSCOPY

Advantages:

1.   See Elemental Analysis introduction, p. 213.

Distinctive Requirements:

1.   Wet ashing.
2.   Flame is air-C_2H_2, 2483A, Slit 2A.

Remarks:

1.   Interferences from Ca and Si.
2.   Detection limit for AAS is 0.004 ppm.
3.   Detection limit for FAAS is 1×10^{-5} ppm.

References:

Billings, C. E. *et al. J. Air Poll. Control Assoc.* 23(9):773-777 (September 1973).
Kim, A. G., and I. J. Douglas. *J. Chromatogr. Sci.* 11(12):615-617 (December 1973).

- -

SAMPLING FOR: LEAD

PREFERRED ANALYTICAL PROCEDURE FOR LEVEL 1 INVESTIGATION: SPARK SOURCE MASS SPECTROMETRY

Advantages:

1.   See Elemental Analysis introduction, p. 213.

Distinctive Requirements:

1. See Elemental Analysis introduction.

References:

Ahearn, A. J., Ed. *Trace Analysis by Mass Spectrometry* (New York: Academic Press, 1972).

Hamersma, J. W., S. L. Reynolds and R. F. Maddolone. "IERL-RTP Procedures Manual: Level of Environmental Assessment," TRW Systems Group EPA-600/2-76-160a (June 1976).

PREFERRED ANALYTICAL PROCEDURE FOR LEVEL 2 INVESTIGATION: ATOMIC ABSORPTION SPECTROSCOPY

Advantages:

1. See Elemental Analysis introduction, p. 213.

Distinctive Requirements:

1. Wet ashing.
2. Flame is air-C_2H_2, 2833A, Slit 20A.

Remarks:

1. Interferences from Al, Th, Zr, Ca, Mg, K and Na.
2. Detection limit for AAS is 0.01 ppm.
3. Detection limit for FAAS is 2×10^{-6} ppm.

References:

Billings, C. E. *et al. J. Air Poll. Control Assoc.* 23(9):773-777 (September 1973).

Kim, A. G., and I. J. Douglas. *J. Chromatogr. Sci.* 11(12):615-617 (December 1973).

SAMPLING FOR: LITHIUM

PREFERRED ANALYTICAL PROCEDURE FOR LEVEL 1 INVESTIGATION: SPARK SOURCE MASS SPECTROMETRY

Advantages:

1.   See Elemental Analysis introduction, p. 213.

Distinctive Requirements:

1.   See Elemental Analysis introduction.

References:

Ahearn, A. J., Ed. *Trace Analysis by Mass Spectrometry* (New York: Academic Press, 1972).

Hamersma, J. W., S. L. Reynolds and R. F. Maddolone. "IERL-RTP Procedures Manual: Level of Environmental Assessment," TRW Systems Group EPA-600/2-76-160a (June 1976).

PREFERRED ANALYTICAL PROCEDURE FOR LEVEL 2 INVESTIGATION: ATOMIC ABSORPTION SPECTROSCOPY

Advantages:

1.   See Elemental Analysis introduction, p. 213.

Distinctive Requirements:

1.   Wet ashing.
2.   Flame is air-C_2H_2, 6708A, Slit 40A.

Remarks:

1.   Interference from Sr.
2.   Detection limit for AAS is 0.001 ppm.
3.   Detection limit for FAAS is 3×10^{-6} ppm.

References:

Billings, C. E. *et al. J. Air Poll. Control Assoc.* 23(9):773-777 (September 1973).

Kim, A. G., and I. J. Douglas. *J. Chromatogr. Sci.* 11(12):615-617 (December 1973).

- -

SAMPLING FOR: LUTETIUM

PREFERRED ANALYTICAL PROCEDURE FOR
LEVEL 1 INVESTIGATION: SPARK SOURCE MASS
SPECTROMETRY

Advantages:

1. See Elemental Analysis introduction, p. 213.

Distinctive Requirements:

1. See Elemental Analysis introduction.

References:

Ahearn, A. J., Ed. *Trace Analysis by Mass Spectrometry* (New York: Academic Press, 1972).

Hamersma, J. W., S. L. Reynolds and R. F. Maddolone. "IERL-RTP Procedures Manual: Level of Environmental Assessment," TRW Systems Group EPA-600/2-76-160a (June 1976).

PREFERRED ANALYTICAL PROCEDURE FOR
LEVEL 2 INVESTIGATION: ATOMIC ABSORPTION
SPECTROSCOPY

Advantages:

1. See Elemental Analysis introduction, p. 213.

Distinctive Requirements:

1. Wet ashing.
2. Flame is $N_2 O-C_2 H_2$, 3312A, Slit 7A.

Remarks:

1. Detection limit for AAS is 3 ppm.

References:

Billings, C. E. *et al. J. Air Poll. Control Assoc.* 23(9):773-777 (September 1973).
Kim, A. G., and I. J. Douglas. *J. Chromatogr. Sci.* 11(12):615-617 (December 1973).

SAMPLING FOR: MAGNESIUM

PREFERRED ANALYTICAL PROCEDURE FOR LEVEL 1 INVESTIGATION: SPARK SOURCE MASS SPECTROMETRY

Advantages:

1.   See Elemental Analysis introduction, p. 213.

Distinctive Requirements:

1.   See Elemental Analysis introduction.

References:

Ahearn, A. J., Ed. *Trace Analysis by Mass Spectrometry* (New York: Academic Press, 1972).
Hamersma, J. W., S. L. Reynolds and R. F. Maddolone. "IERL-RTP Procedures Manual: Level of Environmental Assessment," TRW Systems Group EPA-600/2-76-160a (June 1976).

PREFERRED ANALYTICAL PROCEDURE FOR LEVEL 2 INVESTIGATION: ATOMIC ABSORPTION SPECTROSCOPY

Advantages:

1.   See Elemental Analysis introduction, p. 213.

Distinctive Requirements:

1.   Wet ashing.
2.   Flame is air-C_2H_2, 2852A, Slit 20A.

Remarks:

1. Dry ashing is acceptable.
2. Interferences from Al, Si, P, SO_4^{-2}.
3. Buffers to reduce interferences include Ni, La and Sc.
4. Detection limit for AAS is 0.003 ppm.
5. Detection limit for FAAS is 4×10^{-8} ppm.

References:

Billings, C. E. *et al. J. Air Poll. Control Assoc.* 23(9):773-777 (September 1973).
Kim, A. G., and I. J. Douglas. *J. Chromatogr. Sci.* 11(12):615-617 (December 1973).

_ _

SAMPLING FOR: MANGANESE

PREFERRED ANALYTICAL PROCEDURE FOR LEVEL 1 INVESTIGATION: SPARK SOURCE MASS SPECTROMETRY

Advantages:

1. See Elemental Analysis introduction, p. 213.

Distinctive Requirements:

1. See Elemental Analysis introduction.

References:

Ahearn, A. J., Ed. *Trace Analysis by Mass Spectrometry* (New York: Academic Press, 1972).
Hamersma, J. W., S. L. Reynolds and R. F. Maddolone. "IERL-RTP Procedures Manual: Level of Environmental Assessment," TRW Systems Group EPA-600/2-76-160a (June 1976).

PREFERRED ANALYTICAL PROCEDURE FOR LEVEL 2 INVESTIGATION: ATOMIC ABSORPTION SPECTROSCOPY

Advantages:

1. See Elemental Analysis introduction, p. 213.

Distinctive Requirements:

1. Wet ashing.
2. Flame is air $C_2 H_2$, 2795A, Slit 7A.

Remarks:

1. Interferences from Cr and Si. Ca buffer reduces interference effects.
2. Detection limit for AAS is 0.0008 ppm.
3. Detection limit for FAAS is 2×10^{-7} ppm.

References:

Billings, C. E. *et al. J. Air Poll. Control Assoc.* 23(9):773-777 (September 1973).
Kim, A. G., and I. J. Douglas. *J. Chromatogr. Sci.* 11(12):615-617 (December 1973).

_____

SAMPLING FOR: NEODYMIUM

PREFERRED ANALYTICAL PROCEDURE FOR LEVEL 1 INVESTIGATION: SPARK SOURCE MASS SPECTROMETRY

Advantages:

1. See Elemental Analysis introduction, p. 213.

Distinctive Requirements:

1. See Elemental Analysis introduction.

References:

Ahearn, A. J., Ed. *Trace Analysis by Mass Spectrometry* (New York: Academic Press, 1972).
Hamersma, J. W., S. L. Reynolds and R. F. Maddolone. "IERL-RTP Procedures Manual: Level of Environmental Assessment," TRW Systems Group EPA-600/2-76-160a (June 1976).

PREFERRED ANALYTICAL PROCEDURE FOR LEVEL 2 INVESTIGATION: ATOMIC ABSORPTION SPECTROSCOPY

Advantages:

1. See Elemental Analysis introduction.

Distinctive Requirements:

1. Wet ashing.
2. Flame is $N_2O-C_2H_2$, 4634A, Slit 2A.

Remarks:

1. Spectral interferences from Pr at 4925A.
2. Detection limit for AAS is 1 ppm.

References:

Billings, C. E. *et al. J. Air Poll. Control Assoc.* 23(9):773-777 (September 1973).
Kim, A. G., and I. J. Douglas. *J. Chromatogr. Sci.* 11(12):615-617 (December 1973).

SAMPLING FOR: MERCURY

PREFERRED ANALYTICAL PROCEDURE FOR LEVEL 1 INVESTIGATION: FLAMELESS ATOMIC ABSORPTION SPECTROSCOPY

Advantages:

1. Accurate, while accuracy of SSMS is questionable for Level 1 investigation.

Distinctive Requirements:

1. Samples are processed as follows: Organic samples and XAD-2 sorbent are wet ashed by O_2 combustion in a Paar bomb over HNO_3. Inorganics,

ash, and particulates are processed with aqua regia, Na_2CO_3, and then HCl.

References:

Kalb, G. W., and C. Baldeck. "The Development of the Gold Amalgamation Sampling and Analytical Procedure for Investigation of Mercury in Stack Gases," Trodet, Inc. (June 1972).

TRW Systems Group. "Procedures for Process Measurements—Trace Inorganic Materials," (July 1975).

PREFERRED ANALYTICAL PROCEDURE FOR LEVEL 2 INVESTIGATION: FLAMELESS ATOMIC ABSORPTION SPECTROSCOPY

Advantages:

1.   See Level 1, above.

Distinctive Requirements:

1.   Argon carrier 2795 A.

Remarks:

1.   Detection limit for AAS is 0.5 ppm.
2.   Detection limit for FAAS is 2×10^{-5} ppm.

- -

SAMPLING FOR: MOLYBDENUM

PREFERRED ANALYTICAL PROCEDURE FOR LEVEL 1 INVESTIGATION: SPARK SOURCE MASS SPECTROMETRY

Advantages:

1.   See Elemental Analysis introduction, p. 213.

Distinctive Requirements:

1.   See Elemental Analysis introduction.

References:

Ahearn, A. J., Ed. *Trace Analysis by Mass Spectrometry* (New York: Academic Press, 1972).

Hamersma, J. W., S. L. Reynolds and R. F. Maddolone. "IERL-RTP Procedures Manual: Level of Environmental Assessment," TRW Systems Group EPA-600/2-76-160a (June 1976).

PREFERRED ANALYTICAL PROCEDURE FOR LEVEL 2 INVESTIGATION: ATOMIC ABSORPTION SPECTROSCOPY

Advantages:

1.　See Elemental Analysis introduction, p. 213.

Distinctive Requirements:

1.　Fusion with $Li_2B_4O_7$.
2.　Dissolution HCl-HNO_3.
3.　Flame is N_2O-C_2H_2, 3133A, Slit 2A.

Remarks:

1.　Wet ashing is acceptable.
2.　Interferences from Fe, Sr, Mn, Ca and $Al(NO_3)_3$.
3.　NH_4Cl buffer can reduce interferences from Fe and Mn.
4.　Detection limit for AAS is 0.03 ppm.
5.　Detection limit for FAAS is 3×10^{-6} ppm.

References:

Billings, C. E. *et al. J. Air Poll. Control Assoc.* 23(9):773-777 (September 1973).

Kim, A. G., and I. J. Douglas. *J. Chromatogr. Sci.* 11(12):615-617 (December 1973).

_ _

SAMPLING FOR: NICKEL

PREFERRED ANALYTICAL PROCEDURE FOR LEVEL 1 INVESTIGATION: SPARK SOURCE MASS SPECTROMETRY

Advantages:

1.　See Elemental Analysis introduction, p. 213.

Distinctive Requirements:

1.   See Elemental Analysis introduction.

References:

Ahearn, A. J., Ed. *Trace Analysis by Mass Spectrometry* (New York: Academic Press, 1972).
Hamersma, J. W., S. L. Reynolds and R. F. Maddolone. "IERL-RTP Procedures Manual: Level of Environmental Assessment," TRW Systems Group EPA-600/2-76-160a (June 1976).

PREFERRED ANALYTICAL PROCEDURE FOR LEVEL 2 INVESTIGATION: ATOMIC ABSORPTION SPECTROSCOPY

Advantages:

1.   See Elemental Analysis introduction, p. 213.

Distinctive Requirements:

1.   Wet ashing.
2.   Flame is air-C_2H_2, 2302A, Slit 2A.

Remarks:

1.   Dry ashing is acceptable.
2.   Interferences from Ca.
3.   Detection limit for AAS is 0.005 ppm.
4.   Detection limit for FAAS is 9×10^{-6} ppm.

References:

Billings, C. E. *et al. J. Air Poll. Control Assoc.* 23(9):773-777 (September 1973).
Kim, A. G., and I. J. Douglas. *J. Chromatogr. Sci.* 11(12):615-617 (December 1973).

SAMPLING FOR: NIOBIUM

PREFERRED ANALYTICAL PROCEDURE FOR LEVEL 1 INVESTIGATION: SPARK SOURCE MASS SPECTROMETRY

Advantages:

1. See Elemental Analysis introduction, p. 213.

Distinctive Requirements:

1. See Elemental Analysis introduction.

References:

Ahearn, A. J., Ed. *Trace Analysis by Mass Spectrometry* (New York: Academic Press, 1972).
Hamersma, J. W., S. L. Reynolds and R. F. Maddolone. "IERL-RTP Procedures Manual: Level of Environmental Assessment," TRW Systems Group EPA-600/2-76-160a (June 1976).

PREFERRED ANALYTICAL PROCEDURE FOR LEVEL 2 INVESTIGATION: ATOMIC ABSORPTION SPECTROSCOPY

Advantages:

1. See Elemental Analysis introduction, p. 213.

Distinctive Requirements:

1. Wet ashing.
2. Flame is N_2O-C_2H_2, 3344A, Slit 2A.

Remarks:

1. Detection limit for AAS is 5 ppm.

References:

Billings, C. E. *et al. J. Air Poll. Control Assoc.* 23(9):773-777 (September 1973).
Kim, A. G., and I. J. Douglas. *J. Chromatogr. Sci.* 11(12):615-617 (December 1973).

SAMPLING FOR: NITROGEN

PREFERRED ANALYTICAL PROCEDURE FOR LEVEL 1 INVESTIGATION: COMBUSTION TITRIMETRIC/SPECTROMETRIC

Advantages:

1. More suitable for measuring hydrogen than is SSMS.
2. Quick and low in cost.

PREFERRED ANALYTICAL PROCEDURE FOR LEVEL 2 INVESTIGATION: KJELDAHL DIGESTION FOLLOWED BY SPECTROMETRIC OR TITRIMETRIC PROCEDURE

Advantages:

1. Common, accurate method.
2. Results are reproducible.

Distinctive Requirements:

1. Mercury, sulfuric acid and CrO_3 digestion utilized.

Remarks:

1. Reference outline elaborates on method.

References:

Dee, A., H. H. Martens, C. I. Merrill and J. T. NaKamura. *Anal. Chem.* 45(8): 1477-1481 (July 1973).

- -

SAMPLING FOR: OSMIUM

PREFERRED ANALYTICAL PROCEDURE FOR LEVEL 1 INVESTIGATION: SPARK SOURCE MASS SPECTROMETRY

Advantages:

1. See Elemental Analysis introduction, p. 213.

Distinctive Requirements:

1. See Elemental Analysis introduction.

References:

Ahearn, A. J., Ed. *Trace Analysis by Mass Spectrometry* (New York: Academic Press, 1972).

Hamersma, J. W., S. L. Reynolds and R. F. Maddolone. "IERL-RTP Procedures Manual: Level of Environmental Assessment," TRW Systems Group EPA-600/2-76-160a (June 1976).

PREFERRED ANALYTICAL PROCEDURE FOR LEVEL 2 INVESTIGATION: ATOMIC ABSORPTION SPECTROSCOPY

Advantages:

1. See Elemental Analysis introduction, p. 213.

Distinctive Requirements:

1. Dry ashing.
2. Flame is air-C_2H_2, 2909A, Slit 2A.

Remarks:

1. Detection limit for AAS is 1 ppm.

References:

Billings, C. E. *et al. J. Air Poll. Control Assoc.* 23(9):773-777 (September 1973).

Kim, A. G., and I. J. Douglas. *J. Chromatogr. Sci.* 11(12):615-617 (December 1973).

SAMPLING FOR: OXYGEN

PREFERRED ANALYTICAL PROCEDURE FOR
LEVEL 1 INVESTIGATION: COMBUSTION

Advantages:

1. More suitable for measuring oxygen than is SSMS.
2. Accurate, quick and low in cost.
3. Results are reproducible.

Distinctive Requirements:

1. Combustion and conversion to CO.

References:

Kuch, A. J., A. J. Andreatch and J. P. Mohns. *Anal. Chem.* 39(11):1249-1254 (September 1967).

PREFERRED ANALYTICAL PROCEDURE FOR
LEVEL 2 INVESTIGATION: COMBUSTION

Advantages:

1. See Level 1, above.

Distinctive Requirements:

1. See Level 1, above.

SAMPLING FOR: PALLADIUM

PREFERRED ANALYTICAL PROCEDURE FOR
LEVEL 1 INVESTIGATION: SPARK SOURCE MASS
SPECTROMETRY

Advantages:

1.   See Elemental Analysis introduction, p. 213.

Distinctive Requirements:

1.   See Elemental Analysis introduction.

References:

Ahearn, A. J., Ed. *Trace Analysis by Mass Spectrometry* (New York: Academic Press, 1972).
Hamersma, J. W., S. L. Reynolds and R. F. Maddolone. "IERL-RTP Procedures Manual: Level of Environmental Assessment," TRW Systems Group EPA-600/2-76-160a (June 1976).

PREFERRED ANALYTICAL PROCEDURE FOR
LEVEL 2 INVESTIGATION: ATOMIC ABSORPTION
SPECTROSCOPY

Advantages:

1.   See Elemental Analysis introduction.

Distinctive Requirements:

1.   Wet ashing.
2.   Flame is air-C_2H_2, 2476A, Slit 2A.

Remarks:

1.   Detection limit for AAS is 0.01 ppm.
2.   Detection limit for FAAS is 4×10^{-6} ppm.

References:

Billings, C. E. *et al. J. Air Poll. Control Assoc.* 23(9):773-777 (September 1973).
Kim, A. G., and I. J. Douglas. *J. Chromatogr. Sci.* 11(12):615-617 (December 1973).

SAMPLING FOR: PHOSPHORUS

PREFERRED ANALYTICAL PROCEDURE FOR LEVEL 1 INVESTIGATION: SPARK SOURCE MASS SPECTROMETRY

Advantages:

1.   See Elemental Analysis introduction, p. 213.

Distinctive Requirements:

1.   See Elemental Analysis introduction.

References:

Ahearn, A. J., Ed. *Trace Analysis by Mass Spectrometry* (New York: Academic Press, 1972).
Hamersma, J. W., S. L. Reynolds and R. F. Maddolone. "IERL-RTP Procedures Manual: Level of Environmental Assessment," TRW Systems Group EPA-600/2-76-160a (June 1976).

PREFERRED ANALYTICAL PROCEDURE FOR LEVEL 2 INVESTIGATION: COMBUSTION SPECTROMETRIC/TITRIMETRIC

Advantages:

1.   Accurate and common method.
2.   Results are reproducible.

Distinctive Requirements:

1.   Wet ashing with HF and HNO_3.

2. Fusion with Na_2CO_3.
3. Precipitation with molybate solution.

References:

ASTM Standard D2795.

--------------- -------------------------

SAMPLING FOR: PLATINUM

PREFERRED ANALYTICAL PROCEDURE FOR
LEVEL 1 INVESTIGATION: SPARK SOURCE MASS
SPECTROMETRY

Advantages:

1. See Elemental Analysis introduction, p. 213.

Distinctive Requirements:

1. See Elemental Analysis introduction.

References:

Ahearn, A. J., Ed. *Trace Analysis by Mass Spectrometry* (New York: Academic Press, 1972).
Hamersma, J. W., S. L. Reynolds and R. F. Maddolone. "IERL-RTP Procedures Manual: Level of Environmental Assessment," TRW Systems Group EPA-600/2-76-160a (June 1976).

PREFERRED ANALYTICAL PROCEDURE FOR
LEVEL 2 INVESTIGATION: ATOMIC ABSORPTION
SPECTROSCOPY

Advantages:

1. See Elemental Analysis introduction, p. 213.

Distinctive Requirements:

1. $Li_2B_4O_7$ fusion.
2. Dissolution $HCl-HNO_3$.
3. Flame is air-C_2H_2, 2659A, Slit 2A.

Remarks:

1. Wet ashing is acceptable.
2. Interferences from Pd, Rh, Au, Ir, Ru, Cs and Na. Copper buffer solution may reduce interference.
3. Detection limit for AAS is 0.05 ppm.
4. Detection limit for FAAS is 4×10^{-5} ppm.

References:

Billings, C. E. *et al. J. Air Poll. Control Assoc.* 23(9):773-777 (September 1973).
Kim, A. G., and I. J. Douglas. *J. Chromatogr. Sci.* 11(12):615-617 (December 1973).

SAMPLING FOR: POTASSIUM

PREFERRED ANALYTICAL PROCEDURE FOR LEVEL 1 INVESTIGATION: SPARK SOURCE MASS SPECTROMETRY

Advantages:

1. See Elemental Analysis introduction, p. 213.

Distinctive Requirements:

1. See Elemental Analysis introduction.

References:

Ahearn, A. J., Ed. *Trace Analysis by Mass Spectrometry* (New York: Academic Press, 1972).
Hamersma, J. W., S. L. Reynolds and R. F. Maddolone. "IERL-RTP Procedures Manual: Level of Environmental Assessment," TRW Systems Group EPA-600/2-76-160a (June 1976).

PREFERRED ANALYTICAL PROCEDURE FOR
LEVEL 2 INVESTIGATION: ATOMIC ABSORPTION SPECTROSCOPY

Advantages:

1. See Elemental Analysis introduction, p. 213.

Distinctive Requirements:

1. Wet ashing.
2. Flame is air-$C_2 H_2$, 7655A, Slit 13A.

Remarks:

1. Dry ashing is acceptable.
2. Interference from sodium.
3. Detection limit for AAS is 0.003 ppm.
4. Detection limit for FAAS is 4 x 10^{-5} ppm.

References:

Billings, C. E. *et al. J. Air Poll. Control Assoc.* 23(9):773-777 (September 1973).

Kim, A. G., and I. J. Douglas. *J. Chromatogr. Sci.* 11(12):615-617 (December 1973).

SAMPLING FOR: PRASEODYNIUM

PREFERRED ANALYTICAL PROCEDURE FOR
LEVEL 1 INVESTIGATION: SPARK SOURCE MASS
SPECTROMETRY

Advantages:

1. See Elemental Analysis introduction, p. 213.

Distinctive Requirements:

1. See Elemental Analysis introduction.

References:

Ahearn, A. J., Ed. *Trace Analysis by Mass Spectrometry* (New York: Academic Press, 1972).

Hamersma, J. W., S. L. Reynolds and R. F. Maddolone. "IERL-RTP Procedures Manual: Level of Environmental Assessment," TRW Systems Group EPA-600/2-76-160a (June 1976).

PREFERRED ANALYTICAL PROCEDURE FOR LEVEL 2 INVESTIGATION: ATOMIC ABSORPTION SPECTROSCOPY

Advantages:

1. See Elemental Analysis introduction, p. 213.

Distinctive Requirements:

1. Wet ashing.
2. Flame is $N_2O-C_2H_2$, 4951A, Slit 2A.

Remarks:

1. Detection limit for AAS is 10 ppm.

References:

Billings, C. E. *et al. J. Air Poll. Control Assoc.* 23(9):773-777 (September 1973).
Kim, A. G., and I. J. Douglas. *J. Chromatogr. Sci.* 11(12):615-617 (December 1973).

SAMPLING FOR: RHENIUM

PREFERRED ANALYTICAL PROCEDURE FOR LEVEL 1 INVESTIGATION: SPARK SOURCE MASS SPECTROMETRY

Advantages:

1. See Elemental Analysis introduction, p. 213.

Distinctive Requirements:

1. See Elemental Analysis introduction.

References:

Ahearn, A. J., Ed. *Trace Analysis by Mass Spectrometry* (New York: Academic Press, 1972).

Hamersma, J. W., S. L. Reynolds and R. F. Maddolone. "IERL-RTP Procedures Manual: Level of Environmental Assessment," TRW Systems Group EPA-600/2-76-160a (June 1976).

PREFERRED ANALYTICAL PROCEDURE FOR LEVEL 2 INVESTIGATION: ATOMIC ABSORPTION SPECTROSCOPY

Advantages:

1.   See Elemental Analysis introduction, p. 213.

Distinctive Requirements:

1.   Wet ashing.
2.   Flame is $N_2 O\text{-}C_2 H_2$, 3460A, Slit 2A.

Remarks:

1.   Detection limit for AAS is 1 ppm.

References:

Billings, C. E. *et al. J. Air Poll. Control Assoc.* 23(9):773-777 (September 1973).

Kim, A. G., and I. J. Douglas. *J. Chromatogr. Sci.* 11(12):615-617 (December 1973).

- -

SAMPLING FOR: RHODIUM

PREFERRED ANALYTICAL PROCEDURE FOR LEVEL 1 INVESTIGATION: SPARK SOURCE MASS SPECTROMETRY

Advantages:

1.   See Elemental Analysis introduction, p. 213.

Distinctive Requirements:

1. See Elemental Analysis introduction, p. 213.

References:

Ahearn, A. J., Ed. *Trace Analysis by Mass Spectrometry* (New York: Academic Press, 1972).

Hamersma, J. W., S. L. Reynolds and R. F. Maddolone. "IERL-RTP Procedures Manual: Level of Environmental Assessment," TRW Systems Group EPA-600/2-76-160a (June 1976).

PREFERRED ANALYTICAL PROCEDURE FOR LEVEL 2 INVESTIGATION: ATOMIC ABSORPTION SPECTROSCOPY

Advantages:

1. See Elemental Analysis introduction, p. 213.

Distinctive Requirements:

1. Wet ashing.
2. Flame is air-C_2H_2, 3435A, Slit 2A.

Remarks:

1. Interferences from Na, Pt, Pd, Au, Ir, Ru and Cs.
2. Detection limit for AAS is 0.02 ppm.
3. Detection limit for FAAS is 8×10^{-6} ppm.

References:

Billings, C. E. *et al. J. Air Poll. Control Assoc.* 23(9):773-777 (September 1973).

Kim, A. G., and I. J. Douglas. *J. Chromatogr. Sci.* 11(12):615-617 (December 1973).

_ _

SAMPLING FOR: RUBIDIUM

PREFERRED ANALYTICAL PROCEDURE FOR LEVEL 1 INVESTIGATION: SPARK SOURCE MASS SPECTROMETRY

Advantages:

1. See Elemental Analysis introduction, p. 213.

Distinctive Requirements:

1. See Elemental Analysis introduction.

References:

Ahearn, A. J., Ed. *Trace Analysis by Mass Spectrometry* (New York: Academic Press, 1972).

Hamersma, J. W., S. L. Reynolds and R. F. Maddolone. "IERL-RTP Procedures Manual: Level of Environmental Assessment," TRW Systems Group EPA-600/2-76-160a (June 1976).

PREFERRED ANALYTICAL PROCEDURE FOR LEVEL 2 INVESTIGATION: ATOMIC ABSORPTION SPECTROSCOPY

Advantages:

1. See Elemental Analysis introduction, p. 213.

Distinctive Requirements:

1. Wet ashing.
2. Flame is air-C_2H_2, 7800A, Slit 40A.

Remarks:

1. Dry ashing is acceptable.
2. Interferences from sodium and potassium.
3. Detection limit for AAS is 0.005 ppm.
4. Detection limit for FAAS is 1×10^{-6} ppm.

References:

Billings, C. E. *et al. J. Air Poll. Control Assoc.* 23(9):773-777 (September 1973).

Kim, A. G., and I. J. Douglas. *J. Chromatogr. Sci.* 11(12):615-617 (December 1973).

SAMPLING FOR: RUTHENIUM

PREFERRED ANALYTICAL PROCEDURE FOR LEVEL 1 INVESTIGATION: SPARK SOURCE MASS SPECTROMETRY

Advantages:

1.　See Elemental Analysis introduction, p. 213.

Distinctive Requirements:

1.　See Elemental Analysis introduction.

References:

Ahearn, A. J., Ed. *Trace Analysis by Mass Spectrometry* (New York: Academic Press, 1972).
Hamersma, J. W., S. L. Reynolds and R. F. Maddolone. "IERL-RTP Procedures Manual: Level of Environmental Assessment," TRW Systems Group EPA-600/2-76-160a (June 1976).

PREFERRED ANALYTICAL PROCEDURE FOR LEVEL 2 INVESTIGATION: ATOMIC ABSORPTION SPECTROSCOPY

Advantages:

1.　See Elemental Analysis introduction, p. 213.

Distinctive Requirements:

1.　Dry ashing.
2.　Flame is air-C_2H_2, 3499A, Slit 7A.

Remarks:

1.　Detection limit for AAS is 0.3 ppm.

References:

Billings, C. E. *et al. J. Air Poll. Control Assoc.* 23(9):773-777 (September 1973).
Kim, A. G., and I. J. Douglas. *J. Chromatogr. Sci.* 11(12):615-617 (December 1973).

SAMPLING FOR: SAMARIUM

PREFERRED ANALYTICAL PROCEDURE FOR LEVEL 1 INVESTIGATION: SPARK SOURCE MASS SPECTROMETRY

Advantages:

1. See Elemental Analysis introduction, p. 213.

Distinctive Requirements:

1. See Elemental Analysis introduction.

References:

Ahearn, A. J., Ed. *Trace Analysis by Mass Spectrometry* (New York: Academic Press, 1972).
Hamersma, J. W., S. L. Reynolds and R. F. Maddolone. "IERL-RTP Procedures Manual: Level of Environmental Assessment," TRW Systems Group EPA-600/2-76-160a (June 1976).

PREFERRED ANALYTICAL PROCEDURE FOR LEVEL 2 INVESTIGATION: ATOMIC ABSORPTION SPECTROSCOPY

Advantages:

1. See Elemental Analysis introduction, p. 213.

Distinctive Requirements:

1. Dry ashing.
2. Flame is N_2O-C_2H_2, 4297A, Slit 2A.

Remarks:

1. Detection limit for AAS is 5 ppm.

References:

Billings, C. E. *et al. J. Air Poll. Control Assoc.* 23(9):773-777 (September 1973).
Kim, A. G., and I. J. Douglas. *J. Chromatogr. Sci.* 11(12):615-617 (December 1973).

SAMPLING FOR: SCANDIUM

PREFERRED ANALYTICAL PROCEDURE FOR LEVEL 1 INVESTIGATION: SPARK SOURCE MASS SPECTROMETRY

Advantages:

1. See Elemental Analysis introduction, p. 213.

Distinctive Requirements:

1. See Elemental Analysis introduction.

References:

Ahearn, A. J., Ed. *Trace Analysis by Mass Spectrometry* (New York: Academic Press, 1972).
Hamersma, J. W., S. L. Reynolds and R. F. Maddolone. "IERL-RTP Procedures Manual: Level of Environmental Assessment," TRW Systems Group EPA-600/2-76-160a (June 1976).

PREFERRED ANALYTICAL PROCEDURE FOR LEVEL 2 INVESTIGATION: ATOMIC ABSORPTION SPECTROSCOPY

Advantages:

1. See Elemental Analysis introduction, p. 213.

Distinctive Requirements:

1. Dry ashing.
2. Flame is $N_2 O$-$C_2 H_2$, 3912A, Slit 7A.

Remarks:

1. Detection limit for AAS is 0.02 ppm.

References:

Billings, C. E. *et al. J. Air Poll. Control Assoc.* 23(9):773-777 (September 1973).
Kim, A. G., and I. J. Douglas. *J. Chromatogr. Sci.* 11(12):615-617 (December 1973).

SAMPLING FOR: SELENIUM

PREFERRED ANALYTICAL PROCEDURE FOR LEVEL 1 INVESTIGATION: SPARK SOURCE MASS SPECTROMETRY

Advantages:

1. See Elemental Analysis introduction, p. 213.

Distinctive Requirements:

1. See Elemental Analysis introduction.

References:

Ahearn, A. J., Ed. *Trace Analysis by Mass Spectrometry* (New York: Academic Press, 1972).
Hamersma, J. W., S. L. Reynolds and R. F. Maddolone. "IERL-RTP Procedures Manual: Level of Environmental Assessment," TRW Systems Group EPA-600/2-76-160a (June 1976).

PREFERRED ANALYTICAL PROCEDURE FOR LEVEL 2 INVESTIGATION: ATOMIC ABSORPTION SPECTROSCOPY HYDRIDE EVOLUTION TECHNIQUE

Advantages:

1. See Elemental Analysis introduction, p. 213.

Distinctive Requirements:

1. Parr oxygen bomb combustion at 4 atm by HCl/HNO_3.
2. Hydride is generated by $TiCl_3$-Mg method.
3. Ar-H_2 air flame, 1960A, Slit 20A.

Remarks:

1. Interference from copper.
2. Detection limit for AAS is 0.1 ppm.
3. Detection limit for FAAS is 9×10^{-6} ppm.
4. Detection limit for HE is 0.002 ppm.

References:

Babu, S. P. "Trace Elements in Fuel," Advances in Chemistry Series No. 141 (Washington, D.C.: American Chemical Society, 1975).

- -

SAMPLING FOR: SILICON

PREFERRED ANALYTICAL PROCEDURE FOR LEVEL 1 INVESTIGATION: SPARK SOURCE MASS SPECTROMETRY

Advantages:

1. See Elemental Analysis introduction, p. 213.

Distinctive Requirements:

1. See Elemental Analysis introduction.

References:

Ahearn, A. J., Ed. *Trace Analysis by Mass Spectrometry* (New York: Academic Press, 1972).
Hamersma, J. W., S. L. Reynolds and R. F. Maddolone. "IERL-RTP Procedures Manual: Level of Environmental Assessment," TRW Systems Group EPA-600/2-76-160a (June 1976).

PREFERRED ANALYTICAL PROCEDURE FOR LEVEL 2 INVESTIGATION: ATOMIC ABSORPTION SPECTROSCOPY

Advantages:

1. See Elemental Analysis introduction, p. 213.

Distinctive Requirements:

1. Wet ashing.
2. Flame is $N_2 O\text{-}C_2 H_2$, 2561A, Slit 2A.

Remarks:

1. Detection limit for AAS is 0.1 ppm.
2. Detection limit for FAAS is 5×10^{-8} ppm.

References:

Billings, C. E. *et al. J. Air Poll. Control Assoc.* 23(9):773-777 (September 1973).
Kim, A. G., and I. J. Douglas. *J. Chromatogr. Sci.* 11(12):615-617 (December 1973).

SAMPLING FOR: SILVER

PREFERRED ANALYTICAL PROCEDURE FOR LEVEL 1 INVESTIGATION: SPARK SOURCE MASS SPECTROMETRY

Advantages:

1. See Elemental Analysis introduction, p. 213.

Distinctive Requirements:

1. See Elemental Analysis introduction.

References:

Ahearn, A. J., Ed. *Trace Analysis by Mass Spectrometry* (New York: Academic Press, 1972).

Hamersma, J. W., S. L. Reynolds and R. F. Maddolone. "IERL-RTP Procedures Manual: Level of Environmental Assessment," TRW Systems Group EPA-600/2-76-160a (June 1976).

PREFERRED ANALYTICAL PROCEDURE FOR LEVEL 2 INVESTIGATION: ATOMIC ABSORPTION SPECTROSCOPY

Advantages:

1. See Elemental Analysis introduction, p. 213.

Distinctive Requirements:

1. Wet ashing.
2. Flame is air-C_2H_2, 3281A, Slit 7A.

Remarks:

1. Dry ashing is acceptable.
2. Interference from Th.
3. Detection limit for AAS is 0.001 ppm.
4. Detection limit for FAAS is 1×10^{-7} ppm.

References:

Billings, C. E. *et al. J. Air Poll. Control Assoc.* 23(9):773-777 (September 1973).
Kim, A. G., and I. J. Douglas. *J. Chromatogr. Sci.* 11(12):615-617 (December 1973).

- -

SAMPLING FOR: SODIUM

PREFERRED ANALYTICAL PROCEDURE FOR LEVEL 1 INVESTIGATION: SPARK SOURCE MASS SPECTROMETRY

Advantages:

1. See Elemental Analysis introduction, p. 213.

Distinctive Requirements:

1. See Elemental Analysis introduction, p. 213.
2. Sodium is not measured in particulate, ash and inorganic samples.

References:

Ahearn, A. J., Ed. *Trace Analysis by Mass Spectrometry* (New York: Academic Press, 1972).
Hamersma, J. W., S. L. Reynolds and R. F. Maddolone. "IERL-RTP Procedures Manual: Level of Environmental Assessment," TRW Systems Group EPA-600/2-76-160a (June 1976).

PREFERRED ANALYTICAL PROCEDURE FOR LEVEL 2 INVESTIGATION: ATOMIC ABSORPTION SPECTROSCOPY

Advantages:

1. See Elemental Analysis introduction.

Distinctive Requirements:

1. Wet ashing.
2. Flame is air-C_2H_2, 5890A, Slit 4A.

Remarks:

1. Dry ashing is acceptable.
2. Detection limit for AAS is 0.0008 ppm.
3. Detection limit for FAAS is 1×10^{-7} ppm.

References:

Billings, C. E. *et al. J. Air Poll. Control Assoc.* 23(9):773-777 (September 1973).
Kim, A. G., and I. J. Douglas. *J. Chromatogr. Sci.* 11(12):615-617 (December 1973).

SAMPLING FOR: STRONTIUM

PREFERRED ANALYTICAL PROCEDURE FOR
LEVEL 1 INVESTIGATION: SPARK SOURCE MASS
SPECTROMETRY

Advantages:

1. See Elemental Analysis introduction, p. 213.

Distinctive Requirements:

1. See Elemental Analysis introduction.

References:

Ahearn, A. J., Ed. *Trace Analysis by Mass Spectrometry* (New York: Academic Press, 1972).

Hamersma, J. W., S. L. Reynolds and R. F. Maddolone. "IERL-RTP Procedures Manual: Level of Environmental Assessment," TRW Systems Group EPA-600/2-76-160a (June 1976).

PREFERRED ANALYTICAL PROCEDURE FOR
LEVEL 2 INVESTIGATION: ATOMIC ABSORPTION
SPECTROSCOPY

Advantages:

1. See Elemental Analysis introduction, p. 213.

Distinctive Requirements:

1. Wet ashing.
2. Flame is air-$C_2 H_2$, 4607A, Slit 13A.

Remarks:

1. Dry ashing is acceptable.
2. Interferences from Al, P, Na and K.
3. Detection limit for AAS is 0.005 ppm.
4. Detection limit for FAAS is 1×10^{-6} ppm.

References:

Billings, C. E. *et al. J. Air Poll. Control Assoc.* 23(9):773-777 (September 1973).

Kim, A. G., and I. J. Douglas. *J. Chromatogr. Sci.* 11(12):615-617 (December 1973).

_ _

SAMPLING FOR: SULFUR

PREFERRED ANALYTICAL PROCEDURE FOR LEVEL 1 INVESTIGATION: SPARK SOURCE MASS SPECTROMETRY

Advantages:

1. See Elemental Analysis introduction, p. 213.

Distinctive Requirements:

1. See Elemental Analysis introduction.

References:

Ahearn, A. J., Ed. *Trace Analysis by Mass Spectrometry* (New York: Academic Press, 1972).
Hamersma, J. W., S. L. Reynolds and R. F. Maddolone. "IERL-RTP Procedures Manual: Level of Environmental Assessment," TRW Systems Group EPA-600/2-76-160a (June 1976).

PREFERRED ANALYTICAL PROCEDURE FOR LEVEL 2 INVESTIGATION: COMBUSTION GRAVIMETRIC

Advantages:

1. Accurate common technique.
2. Results are reproducible.

Distinctive Requirements:

1. Sample is combusted under 20-30 atm.
2. Sulfur is determined as $BaSO_4$ gravimetrically.

Remarks:

1. Eshka fusion method is also acceptable.

_ _

SAMPLING FOR: TANTALUM

PREFERRED ANALYTICAL PROCEDURE FOR
LEVEL 1 INVESTIGATION: SPARK SOURCE MASS
SPECTROMETRY

Advantages:

1. See Elemental Analysis introduction.

Distinctive Requirements:

1. See Elemental Analysis introduction.

References:

Ahearn, A. J., Ed. *Trace Analysis by Mass Spectrometry* (New York: Academic Press, 1972).

Hamersma, J. W., S. L. Reynolds and R. F. Maddolone. "IERL-RTP Procedures Manual: Level of Environmental Assessment," TRW Systems Group EPA-600/2-76-160a (June 1976).

PREFERRED ANALYTICAL PROCEDURE FOR
LEVEL 2 INVESTIGATION: ATOMIC ABSORPTION
SPECTROSCOPY

Advantages:

1. See Elemental Analysis introduction, p. 213.

Distinctive Requirements:

1. Wet ashing.
2. Flame is $N_2O-C_2H_2$, 2715A, Slit 2A.

Remarks:

1. Dry ashing is acceptable.
2. Detection limit for AAS is 3 ppm.

References:

Billings, C. E. *et al. J. Air Poll. Control Assoc.* 23(9):773-777 (September 1973).
Kim, A. G., and I. J. Douglas. *J. Chromatogr. Sci.* 11(12):615-617 (December 1973).

SAMPLING FOR: TELLURIUM

PREFERRED ANALYTICAL PROCEDURE FOR LEVEL 1 INVESTIGATION: SPARK SOURCE MASS SPECTROMETRY

Advantages:

1.   See Elemental Analysis introduction, p. 213.

Distinctive Requirements:

1.   See Elemental Analysis introduction.

References:

Ahearn, A. J., Ed. *Trace Analysis by Mass Spectrometry* (New York: Academic Press, 1972).
Hamersma, J. W., S. L. Reynolds and R. F. Maddolone. "IERL-RTP Procedures Manual: Level of Environmental Assessment," TRW Systems Group EPA-600/2-76-160a (June 1976).

PREFERRED ANALYTICAL PROCEDURE FOR LEVEL 2 INVESTIGATION: ATOMIC ABSORPTION SPECTROSCOPY HYDRIDE EVOLUTION TECHNIQUE

Distinctive Requirements:

1.   Wet ashing: hydride is generated by $NaBH_4$.
2.   Flame is air-C_2H_2, 1960A, Slit 2A.

Remarks:

1.   Interferences from Cu.

2.   Detection limit for AAS is 0.05 ppm.
3.   Detection limit for FAAS is 0.003 ppm.
4.   Detection limit for HE is 0.004 ppm.

References:

Babu, S. P. "Trace Elements in Fuel," Advances in Chemistry Series No. 141 (Washington, D.C.: American Chemical Society, 1975).

SAMPLING FOR: TERBIUM

PREFERRED ANALYTICAL PROCEDURE FOR LEVEL 1 INVESTIGATION: SPARK SOURCE MASS SPECTROMETRY

Advantages:

1.   See Elemental Analysis introduction, p. 213.

Distinctive Requirements:

1.   See Elemental Analysis introduction.

References:

Ahearn, A. J., Ed. *Trace Analysis by Mass Spectrometry* (New York: Academic Press, 1972).
Hamersma, J. W., S. L. Reynolds and R. F. Maddolone. "IERL-RTP Procedures Manual: Level of Environmental Assessment," TRW Systems Group EPA-600/2-76-160a (June 1976).

PREFERRED ANALYTICAL PROCEDURE FOR LEVEL 2 INVESTIGATION: ATOMIC ABSORPTION SPECTROSCOPY

Advantages:

1.   See Elemental Analysis introduction, p. 213.

Distinctive Requirements:

1. Wet ashing.
2. Flame is air-$C_2 H_2$, 4326A, Slit 2A.

Remarks:

1. Detection limit for AAS is 0.05 ppm.
2. Detection limit for FAAS is 1×10^{-6} ppm.

References:

Billings, C. E. *et al. J. Air Poll. Control Assoc.* 23(9):773-777 (September 1973).
Kim, A. G., and I. J. Douglas. *J. Chromatogr. Sci.* 11(12):615-617 (December 1973).

— —

SAMPLING FOR: THALLIUM

PREFERRED ANALYTICAL PROCEDURE FOR LEVEL 1 INVESTIGATION: SPARK SOURCE MASS SPECTROMETRY

Advantages:

1. See Elemental Analysis introduction, p. 213.

Distinctive Requirements:

1. See Elemental Analysis introduction.

References:

Ahearn, A. J., Ed. *Trace Analysis by Mass Spectrometry* (New York: Academic Press, 1972).
Hamersma, J. W., S. L. Reynolds and R. F. Maddolone. "IERL-RTP Procedures Manual: Level of Environmental Assessment," TRW Systems Group EPA-600/2-76-160a (June 1976).

PREFERRED ANALYTICAL PROCEDURE FOR LEVEL 2 INVESTIGATION: ATOMIC ABSORPTION SPECTROSCOPY

Advantages:

1.   See Elemental Analysis introduction, p. 213.

Distinctive Requirements:

1.   Wet ashing.
2.   Flame is air-C_2H_2, 2768A, Slit 20A.

Remarks:

1.   Detection limit for AAS is 0.02 ppm.
2.   Detection limit for FAAS is 1×10^{-6} ppm.

References:

Billings, C. E. *et al. J. Air Poll. Control Assoc.* 23(9):773-777 (September 1973).
Kim, A. G., and I. J. Douglas. *J. Chromatogr. Sci.* 11(12):615-617 (December 1973).

- -

SAMPLING FOR: THULIUM

PREFERRED ANALYTICAL PROCEDURE FOR LEVEL 1 INVESTIGATION: SPARK SOURCE MASS SPECTROMETRY

Advantages:

1.   See Elemental Analysis introduction, p. 213.

Distinctive Requirements:

1.   See Elemental Analysis introduction.

References:

Ahearn, A. J., Ed. *Trace Analysis by Mass Spectrometry* (New York: Academic Press, 1972).

Hamersma, J. W., S. L. Reynolds and R. F. Maddolone. "IERL-RTP Procedures Manual: Level of Environmental Assessment," TRW Systems Group EPA-600/2-76-160a (June 1976).

PREFERRED ANALYTICAL PROCEDURE FOR LEVEL 2 INVESTIGATION: ATOMIC ABSORPTION SPECTROSCOPY

Advantages:

1. See Elemental Analysis introduction, p. 213.

Distinctive Requirements:

1. Wet ashing.
2. Flame is air-C_2H_2, 4106A.

Remarks:

1. Detection limit for AAS is 1 ppm.

References:

Billings, C. E. *et al. J. Air Poll. Control Assoc.* 23(9):773-777 (September 1973).
Kim, A. G., and I. J. Douglas. *J. Chromatogr. Sci.* 11(12):615-617 (December 1973).

SAMPLING FOR: TIN

PREFERRED ANALYTICAL PROCEDURE FOR LEVEL 1 INVESTIGATION: SPARK SOURCE MASS SPECTROMETRY

Advantages:

1. See Elemental Analysis introduction, p. 213.

Distinctive Requirements:

1. See Elemental Analysis introduction.

References:

Ahearn, A. J., Ed. *Trace Analysis by Mass Spectrometry* (New York: Academic Press, 1972).

Hamersma, J. W., S. L. Reynolds and R. F. Maddolone. "IERL-RTP Procedures Manual: Level of Environmental Assessment," TRW Systems Group EPA-600/2-76-160a (June 1976).

PREFERRED ANALYTICAL PROCEDURE FOR LEVEL 2 INVESTIGATION: ATOMIC ABSORPTION SPECTROSCOPY HYDRIDE EVOLUTION TECHNIQUE

Distinctive Requirements:

1. Wet ashing.
2. Hydride is generated by $TiCl_3$-Mg method.
3. Flame is air-C_2H_2, 2246A, Slit 7A.

Remarks:

1. Dry ashing is acceptable.
2. Interferences from Na and PO_4.
3. Detection limit for AAS is 0.05 ppm.
4. Detection limit for FAAS is 0.0003 ppm.
5. Detection limit for HE is 0.004 ppm.

References:

Babu, S. P. "Trace Elements in Fuel," Advances in Chemistry Series No. 141 (Washington, D.C.: American Chemical Society, 1975).

- -

SAMPLING FOR: TITANIUM

PREFERRED ANALYTICAL PROCEDURE FOR LEVEL 1 INVESTIGATION: SPARK SOURCE MASS SPECTROMETRY

Advantages:

1. See Elemental Analysis introduction, p. 213.

Distinctive Requirements:

1. See Elemental Analysis introduction, p. 213.

References:

Ahearn, A. J., Ed. *Trace Analysis by Mass Spectrometry* (New York: Academic Press, 1972).
Hamersma, J. W., S. L. Reynolds and R. F. Maddolone. "IERL-RTP Procedures Manual: Level of Environmental Assessment," TRW Systems Group EPA-600/2-76-160a (June 1976).

PREFERRED ANALYTICAL PROCEDURE FOR LEVEL 2 INVESTIGATION: ATOMIC ABSORPTION SPECTROSCOPY

Advantages:

1. See Elemental Analysis introduction, p. 213.

Distinctive Requirements:

1. Wet ashing.
2. Flame is $N_2 O$-$C_2 H_2$, 3643A, Slit 2A.

Remarks:

1. Detection limit for AAS is 0.1 ppm.
2. Detection limit for FAAS is 4×10^{-5} ppm.

References:

Billings, C. E. *et al. J. Air Poll. Control Assoc.* 23(9):773-777 (September 1973).
Kim, A. G., and I. J. Douglas. *J. Chromatogr. Sci.* 11(12):615-617 (December 1973).

SAMPLING FOR: TUNGSTEN

PREFERRED ANALYTICAL PROCEDURE FOR LEVEL 1 INVESTIGATION: SPARK SOURCE MASS SPECTROMETRY

Advantages:

1.   See Elemental Analysis introduction, p. 213.

Distinctive Requirements:

1.   See Elemental Analysis introduction.

References:

Ahearn, A. J., Ed. *Trace Analysis by Mass Spectrometry* (New York: Academic Press, 1972).

Hamersma, J. W., S. L. Reynolds and R. F. Maddolone. "IERL-RTP Procedures Manual: Level of Environmental Assessment," TRW Systems Group EPA-600/2-76-160a (June 1976).

PREFERRED ANALYTICAL PROCEDURE FOR LEVEL 2 INVESTIGATION: ATOMIC ABSORPTION SPECTROSCOPY

Advantages:

1   See Elemental Analysis introduction, p. 213.

Distinctive Requirements:

1.   Wet ashing.
2.   Flame is $N_2O-C_2H_2$, 4009A, Slit 2A.

Remarks:

1.   Detection limit for AAS is 3 ppm.

References:

Billings, C. E. *et al. J. Air Poll. Control Assoc.* 23(9):773-777 (September 1973).

Kim, A. G., and I. J. Douglas. *J. Chromatogr. Sci.* 11(12):615-617 (December 1973).

- -

SAMPLING FOR: URANIUM

PREFERRED ANALYTICAL PROCEDURE FOR LEVEL 1 INVESTIGATION: SPARK SOURCE MASS SPECTROMETRY

Advantages:

1. See Elemental Analysis introduction, p. 213.

Distinctive Requirements:

1. See Elemental Analysis introduction.

References:

Ahearn, A. J., Ed. *Trace Analysis by Mass Spectrometry* (New York: Academic Press, 1972).

Hamersma, J. W., S. L. Reynolds and R. F. Maddolone. "IERL-RTP Procedures Manual: Level of Environmental Assessment," TRW Systems Group EPA-600/2-76-160a (June 1976).

PREFERRED ANALYTICAL PROCEDURE FOR LEVEL 2 INVESTIGATION: ATOMIC ABSORPTION SPECTROSCOPY

Advantages:

1. See Elemental Analysis introduction, p. 213.

Distinctive Requirements:

1. Wet ashing.
2. Flame is N_2O-C_2H_2, 3514A, Slit 21A.

Remarks:

1. Detection limit for AAS is 30 ppm.

References:

Billings, C. E. *et al. J. Air Poll. Control Assoc.* 23(9):773-777 (September 1973).

Kim, A. G., and I. J. Douglas. *J. Chromatogr. Sci.* 11(12):615-617 (December 1973).

- -

SAMPLING FOR: VANADIUM

PREFERRED ANALYTICAL PROCEDURE FOR LEVEL 1 INVESTIGATION: SPARK SOURCE MASS SPECTROMETRY

Advantages:

1.   See Elemental Analysis introduction, p. 213.

Distinctive Requirements:

1.   See Elemental Analysis introduction.

References:

Ahearn, A. J., Ed. *Trace Analysis by Mass Spectrometry* (New York: Academic Press, 1972).
Hamersma, J. W., S. L. Reynolds and R. F. Maddolone. "IERL-RTP Procedures Manual: Level of Environmental Assessment," TRW Systems Group EPA-600/2-76-160a (June 1976).

PREFERRED ANALYTICAL PROCEDURE FOR LEVEL 2 INVESTIGATION: ATOMIC ABSORPTION SPECTROSCOPY

Advantages:

1.   See Elemental Analysis introduction, p. 213.

Distinctive Requirements:

1.   Wet ashing.
2.   Flame is $N_2 O\text{-}C_2 H_2$, 3184A, Slit 2A.

Remarks:

1.   Detection limit for AAS is 0.02 ppm.
2.   Detection limit for FAAS is 3×10^{-6} ppm.

References:

Billings, C. E. *et al.* *J. Air Poll. Control Assoc.* 23(9):773-777 (September 1973).
Kim, A. G., and I. J. Douglas. *J. Chromatogr. Sci.* 11(12):615-617 (December 1973).

SAMPLING FOR: YTTERBIUM

PREFERRED ANALYTICAL PROCEDURE FOR LEVEL 1 INVESTIGATION: SPARK SOURCE MASS SPECTROMETRY

Advantages:

1.   See Elemental Analysis introduction, p. 213.

Distinctive Requirements:

1.   See Elemental Analysis introduction.

References:

Ahearn, A. J., Ed. *Trace Analysis by Mass Spectrometry* (New York: Academic Press, 1972).
Hamersma, J. W., S. L. Reynolds and R. F. Maddolone. "IERL-RTP Procedures Manual: Level of Environmental Assessment," TRW Systems Group EPA-600/2-76-160a (June 1976).

PREFERRED ANALYTICAL PROCEDURE FOR LEVEL 2 INVESTIGATION: ATOMIC ABSORPTION SPECTROSCOPY

Advantages:

1.   See Elemental Analysis introduction, p. 213.

Distinctive Requirements:

1.   Wet ashing.
2.   Flame is N_2O-C_2H_2, 3184A, Slit 2A.

Remarks:

1.   Detection limit for AAS is 0.04 ppm.

References:

Billings, C. E. *et al. J. Air Poll. Control Assoc.* 23(9):773-777 (September 1973).
Kim, A. G., and I. J. Douglas. *J. Chromatogr. Sci.* 11(12):615-617 (December 1973).

SAMPLING FOR: YTTRIUM

PREFERRED ANALYTICAL PROCEDURE FOR LEVEL 1 INVESTIGATION: SPARK SOURCE MASS SPECTROMETRY

Advantages:

1.   See Elemental Analysis introduction, p. 213.

Distinctive Requirements:

1.   See Elemental Analysis introduction.

References:

Ahearn, A. J., Ed. *Trace Analysis by Mass Spectrometry* (New York: Academic Press, 1972).
Hamersma, J. W., S. L. Reynolds and R. F. Maddolone. "IERL-RTP Procedures Manual: Level of Environmental Assessment," TRW Systems Group EPA-600/2-76-160a (June 1976).

PREFERRED ANALYTICAL PROCEDURE FOR LEVEL 2 INVESTIGATION: ATOMIC ABSORPTION SPECTROSCOPY

Advantages:

1.   See Elemental Analysis introduction, p. 213.

Distinctive Requirements:

1. Wet ashing.
2. Flame is $N_2 O$-$C_2 H_2$, 4077A, Slit 2A.

Remarks:

1. Detection limit for AAS is 0.3 ppm.

References:

Billings, C. E. *et al. J. Air Poll. Control Assoc.* 23(9):773-777 (September 1973).
Kim, A. G., and I. J. Douglas. *J. Chromatogr. Sci.* 11(12):615-617 (December 1973).

- -

SAMPLING FOR: ZINC

**PREFERRED ANALYTICAL PROCEDURE FOR
LEVEL 1 INVESTIGATION: SPARK SOURCE MASS
SPECTROMETRY**

Advantages:

1. See Elemental Analysis introduction, p. 213.

Distinctive Requirements:

1. See Elemental Analysis introduction.

References:

Ahearn, A. J., Ed. *Trace Analysis by Mass Spectrometry* (New York: Academic Press, 1972).
Hamersma, J. W., S. L. Reynolds and R. F. Maddolone. "IERL-RTP Procedures Manual: Level of Environmental Assessment," TRW Systems Group EPA-600/2-76-160a (June 1976).

PREFERRED ANALYTICAL PROCEDURE FOR
LEVEL 2 INVESTIGATION: ATOMIC ABSORPTION SPECTROSCOPY

Advantages:

1. See Elemental Analysis introduction, p. 213.

Distinctive Requirements:

1. Wet ashing.
2. Flame is air-C_2H_2, 2138A, Slit 20A.

Remarks:

1. Interferences from Fe, Al, Na, K, Mg and Ca.
2. Detection limit for AAS is 0.001 ppm.
3. Detection limit for FAAS is 3×10^{-8} ppm.

References:

Billings, C. E. *et al. J. Air Poll. Control Assoc.* 23(9):773-777 (September 1973).

Kim, A. G., and I. J. Douglas. *J. Chromatogr. Sci.* 11(12):615-617 (December 1973).

SAMPLING FOR: ZIRCONIUM

PREFERRED ANALYTICAL PROCEDURE FOR
LEVEL 1 INVESTIGATION: SPARK SOURCE MASS
SPECTROMETRY

Advantages:

1. See Elemental Analysis introduction, p. 213.

Distinctive Requirements:

1. See Elemental Analysis introduction.

References:

Ahearn, A. J., Ed. *Trace Analysis by Mass Spectrometry* (New York: Academic Press, 1972).

Hamersma, J. W., S. L. Reynolds and R. F. Maddolone. "IERL-RTP Procedures Manual: Level of Environmental Assessment," TRW Systems Group EPA-600/2-76-160a (June 1976).

PREFERRED ANALYTICAL PROCEDURE FOR LEVEL 2 INVESTIGATION: ATOMIC ABSORPTION SPECTROSCOPY

Advantages:

1. See Elemental Analysis introduction.

Distinctive Requirements:

1. Dry ashing.
2. Flame is $N_2O-C_2H_2$, 3601A, Slit 2A.

Remarks:

1. Detection limit for AAS is 5 ppm.

References:

Billings, C. E. *et al. J. Air Poll. Control Assoc.* 23(9):773-777 (September 1973).
Kim, A. G., and I. J. Douglas. *J. Chromatogr. Sci.* 11(12):615-617 (December 1973).

- -

OTHER PROCEDURES FOR ELEMENTAL ANALYSIS

METHOD: X-RAY FLUORESCENCE (XRF)

There are five common methods used for quantification in XRF:

1. internal standard;
2. standard dilution;
3. standard addition;
4. absolute calculation; and
5. direct comparison with suitable standards.

Intensities of X-rays and fluorescence are dependent on matrix composition. Varying particle sizes can alter fluorescence intensity. Careful sample handling is important to minimize differences in chemical and physical forms. The resolving power of the instrument directly affects spectral interferences.

References:

Dulka, J. J., and T. H. Risby. *Anal. Chem.* 48:640A (1976).
Winefordner, J. D., Ed. *Trace Analysis-Spectroscopic Methods for Elements* (New York: John Wiley and Sons, Inc., 1976).

CHAPTER 10

ANION ANALYSIS

INTRODUCTION

This section gives the methods for anion analysis. Obviously not all possible anions are defined, but most of the available analytical methods are identified.

The Level 1 investigation in most cases utilizes spark source mass spectrometry (SSMS). The level of accuracy of this method for anions is relatively low. The Level 2 investigation utilizes the many elaborate inorganic compound characterization instruments. These may include electron microscopy, X-ray diffraction, etc.

- -

SAMPLING FOR: ANION ANALYSIS (AMMONIA)

PREFERRED ANALYTICAL PROCEDURE FOR LEVEL 1 INVESTIGATION: COLORIMETRIC

Advantages:

1.   Simple and fast.

Distinctive Requirements:

1.   As specified by reagent test kit.

**PREFERRED ANALYTICAL PROCEDURE FOR
LEVEL 2 INVESTIGATION: COLORIMETRIC—DL -
0.005-1.0 ppm; TITRIMETRIC—DL - 1.0 - 0.25 ppm**

Advantages:

1. EPA Method; accurate for analyzing ammonia in water.

Distinctive Requirements:

1. Sample buffered at pH 9.5 and distilled into a solution of boric acid.
2. The ammonia in the distillate can be determined colorimetrically by nesslerization or titrimetrically with H_2SO_4 using a mixed indicator (methyl red/methylene blue).

Remarks:

1. Volatile organic alkaline compounds may cause an off color in the nesslerization procedure.

References:

ASTM Standard D431.

- -

SAMPLING FOR: ARSENATE/ARSENITE

**PREFERRED ANALYTICAL PROCEDURE FOR
LEVEL 1 INVESTIGATION: SSMS**

Advantages:

1. Elemental analysis provides upper concentration limit for anions.
2. SSMS has multicomponent capability.

Distinctive Requirements:

1. Aqueous samples: slurry with graphite and briquette.
2. Solid samples: see Elemental Analysis, p. 213.

PREFERRED ANALYTICAL PROCEDURE FOR
LEVEL 2 INVESTIGATION: SPECTROMETRIC

Advantages:

1. Accurate, fast technique for measuring arsenic.
2. ASTM Method.

Distinctive Requirements:

1. 25 ml of sample is acidified with HCl, mixed with KI and $SnCl_2$.
2. 3 g of zinc are added and the arsine generated is bubbled through a silver diethyldithiocarbonate-pyridine solution. Absorbance is measured at 540 nm within 30 minutes.

Remarks:

1. Arsenic measurement by AAS is also an acceptable technique and may be subject to fewer interferences.

References:

ASTM Standard D2972.

_ _

SAMPLING FOR: BROMIDE

PREFERRED ANALYTICAL PROCEDURE FOR
LEVEL 1 INVESTIGATION: SSMS

Advantages:

1. Same as for arsenate.

Distinctive Requirements:

1. Same as for arsenate.

PREFERRED ANALYTICAL PROCEDURE FOR
LEVEL 2 INVESTIGATION: TITRIMETRIC

Advantages:

1. Accurate and fast technique for measuring bromide.
2. ASTM method.

Distinctive Requirements:

1. Sufficient NaCl is added to 100 ml of the sample to produce a 3-g chloride content.
2. KClO is added to oxidize bromide to bromine (excess is destroyed with $NaClO_2$).
3. KI is added and liberated I_2 is titrated with 0.01 N $Na_2 S_2 O_3$.

Remarks:

1. Measures bromide and iodide; thus, this method is to be used on conjunction with iodide determination.
2. Fe^{+2}, Mn^{+2} interfere, but may be removed by treatment with CaO.

- -

SAMPLING FOR: CARBONATE (BICARBONATE)

PREFERRED ANALYTICAL PROCEDURE FOR
LEVEL 1 INVESTIGATION: TITRIMETRIC

Advantages:

1. Rapid, simple analysis technique.

Distinctive Requirements:

1. As specified by reagent test kit.

PREFERRED ANALYTICAL PROCEDURE FOR
LEVEL 2 INVESTIGATION: TITRIMETRIC

Advantages:

1. Accurate technique for measuring total carbonate present.

Distinctive Requirements:

1. CO_2 is liberated by acidifying and heating the sample in a closed system.
2. CO_2 is absorbed in a barium hydroxide solution.
3. Excess barium hydroxide is titrated with $0.04\ N$ HCl.

Remarks:

1. Sulfides (H_2S) interfered but are removed by scrubbing with an iodine solution. Other interferences are removed by scrubbing with chromic acid.
2. From pH measurement H_2CO_3, HCO_3^-, CO_3^{2-} concentrations may be estimated.

References:

ASTM Standard D513.

SAMPLING FOR: CHLORIDE

**PREFERRED ANALYTICAL PROCEDURE FOR
LEVEL 1 INVESTIGATION: SSMS**

Advantages:

1. As for arsenate.

Distinctive Requirements:

1. As for arsenate.

**PREFERRED ANALYTICAL PROCEDURE FOR
LEVEL 2 INVESTIGATION: TITRIMETRIC**

Advantages:

1. Accurate technique for measuring chloride content of industrial wastewater.

Distinctive Requirements:

1.   50-ml sample is titrated with 0.025 N silver nitrate to a potassium chromate endpoint.
2.   Sulfites are oxidized to sulfates by H_2O_2 addition.

Remarks:

1.   Phosphates (> 250 ppm) interfere.
2.   Iodine and bromide may also interfere with visual endpoint; potentiometric titration may solve this problem.

References:

ASTM Standard D512.

_ _ _ _ _'_ _

SAMPLING FOR: CYANIDE

PREFERRED ANALYTICAL PROCEDURE FOR LEVEL 1 INVESTIGATION: COLORIMETRIC

Advantages:

1.   Rapid, simple analysis technique.

Distinctive Requirements:

1.   As specified by reagent test kit.

PREFERRED ANALYTICAL PROCEDURE FOR LEVEL 2 INVESTIGATION: TITRIMETRIC/SPECTROMETRIC

Advantages:

1.   Accurate technique for measuring cyanide.

Distinctive Requirements:

1.   Titration: titration with $AgNO_3$ to rhodamine endpoint.
2.   Spectrometric: neutralize absorption solution with acetic acid to pH 6.5-8.0. 0.2 ml of chloramine T solution is added. Absorbance measured at 620 nm after 20 minutes.

Remarks:

1. Titration method applies when cyanide concentration is $>$ 1 ppm; spectrometric method for $<$ 1 ppm.

References:

ASTM Standard D2036.

SAMPLING FOR: FLUORIDE

**PREFERRED ANALYTICAL PROCEDURE FOR
LEVEL 1 INVESTIGATION: SSMS**

Advantages:

1. As for arsenate.

Distinctive Requirements:

1. As for arsenate.

**PREFERRED ANALYTICAL PROCEDURE FOR
LEVEL 2 INVESTIGATION: SPECIFIC ION ELECTRODE (SIE)**

Advantages:

1. Accurate, simple and rapid method for analyzing fluoride.

Distinctive Requirements:

1. pH is adjusted to 5.2-5.5 with 0.5 N H_2SO_4.
2. CO_2 is removed by heating on a hot water bath.
3. Buffer is added and fluoride is measured by known addition method.

Remarks:

1. SIE is more accurate and simpler than distillation–spectrometric method (SPADNS).

References:

Ruch, R. R., J. H. Gluskoter and N. F. Shimp. "Occurrence and Distribution of Potentially Volatile Trace Elements in Coal," *Environmental Geology Notes* 72 (1974).

Thomas, Josephus, Jr., and Harold J. Gluskoter. "Determination of Fluoride in Coal with the Fluoride Ion-Selective Electrode," *Analytical Chemistry* 46(a): 1321-1323 (1974).

SAMPLING FOR: IODIDE

___

PREFERRED ANALYTICAL PROCEDURE FOR LEVEL 1 INVESTIGATION: SSMS

Advantages:

1.   As for arsenate.

Distinctive Requirements:

1.   As for arsenate.

PREFERRED ANALYTICAL PROCEDURE FOR LEVEL 2 INVESTIGATION: SPECTROMETRIC

Advantages:

1.   Accurate method for analysis of iodide.

Distinctive Requirements:

1.   Iodide is determined by oxidation to iodate with saturated bromine water in acid solution.
2.   Excess bromine is destroyed by addition of sodium formate.
3.   Sample is titrated with 0.01 N sodium thio-sulfate solution.

Remarks:

1.   Effects of Fe^{+3}, Mn^{+2}, and organic matter are removed by treatment with CaO. This method is used in conjunction with bromide determination.

References:

ASTM Standard D1246.

SAMPLING FOR: NITRATE

**PREFERRED ANALYTICAL PROCEDURE FOR
LEVEL 1 INVESTIGATION: COLORIMETRIC**

Advantages:

1.   Rapid, simple analysis technique.

Distinctive Requirements:

1.   As specified by reagent test kit.

**PREFERRED ANALYTICAL PROCEDURE FOR
LEVEL 2 INVESTIGATION: SPECTROMETRIC**

Advantages:

1.   Accurate technique for nitrate analysis.

Distinctive Requirements:

1.   5 ml of sample is mixed with brusine-sulfanilic acid solution then mixed with 10 ml of 15.6 M H_2SO_4.
2.   Color is developed for 10 # 1 minutes in a dark area; absorbance is measured at 410 nm.

Remarks:

1.   Color does not follow. Beer-Lambert relation, however, plotting absorbance vs concentration yields a smooth curve.
2.   Turbid or colored samples interfere, but may be removed by filtration and treatment with Al_2O_3 and activated carbon.

References:

ASTM Standard D992.

SAMPLING FOR: NITRATE

PREFERRED ANALYTICAL PROCEDURE FOR LEVEL 1 INVESTIGATION: COLORIMETRIC

Advantages:

1.   Rapid, simple analysis method.

Distinctive Requirements:

1.   As specified by reagent test kit.

PREFERRED ANALYTICAL PROCEDURE FOR LEVEL 2 INVESTIGATION: SPECTROMETRIC

Advantages:

1.   Accurate technique for analysis of nitrate.

Distinctive Requirements:

1.   pH is adjusted to 7 with CH_3COOH.
2.   If sample has appreciable color, filter with $Al(OH)_3$ gel.
3.   EDTA is added to complex cations.
4.   2 ml of sulfanilic acid solution and 2 ml of naphthylamine hydrochloride are added to the sample, solution is buffered at pH 2.0-2.5 with $Na_2C_2H_3O_2$ solution, allowed to stand 30 minutes, and absorbance measured at 515 nm.

Remarks:

1.   Mercury II causes high results while copper (II) catalyzes the decomposition of the diazonium salt and thus leads to low results.
2.   Certain bacteria utilize nitrites in their metabolism. Storage at low temperature minimizes this effect.

References:

ASTM Standard D1254.

SAMPLING FOR: ORTHO-PHOSPHATE

PREFERRED ANALYTICAL PROCEDURE FOR
LEVEL 1 INVESTIGATION: COLORIMETRIC

Advantages:

1. Rapid, simple analysis technique.

Distinctive Requirements:

1. As specified by reagent test kit.

PREFERRED ANALYTICAL PROCEDURE FOR
LEVEL 2 INVESTIGATION: SPECTROMETRIC

Advantages:

1. Accurate technique for analyzing phosphate.

Distinctive Requirements:

1. If pH is greater than 7, sample is neutralized with H_2SO_4.
2. Molybdate reagent and stannous chloride reagent are added.
3. Absorption is measured at 690 nm between 10 and 12 minutes after reagent addition.

Remarks:

1. Color intensity is time and temperature dependent.
2. Solution may be extracted with benzene-isobutanol solvent to remove interferences and increase sensitivity.

References:

ASTM Standard D515.

SAMPLING FOR: SULFIDE

PREFERRED ANALYTICAL PROCEDURE FOR
LEVEL 1 INVESTIGATION: SSMS

Advantages:

1. As for arsenate.

Distinctive Requirements:

1. As for arsenate.

PREFERRED ANALYTICAL PROCEDURE FOR
LEVEL 2 INVESTIGATION: TITRIMETRIC

Advantages:

1. Rapid and accurate technique for measuring sulfide.

Distinctive Requirements:

1. Sample is acidified and stripped with an inert gas and collection in a zinc acetate solution.
2. Iodine solution is added to collection vessels, acidified with HCl and back-titrated with 0.025 N sodium thiosulfate solution.

References:

ASTM Standard D2579.

SAMPLING FOR: SULFITE

PREFERRED ANALYTICAL PROCEDURE FOR
LEVEL 1 INVESTIGATION: COLORIMETRIC

Advantages:

1. Method provides a simple and rapid analysis technique.

Distinctive Requirements:

1. As specified by reagent test kit.

PREFERRED ANALYTICAL PROCEDURE FOR
LEVEL 2 INVESTIGATION: TITRIMETRIC

Advantages:

1. Rapid and accurate technique for measuring sulfite.

Distinctive Requirements:

1. Air is excluded while sample is being taken by use of apparatus described in reference.
2. HCl, KI, and KIO_3 are added.
3. Excess iodine chloride formed is titrated with $0.01\ N\ Na_2S_2O_3$ using a dead stop endpoint-indicating apparatus.

Remarks:

1. Starch indicator may be used.

References:

ASTM Standard D1339.

SAMPLING FOR: SULFATE

PREFERRED ANALYTICAL PROCEDURE FOR
LEVEL 1 INVESTIGATION: TURBIDIMETRIC/COLORIMETRIC

Advantages:

1. Provides a rapid, simple analysis technique.

Distinctive Requirements:

1. As specified by reagent test kit.

PREFERRED ANALYTICAL PROCEDURE FOR
LEVEL 2 INVESTIGATION: GRAVIMETRIC

Advantages:

1.   Accurate technique for measuring sulfate.

Distinctive Requirements:

1.   Sample is filtered, pH adjusted to 4.5 with HCl, hot $BaCl_2$ added, allowed to stand for 2 hr, filtered and ignited at 800°C.

Remarks:

1.   A titrimetric method may also be used for $BaCl_2$, titrating in an alcohol solution to a thorin endpoint.

Reference:

ASTM Standard D516.

CHAPTER 11

FUEL ANALYSIS

INTRODUCTION

In this section the methods for fuel analysis are given. These are ASTM methods. They are presented here because of the extraordinary role fuel and its combustion play in air pollution. Generally, a complete fuel analysis prior to combustion and use will result in knowledge of the emissions to be effected. Obviously, this can lead to the selection of the proper fuel to meet air emission standards and to the establishment of a workable fuel-switching program.

-- -- -- -- -- -- -- --- -- -- -- -- -- -- -- -- -- -- -- -- --

SAMPLING FOR: COAL: MOISTURE

PREFERRED ANALYTICAL PROCEDURE FOR LEVEL 1 INVESTIGATION: GRAVIMETRIC

Advantages:

1. Rapid, simple and accurate.
2. ASTM Method.

Distinctive Requirements:

1. 1 g of sample is dried at 104-110°C for 1 hr and weighed.

References:

ASTM Standard D271.

PREFERRED ANALYTICAL PROCEDURE FOR LEVEL 2 INVESTIGATION: GRAVIMETRIC

Advantages:

1. See above.

Distinctive Requirements:

1. See Level 1, above.

Remarks:

1. Moisture, ash, volatile matter, fixed carbon, heating value and sulfur comprise a proximate analysis.

_ _

SAMPLING FOR: COAL: ASH

PREFERRED ANALYTICAL PROCEDURE FOR LEVEL 1 INVESTIGATION: COMBUSTION

Advantages:

1. Rapid, simple and accurate.
2. ASTM Method.

Distinctive Requirements:

1. Dried coal sample heated at 750°C to a constant weight.

PREFERRED ANALYTICAL PROCEDURE FOR LEVEL 2 INVESTIGATION: COMBUSTION

Advantages:

1. See Level 1, above.

Distinctive Requirements:

1. See Level 1, above.

_ _

SAMPLING FOR: COAL: VOLATILE MATTER

PREFERRED ANALYTICAL PROCEDURE FOR
LEVEL 1 INVESTIGATION: COMBUSTION (CONTROLLED)

Advantages:

1. Simple, rapid and accurate.
2. ASTM Method.

Distinctive Requirements:

1. 1 g of sample is placed in Pt crucible with cover in oven at 950° ± 20°C.
2. After rapid initial discharge of volatiles has subsided, crucible is heated 7 more minutes, cooled and weighed.

References:

ASTM Standard D271.

PREFERRED ANALYTICAL PROCEDURE FOR
LEVEL 2 INVESTIGATION: COMBUSTION

Advantages:

1. See Level 1, above.

Distinctive Requirements:

1. See Level 1, above.

Remarks:

1. Crucible cover should fit tightly.

- -

SAMPLING FOR: COAL: FIXED CARBON

PREFERRED ANALYTICAL PROCEDURE FOR
LEVEL 1 INVESTIGATION: DIFFERENCE

Advantages:

1. Rapid, simple and accurate.
2. ASTM Method.

Distinctive Requirements:

1. Fixed carbon, % = 100−(moisture % + ash % + volatile matter %).

References:

ASTM Standard D271.

PREFERRED ANALYTICAL PROCEDURE FOR
LEVEL 2 INVESTIGATION: COMBUSTION

Advantages:

1. See Level 1, above.

Distinctive Requirements:

1. See Level 1, above.

- -

SAMPLING FOR: COAL: SULFUR

PREFERRED ANALYTICAL PROCEDURE FOR
LEVEL 1 INVESTIGATION: COMBUSTION

Advantages:

1. Rapid, accurate and simple.
2. ASTM Method.

Distinctive Requirements:

1, Sulfur is determined from bomb washings from oxygen bomb colorimeter following colorimetric determination.
2. NH_4OH is added, and solution boiled and filtered.
3. Saturated bromine water is added, solution is acidified with HCl.
4. $BaCl_2$ is added and Pt ignited at 925°C to a constant weight.

References:

ASTM Standard D271.

PREFERRED ANALYTICAL PROCEDURE FOR LEVEL 2 INVESTIGATION: COMBUSTION

Advantages:

1. See Level 1, above.

Distinctive Requirements:

1. See Level 1, above.

Remarks:

1. Bomb washing method.

_ _

SAMPLING FOR: COAL: HEATING VALUE

PREFERRED ANALYTICAL PROCEDURE FOR LEVEL 1 INVESTIGATION: COMBUSTION

Advantages:

1. Fast, simple and accurate.
2. ASTM Method.

Distinctive Requirements:

1. 1 g of coal is combusted in a bomb under 20-30 atm. of oxygen.
2. Bomb washings are titrated with standard base to determine correction factor.

References:

ASTM Standard D271.

PREFERRED ANALYTICAL PROCEDURE FOR LEVEL 2 INVESTIGATION: COMBUSTION

Advantages:

1. See Level 1, above.

Distinctive Requirements:

1. See Level 1, above.

Remarks:

1. Washings are used for sulfur analysis.
2. Corrections are made for nitrogen and sulfur content.

- -

SAMPLING FOR: COAL CARBON

PREFERRED ANALYTICAL PROCEDURE FOR LEVEL 1 INVESTIGATION: COMBUSTION

Advantages:

1. Accurate, simple and fast.
2. ASTM Method.

Distinctive Requirements:

1. 0.2 g of coal is combusted in stream of oxygen at 850-900°C in a combustion train.
2. CO_2 formed is absorbed in soda lime and weighed.

References:

ASTM Standard D271.

PREFERRED ANALYTICAL PROCEDURE FOR
LEVEL 2 INVESTIGATION: COMBUSTION

Advantages:

1. Same as Level 1, above.

Distinctive Requirements:

1. Same as Level 1, above.

Remarks:

1. Carbon, hydrogen, nitrogen and oxygen comprise an ultimate analysis.
2. SO_2, which is formed, is absorbed on CaO, $PbCrO_4$ or silver.

_ _

SAMPLING FOR: COAL: HYDROGEN

PREFERRED ANALYTICAL PROCEDURE FOR
LEVEL 1 INVESTIGATION: COMBUSTION

Advantages:

1. Fast, simple and accurate.
2. ASTM Standard.

Distinctive Requirements:

1. 0.2 g of coal is combusted in stream of oxygen at 850-900°C in a combustion train.
2. H_2O formed is adsorbed on $Mg(ClO_4)$ and weighed.

References:

ASTM Standard D271.

PREFERRED ANALYTICAL PROCEDURE FOR
LEVEL 2 INVESTIGATION: COMBUSTION

Advantages:

1. See Level 1, above.

Distinctive Requirements:

1. See Level 1, above.

- -

SAMPLING FOR: COAL: NITROGEN

PREFERRED ANALYTICAL PROCEDURE FOR
LEVEL 1 INVESTIGATION: KJELDAHL-GUNNING METHOD

Advantages:

1. Accurate in determining nitrogen.
2. ASTM Method.

Distinctive Requirements:

1. Nitrogen is converted to ammonium salts by digestion with hot H_2SO_4 (and a catalytic amount of Hg) for several hours.
2. Zinc is added and a solution of K_2S in NaOH is added.
3. The mixture is distilled into 0.2 $(NH_4)_2SO_4$ and backtitrated to a methyl red end point with 0.02 NaOH.

References:

ASTM Standard D271.

PREFERRED ANALYTICAL PROCEDURE FOR
LEVEL 2 INVESTIGATION: KJELDAHL-GUNNING METHOD

Advantages:

1. See Level 1, above.

Distinctive Requirements:

1. See Level 1, above.

- -

SAMPLING FOR: COAL: OXYGEN

**PREFERRED ANALYTICAL PROCEDURE FOR
LEVEL 1 INVESTIGATION: DIFFERENCE**

Distinctive Requirements:

1. Oxygen, % = 100 (% hydrogen, % carbon, % nitrogen, % sulfur, % moisture, % ash).

References:

ASTM Standard D271.

**PREFERRED ANALYTICAL PROCEDURE FOR
LEVEL 2 INVESTIGATION: DIFFERENCE**

Advantages:

1. See Level 1, above.

Distinctive Requirements:

1. See Level 1, above.

- -

SAMPLING FOR: COAL TRACE ELEMENTS

**PREFERRED ANALYTICAL PROCEDURE FOR
LEVEL 1 INVESTIGATION: SSMS**

Advantages:

1. See Elemental Analysis, p. 213.

Distinctive Requirements:

1.   See Elemental Analysis.

PREFERRED ANALYTICAL PROCEDURE FOR LEVEL 2 INVESTIGATION: FAAS/AAS

Advantages:

1.   See Elemental Analysis.

Distinctive Requirements:

1.   See Elemental Analysis.

- -

SAMPLING FOR: RESIDUAL OIL H_2O BITUMINOUS MATERIAL

PREFERRED ANALYTICAL PROCEDURE FOR LEVEL 1 INVESTIGATION: DISTILLATION

Advantages:

1.   Good technique for determination of H_2O and bituminous material in residual oil.
2.   ASTM Method.

Distinctive Requirements:

1.   Sample weighed to ± 1% transferred to a Dean-Stark apparatus.
2.   Sample is distilled and water measured in trap.

References:

ASTM Standard D1246.

PREFERRED ANALYTICAL PROCEDURE FOR
LEVEL 2 INVESTIGATION: DISTILLATION

Advantages:

1. See Level 1, above.

Distinctive Requirements:

1. See Level 1, above.

- -

SAMPLING FOR: RESIDUAL OIL: ASH

PREFERRED ANALYTICAL PROCEDURE FOR
LEVEL 1 INVESTIGATION: COMBUSTION

Advantages:

1. Excellent method for measuring ash content of residual oil.
2. ASTM Method.

Distinctive Requirements:

1. 100 g of sample is combusted in an oven crucible or evaporating dish.
2. Carbonaceous residue is ashed at $775°C$, cooled and weighed.

References:

ASTM Standard D482.

PREFERRED ANALYTICAL PROCEDURE FOR
LEVEL 2 INVESTIGATION: COMBUSTION

Advantages:

1. See Level 1, above.

Distinctive Requirements:

1. See Level 1, above.

Remarks:

1. Ash sample can be analyzed for trace elements.

SAMPLING FOR: RESIDUAL OIL HEATING VALUE

PREFERRED ANALYTICAL PROCEDURE FOR
LEVEL 1 INVESTIGATION: COMBUSTION

Advantages:

1. Good method for determining the heating value of residual oil.
2. ASTM Method.

Distinctive Requirements:

1. Sample is combusted in O_2 bomb with 1 ml of H_2O and under 30 atm.
2. Contents of bomb are titrated to determine a correction factor.

References:

ASTM Standard D280.

PREFERRED ANALYTICAL PROCEDURE FOR
LEVEL 2 INVESTIGATION: COMBUSTION

Advantages:

1. See Level 1, above.

Distinctive Requirements:

1. See Level 1, above.

SAMPLING FOR: RESIDUAL OIL: SULFUR

PREFERRED ANALYTICAL PROCEDURE FOR LEVEL 1 INVESTIGATION: BOMB WASHING METHOD

Advantages:

1. Provides a good determination of sulfur content of residual oil.

Distinctive Requirements:

1. Bomb washings from heating value determination are acidified with HCl.
2. Saturated bromine water is added and the solution is boiled.
3. $BaCl_2$ is added, the solution filtered and the precipitate ignited at $925°C$.

References:

ASTM Standard D280, D1252.

PREFERRED ANALYTICAL PROCEDURE FOR LEVEL 2 INVESTIGATION: BOMB WASHING METHOD

Advantages:

1. See Level 1, above.

Distinctive Requirements:

1. See Level 1, above.

SAMPLING FOR: RESIDUAL OIL: TRACE ELEMENTS

**PREFERRED ANALYTICAL PROCEDURE FOR
LEVEL 1 INVESTIGATION: SSMS**

Advantages:

1.   See Elemental Analysis, p. 213.

Distinctive Requirements:

1.   See Elemental Analysis.

**PREFERRED ANALYTICAL PROCEDURE FOR
LEVEL 2 INVESTIGATION: FAAS/AAS**

Advantages:

1.   See Elemental Analysis.

Distinctive Requirements:

1.   See Elemental Analysis.

CHAPTER 12

ATMOSPHERIC AIR QUALITY

INTRODUCTION

The understanding of the atmospheric interractions of the physical, chemical and biological processes as related to regional air pollution is somewhat vague, because of the relatively small amount of work in sampling and experimental analysis that has been accomplished. Mathematical modeling work has also been limited due to a general lack of data. This chapter examines the experiments and models that have been carried out as they are related to man and his effect on the urban environment.

REGIONAL PROBLEMS OF AIR POLLUTION

Urban air pollution historically has been the focus of attention of those concerned with the health and welfare of city dwellers. This is due, primarily, to the magnitude of the urban pollution source and its close proximity to densely populated areas. Current considerations have been given to areas such as rural, suburban and other urban, which are a relatively long distance from the urban pollution source. For example, many believe much of the pollution of the Eastern states is generated by the Midwestern industrial states, and that atmospheric conditions transport the pollutants east.

The basis of air pollution studies is the understanding of the interrelationships of source emission parameters, ambient air quality and meteorological conditions. These vary from the emission source location to downwind locations. Further, during the atmospheric transport of the pollutants, various chemical, physical and biological processes may effect the pollutants' physical nature and chemical form. The complexities of these many interactions effects a rather limited understanding of regional or long-distance pollutant atmospheric transport.

As a result, we must become cognizant of the rural environment as it is affected by air pollution. The rural environment is much different than the urban environment and, thus, meteorological characteristics differ greatly. Atmospheric transport, diffusion and chemical decay processes can vary between rural and urban areas. Further, the complexity of the problem is increased by the following:

1. Lower pollution concentrations; pollutant measurements and analyses are made more difficult.
2. Natural pollutant sources, because of their greater proportion of the overall pollution concentration, are of greater significance.
3. The variation of background pollution levels with location (ocean, marsh, desert, swamp, etc.).
4. The scarcity of rural monitoring stations.
5. The allowance of certain chemical reactions and the prolongation of others due to the longer residence time of the pollutants being transported greater distances.

In a later section of this chapter, we will discuss the modifications of the mathematical air pollution models to include these variations in the analysis of the rural environment.

The mechanisms of air pollution formation and transport include:

1. the pollution emission source;
2. meteorological conditions to transport and disperse the pollutants;
3. atmospheric chemical reactions that occur during the transport and dispersion   process and that effect secondary pollutants (these secondary pollutants cause the gravest concern to downwind locations);
4. the reduction of both primary and secondary pollutant concentrations due to the various atmoshperic sinks (*i.e.*, dispersion, washout by rain, etc.).

Therefore, on a regional basis, which is becoming increasingly important in air pollution analysis, the control, measurement and prediction of air pollutant levels is becoming more and more difficult.

For example, from the mid-1960s, when air pollution regulations were born, to the early 1970s, ambient SO_2 concentrations over the eastern portion of the United States declined.[1]   Sulfate levels, on the other hand, did not,[2,3] and it is believed that nearly half of these levels is due to the transformation of sulfur dioxide into sulfate as a result of long-term atmospheric decay. In this same sector of the U.S., in the nonurban areas, it is estimated that over 75% of the ambient sulfates present are due to distant pollutant sources.[1]   As a result of this type of data, the use of tall stacks as a viable pollution control method is being scrutinized. Tall stacks which disperse pollutants, and sulfur dioxide in particular, over wider areas, may effectively limit localized SO_2 impacts but are creating sulfate problems at distant locations.

Additional evidence for the regional importance of air pollution is that oxidant levels of rural areas often exceed regulatory standards.[4,5] In the past, oxidant levels were believed to have been derived solely from the stratosphere. Present studies[6] have shown the source of these oxidants to be hydrocarbons emitted in distant urban areas. Based on an examination of the data, meteorologically effected ozone levels were shown to be incapable of producing the oxidant levels measured at the specified sites.

Of primary concern to a regional air quality plan, therefore, is sulfur dioxide-effected particulates (sulfates) and oxidants; and most of these pollutants are of regional importance because of their long-range transport and chemical transformation.

Secondary pollutant control requires control strategies encompassing large areas. Air pollution impact assessments of tall stacks must begin to study not only the deterioration of air quality on a local basis, but also on a regional basis up to about a 60-mile radius.

Thus, regional management of air quality has become imperative. State agencies must begin to work together using comparable standards. Federal agencies should coordinate the activities of the states with regulations that do not allow disproportionate pollution in some areas which then affects air quality in clean areas tens of miles away.

AIR POLLUTION REGULATIONS

The Clean Air Act of 1970 and its amendments established national air quality standards for ambient pollutant concentrations. These regulations include standards for both short- and long-term averaging periods. Table 12-1 shows these national air quality standards set for sulfur dioxide, total suspended particulate (TSP), carbon monoxide, nitrogen dioxide and photochemical oxidants.

Under these regulations the Environmental Protection Agency (EPA) can enforce regional air quality plans so that these standards can be met. For example, the EPA edict prohibiting parking in mid-town Manhattan to reduce automobile traffic was devised to allow New York City to meet national air quality standards.

Air quality in the United States is regulated by another method. The recent Clean Air Act Amendments of 1977 and the Prevention of Significant Deterioration provisions were promulgated with the intent of protecting and enhancing air quality. Although these regulations are in the process of being

Table 12-1. United States Environmental Protection Agency
Ambient Air Quality Standards

Pollutant	During any 12 consecutive months the average pollutant levels in ambient air shall not exceed:	
	Primary Standard ($\mu g/m^3$)	Secondary Standard ($\mu g/m^3$)
Particulate Matter	75 (annual average)[a]	60 (annual average)[a]
	260 (24-hr average)[b]	150 (24-hr average)[b]
Sulfur Dioxide	80 (annual average)[c]	60 (annual average)[c]
	365 (24-hr average)[b]	260 (24-hr average)[b]
		1300 (3-hr average)[b]
Carbon Monoxide	10 (8-hr average)[b]	Same as primary
	40 (1-hr average)[b]	Same as primary

No distinctions for primary and secondary standards are made for the following:

Photochemical Oxidants	160 (1-hr average)[b]
Hydrocarbons (except methane)	160 (1-hr average)[b]
Nitrogen Dioxide	100 (annual average)[c]

[a]Geometric mean.
[b]This concentration may be exceeded no more than once in any 12 consecutive months.
[c]Arithmetic mean.

interpreted, their intent is to allow no significant deterioration of air quality due to sulfur dioxide and particulates. Future standards will be established for carbon monoxide, nitrogen oxides and hydrocarbons.

Three different classes of allowable increases in concentrations of SO_2 and TSP were established. These and the increments for each class are listed in Table 12-2. For the present, the states will have the authority to reclassify their land with the exceptions of 1) federal lands, and 2) Indian territories. The EPA is to arbitrate disputes between states over the classification of adjacent areas. Further, new emission sources cannot violate the air quality of surrounding areas.

Based on these regulations, the significance of regional air quality planning becomes very apparent.[7,8]

METEOROLOGY

Meteorology plays an important role in air quality, both in the dilution of pollutant concentrations through transport and diffusion and in the removal of pollutants through precipitation scavenging.

Table 12-2. Significant Deterioration Regulations Summary

Three different classes of allowable increases in concentrations of SO_2 and TSP were established.

Class I—areas in which practically any change in air quality would be considered significant

Class II—areas in which deterioration normally accompanying moderate well-controlled growth would be considered insignificant

Class III—areas in which larger deterioration would be allowed

The increments for each class are listed below:

	Class I ($\mu g/m^3$)	Class II ($\mu g/m^3$)	Class III ($\mu g/m^3$)
Sulfur Dioxide			
Annual	2	20	40
24-hr	5	91	182
3-hr	25	512	700
Particulates			
Annual	5	19	37
24-hr	10	37	75

Meteorology is the science of the atmosphere and its phenomenon. The earth's atmosphere is stratified, made up of several layers, each with its own distinct characteristics. However, the troposphere (the layer closest to the earth) is of primary concern with regards to the dispersion and transport of air pollutants. In fact, the lower troposphere, consisting of the first 2000 meters above ground, is the most important in air pollution meteorology. The atmospheric characteristics and changes in this layer are known as micrometeorology and are the result of the interaction of four elements: 1) the sun, 2) the earth, 3) the atmosphere and 4) the natural and anthropogenic surface features. An important factor that will be discussed is the effect of these four interacting atmospheric elements on the micrometeorology of the troposphere and the transport and dispersion of air pollutants.

Wind Speed and Direction

The mechanisms of air pollution involve pollutant emission, transport and diffusion, and effects upon receptors (man, plants, or animals). Although meteorology affects all three phases, of primary importance are transport and diffusion, which are dependent on wind speed and direction.

The effect of wind speed on the transport and diffusion of air pollutants is twofold. The wind speed determines the velocity of the air pollutant as it traverses the sky and contributes to the pollutant dilution rate. That is, the greater the wind velocity, the greater the pollutant velocity and the greater volumes of air introduced to the pollutant, thus, increasing the dilution rate. However, in the case of a stack emission (plume), the higher wind velocity tends to "bend" the plume over, inhibiting its rise potential and forcing pollutants to the ground sooner. The time for dispersion has been reduced and pollutant concentrations at ground level are greater.

Wind direction is the direction from which the wind originates. The wind direction is affected by various meteorological conditions and topographical features. This, in turn, influences the manner in which the pollutants are dispersed. On the local-level eddy currents and other turbulence greatly affect pollutant transport and dispersion. On the regional level, macrometeorology plays an important role as wind flow patterns and climatology must be examined.

Different types of winds develop under different conditions. On clear summer days, light winds are effected when heating of the land surface adjacent to a large body of water such as a lake or ocean is more rapid than the heating of the water. This creates a temperature difference and a density and pressure difference between the air above the land surface and the air above the water. As a result, this difference in air buoyancies produces a local circulation pattern with wind movement from the water towards the land.

Another type of air movement is known as a drainage or slope wind. These usually occur under clear skies when the differences in the rates of heating and cooling of valley floors and sides cause slight density and pressure differences. These result in small-scale circulation patterns characteristic of the valley surface. For example, the steep valley slopes generate the strong downslope winds. The velocity of drainage winds is usually greatest during the winter months. Complex flow patterns are characteristic of valley winds due to valley channeling effects.

Local winds are those wind flow systems caused by large-scale pressure systems. Under these meteorological conditions, a high-pressure area exists in the highland plateaus while a low-pressure area exists in the lowlands, creating a strong pressure-gradient force and high wind speeds of between 20 and 50 mph.

Land breezes occur at night as rapid cooling of the land mass creates a temperature and pressure difference between land and water air masses. These winds are not generally high in velocity but do become rather strong during the winter months.

In general, air motion requires a force to generate the movement and this is a result of many contributing factors. Solar energy is unevenly distributed due to insolation variations and differences in the absorptivity of the earth's surface. This results in large-scale air motions of the earth. In general, the motion is from the tropics towards the polar regions, as an equilibrium state of the earth's energy is attempted. Warm tropical air is more buoyant and it rises; it then travels to the colder polar regions where it is cooled and becomes more dense and drops to the surface. These air masses move along the surface towards the tropics to replace the rising tropical air masses. This is known as the thermal circulation pattern of winds.

Air motion is also affected by the motion of the earth or the Coriolis effect. The earth's rotation is from west to east causing a westward direction of air. The combined effect of the thermal and Coriolis forces on the air masses results in the gradient wind. These winds are located in the upper atmosphere (approximately 300 meters) where frictional effects of the earth's surface are minimal. Frictional forces can effect wind movement up to 100 meters in altitude. This results in a wind velocity gradient with height.

This wind variation contributes to atmospheric turbulence. On the more localized level, surface characteristics such as trees, hills and buildings, produce further atmospheric turbulence. Varying temperatures generate thermal turbulence which contributes to wind variability and its effect on atmospheric stability. Diurnal variations are another factor contributing to wind variability. During the day, turbulence and vertical motions are at a maximum due to solar heating. This results in the maximum amount of energy transfer between the various atmospheric levels. Thus, variation of wind speed with height is minimal during the day. At night vertical motions are minimal.

Based on air motion, the atmosphere can be divided into two layers. From the surface to about 500 meters is the atmospheric mixing layer. In this layer atmospheric motion is affected by surface roughness, horizontal pressure gradients, shear stresses and Coriolis forces. Above this layer is the geostrophic layer where flow is influenced only by horizontal pressure gradients and Coriolis forces. Figure 12-1 shows these two atmospheric zones. Air pollution transport and diffusion is primarily concerned with the atmospheric mixing layer.

In the Ekman layer (Figure 12-1), the wind direction flows clockwise in the Northern Hemisphere and counterclockwise in the Southern Hemisphere. The wind speed in the Ekman layer increases rapidly with height but the wind velocity gradient declines as the free atmosphere layer is approached.

In the surface layer, the vertical turbulent momentum and heat are assumed constant with respect to height to simplify the region's definition. The variation of wind speed with height is critical in this layer because pollutant discharge usually occurs in this layer. This variation is dependent on surface topography.

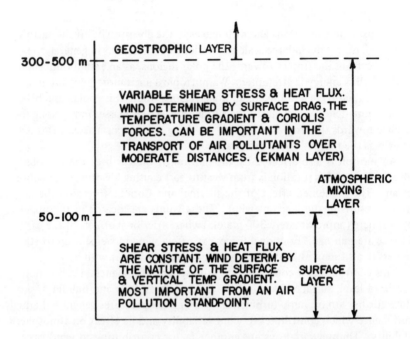

Figure 12-1. Regions of the lower atmosphere according to air flow.

Studies have shown that the atmospheric capacity for diffusing heat and matter is related to both the standard deviation of the wind direction and the range of frequencies of fluctuations that contribute to the standard deviation. These factors, as affected by turbulence, were found to be characteristic of given atmospheric stability conditions and directly related to the dispersion potential of an air mass. Figure 12-2 illustrates the vertical variation of the lateral wind direction (standard deviation) as a function of atmospheric stability. Figure 12-3 shows the vertical variation of the vertical wind direction (standard deviation) as a function of atmospheric stability.

The wind speed variation with height for the entire atmospheric mixing layer can be described by the empirical-power-law profile.[10]

$$\frac{\bar{u}_1(x_3)}{u_G} = \frac{x_3}{h_G}^{\alpha}$$

where u_G = the geostrophic wind speed
        h_G = the thickness of the planetary boundary layer
        $\bar{u}_1$ = the upward displacement of the parcel.

The exponent α is a number which varies from 0.1-0.4 depending on the surface roughness and atmospheric stability. Although applicable to a range of atmospheric stabilities, the equation was developed from data from neutral

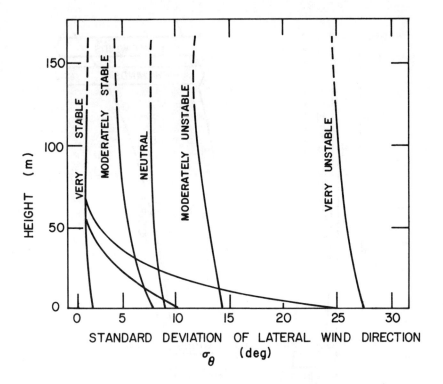

Figure 12-2. Variation of wind direction vs atmospheric stability.[9]

atmospheres. Figure 12-4 gives the surface layer wind profiles under adiabatic conditions and note Table 12-3 for the relationship between roughness (α) factors, topography and atmospheric stability.

The ground plays an important role in determining atmospheric stability. By affecting wind speeds and air temperatures it affects the ability of an air mass to disperse pollutants. This will be discussed further in subsequent sections.

Temperature Stratification

The atmospheric layers are identified as follows:[13]

1) *The Troposphere:* This is the layer closest to the ground extending to an altitude of 15 km over the equator and 10 km over the poles. Temperature decreases with height at a rate of 6.5° C/km. Vertical convection keeps the air relatively well mixed.

2) *The Stratosphere:* This extends from the troposphere to about 50 km in altitude. Temperature is constant in the lower stratosphere and then

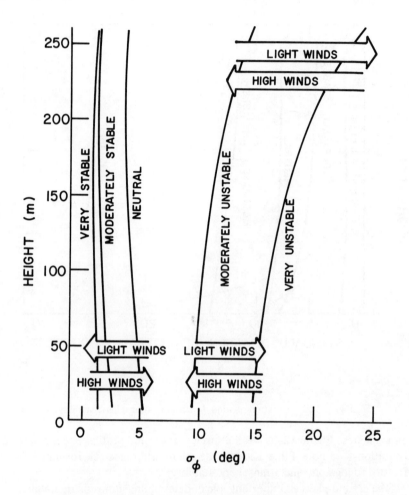

Figure 12-3. Standard deviation of the vertical wind direction.[9]

increases with altitude owing to the absorption of short wave radiation by ozone. At the stratosphere (top of the atmosphere) the temperature reaches $270°K$. There is little vertical mixing in the stratosphere.

3) *The Thermosphere:* This is the uppermost layer. Molecular densities are of the order of 10^{13} mol/cm³ as compared to 5×10^{19} at sea level. Intense ultraviolet radiation dissociates N_2 and O_2. Temperatures exceed $1000°K$.

The troposphere is of primary interest with regards to pollution. Generally in the troposphere, a decrease in temperature accompanies a rise with altitude. Temperatures close to ground level are characterized by the changes in both direction and magnitude. A frequency distribution of a common temperature

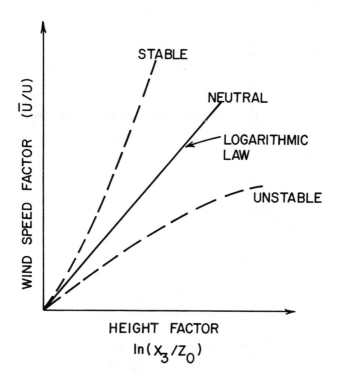

Figure 12-4. True surface layer wind profiles under adiabatic conditions.

Table 12-3. Relationship Between Surface Roughness, Topography, and
Atmospheric Stability[10,11]

Topography	Open country	Suburbs	Urban
Surface Roughness Factor (α)	0.16	0.28	0.40
Atmospheric Stability	**Very stable**	**Neutral**	**Very unstable**
Surface Roughness Factor	0.83	0.14	0.02

gradient is given in Figure 12-5. Negative temperatures represent a normal temperature decrease with altitude and positive values represent inversions.[14]

In the region between 46.4 and 61.0 meters, the temperature gradient most frequently reflects isothermal and/or adiabatic conditions. A significant

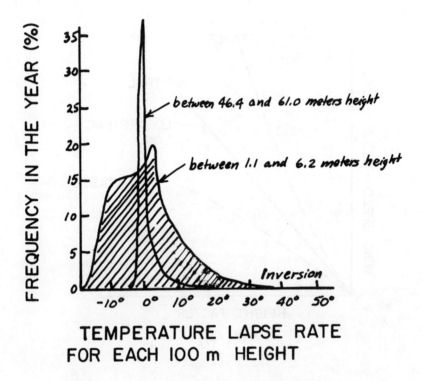

Figure 12-5. Frequency distribution of temperature gradients.[14]

difference in the temperature gradients exists between morning and evening hours. In the morning, the air near the ground is very stable until incoming solar energy starts a mixing action disrupting the stratification. In the evening, the relatively unstable conditions of the day are eventually stabilized as a temperature equilibrium process occurs, thus stratifying the atmosphere.

Temperatures in the lower atmosphere decline with altitude. Temperature differences in air masses cause pressure imbalances, resulting in air flowing from areas of high pressure (near the earth) to areas of low pressure (with increased altitude from earth). On the other hand, surface and heat radiation conditions can cause an increase in ambient temperature with altitude resulting in an inversion that inhibits the dispersion of pollutants.

Figure 12-6 gives the relationship between isothermal conditions and the effect of wind speed and humidity. It shows that the less humid the air the sooner isothermal conditions occur. This corresponds to the fact that dry air is associated with strong outgoing radiation. Therefore, as the humidity increases, the greater the time after sunset before isothermal conditions exist. Wind speed is also shown to influence atmospheric stability.

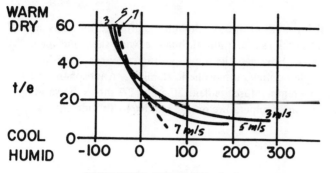

MINUTES BEFORE (-) OR AFTER (+)SUNSET

Figure 12-6. Time of beginning of evening isothermal conditions in relation to temperature, humidity and wind for a layer 1-16 meters in height.[14]

The factors that contribute to the temperature of the earth and the atmosphere at any location include absorption and reflection of solar radiation by atmospheric gas molecules and particles and the absorption, reflection and emission of solar radiation by the earth's surface. The average reflectivity (albedo) of the earth and atmosphere together varies from 30-50% of the incoming radiation with an average value of 34%. Most of this reflection is due to clouds. Table 12-4 shows how incoming solar radiation is reflected and absorbed.

Table 12-4. Reflection and Absorption of Solar Energy

50%	Intercepted by cloud (25% back to space, 23% to the earth and 2% absorbed by clouds)
17%	Absorbed by gases and dust in the atmosphere
12%	Scattered by the air (7% back to space, 5% to earth)
19%	Absorbed by the earth
2%	Reflected by the earth back to space
100%	

These relationships are an integral part of temperature stratification, atmospheric stability and pollutant dispersal.

Atmospheric Stability

The atmospheric stability plays an important role in the atmosphere's pollutant-dispersal capability. A stable atmosphere tends to inhibit pollutant dispersion, increasing pollutant concentrations. An unstable atmosphere aids dispersion, minimizing concentrations. Stability is affected by many meteorological parameters such as insolation, turbulence, wind shear and vertical temperature gradients. However, the temperature gradient is the primary consideration when determining atmospheric stability. Atmospheric lapse rates are based on the temperature gradient. Figure 12-7 shows the varying lapse rates combined on a temperature profile relative to altitude.

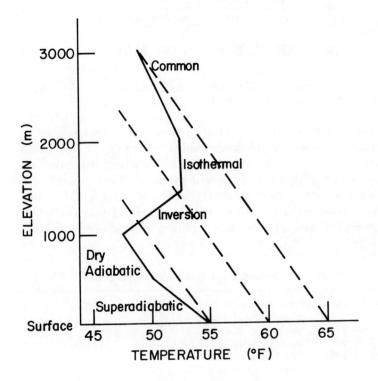

Figure 12-7. Varying lapse rates on a temperature profile relative to altitude.[15]

A dry adiabatic lapse rate is defined as the rate of cooling of a parcel of air for which there is zero heat exchange between the environment. Theoretically, when a small volume of air is forced upward in the atmosphere it will encounter a lower pressure, expand and cool. A temperature lapse rate of -1.0°C/100 meters defines a dry adiabatic state of ascent. A perfectly mixed layer will also exhibit these characteristics, but such a condition actually never occurs.

A superadiabatic lapse rate, or a rate of temperature decrease with height exceeding 1°C/100 meters, usually occurs on days with strong insolation or when a cold air mass is brought over a much warmer surface. A superadiabatic condition favors strong convection, instability and turbulence, and generally is confined to the lowest 10 meters of the atmosphere.

A neutral atmospheric lapse rate occurs when the environmental (actual) lapse rate nearly coincides with the dry adiabatic lapse rate (dashed line). Under these conditions there is a slight tendency for a displaced particle to gain or lose buoyancy. Neutral atmospheres are characterized by overcast skies and moderate-to-strong wind speeds.

A subadiabatic lapse rate defines an atmosphere in which the temperature decreases with height more gradually than the rate of $1°C/100$ meters. This condition is slightly stable and tends to inhibit air displacement.

An isothermal lapse rate exists when the ambient temperature remains constant with height. This lapse rate tends to suppress vertical motion and is normally short-lived.[7,8]

Thus, as noted, the lapse rate in the lower portion of the atmosphere has a great influence on the vertical motion of air. Pollutants in a neutral atmosphere are not affected by buoyant forces. Unstable atmospheres enhance vertical motion. Stable atmospheres oppose vertical motion and pollutant dispersal. Figure 12-8 summarizes this. Some definitions are:

1) *Adiabatic lapse rate:* Temperature decreases with height such that any vertical movement imparted to an air parcel will result in the parcel maintaining the same temperature or density as the surrounding air (neutral stability— $1°C/100$ meters).

2) *Superadiabatic:* A rising parcel of air will be warmer than its environment so that it becomes more buoyant and continues rising (unstable).

3) *Subadiabatic:* A rising air parcel is cooler than its surroundings so it becomes less buoyant and returns (stable).

4) *Isothermal:* The temperature remains constant with height (stable).

5) *Inversion:* The temperature increases with height (extremely stable).

Inversions occur when a layer of warm air covers a layer of cold air. Inversions result in little mixing or vertical dispersion of pollutants takes place. Inversions can form either through cooling from below or a heating from above process. They often form, particularly at night, because of radiation cooling on the ground. Surface inversions result when a warm air mass moves with a horizontal motion over a cool surface. Inversions that result from heating from above and involve the spreading, sinking and compresssion of an air mass as it moves horizontally are termed subsidence inversions. The upper layers of air undergoing the greatest elevation change will experience the greatest degree of compression and temperature increase; if this temperature increase with height is sufficient, an inversion will occur. The types of temperature inversions common in the lower atmosphere are as follows:[7,8,16]

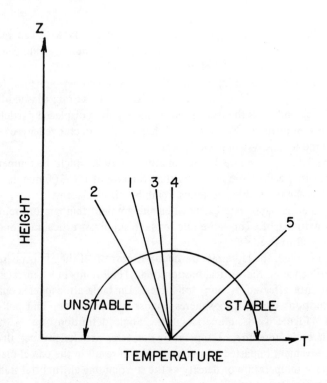

Figure 12-8.   Temperature profiles found in the lower atmosphere relative to stability class.

1) *Surface or Radiation Inversion:* One form of inversion is shown by the examples of plume dispersion relative to stability in Figure 12-9. Structure (e) is usually found at night with light winds and clear skies when the loss of heat by long-wave radiation from the ground surface cools the surface and the air adjacent to it. This condition is usually found in the open country and, with the proper humidity, will lead to the formation of radiation fog.

2) *Elevated Inversions:* This type of inversion exists above the ground surface acting as a barrier between the mixed layer close to the surface and that aloft. Figure 12-10 shows this condition exhibiting a temperature decrease with height capped by an inversion layer. Above the inversion there is a more adiabatic decrease in temperature with height. The cause of this type of inversion may be:

   a. Subsidence—the gradual decent of air aloft accompanied by adiabatic warming of the descending layer resulting from the increase in pressure.

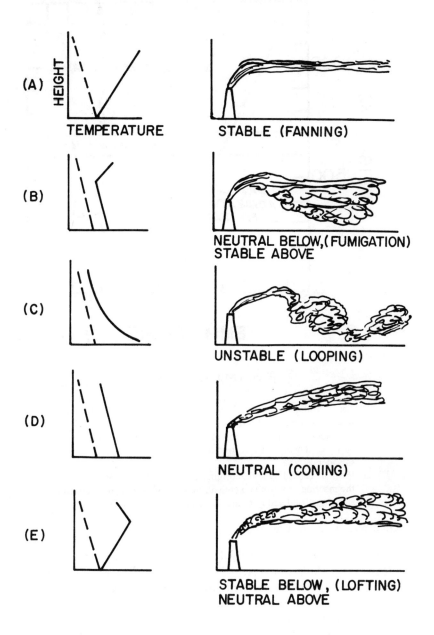

Figure 12-9. Plume behavior as a function of atmospheric stability.

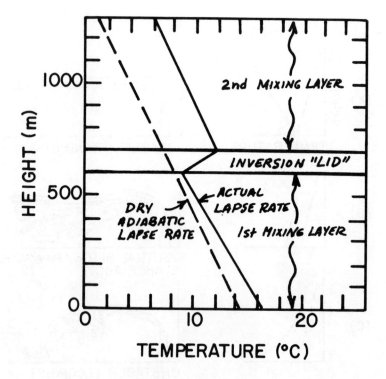

Figure 12-10. Elevated inversion.[16]

This subsidence inversion is most persistent during the fall and summer months and can last for days.

b. Sea Breeze—the introduction of a layer of cool air beneath a warmer air mass. These are short-lived inversions and not as persistent as the subsidence type.

c. Frontal—caused by a meteorological front in the atmosphere due to the meeting of warm and cold air masses. This type of inversion is also short-lived and not quite as persistent as the subsidence type.

Atmospheric Turbulence

Turbulence refers to the irregular and apparently random atmospheric fluctuations. The primary factors contributing to atmospheric turbulence are: 1) surface roughness elements, and 2) surface and ambient air temperature differences. Generally, a gradient wind flow or air temperature increase near the earth's surface effects an increase in turbulence. The turbulent motion of the atmosphere affects vertical motion and in turn effects the dispersion of pollutants. Some of the meteorological parameters influencing atmospheric turbulence will be discussed.

Solar Radiation

1) *Mid-day Heat Exchange:* The earth's atmosphere receives about 2 cal/ cm² of vertical solar radiation each minute during mid-day. This is known as the "solar constant." Figure 12-11 illustrates the solar heat exchange on a typical summer day. The subsequent paragraphs examine the effect this heat exchange on the dispersion of pollutants.

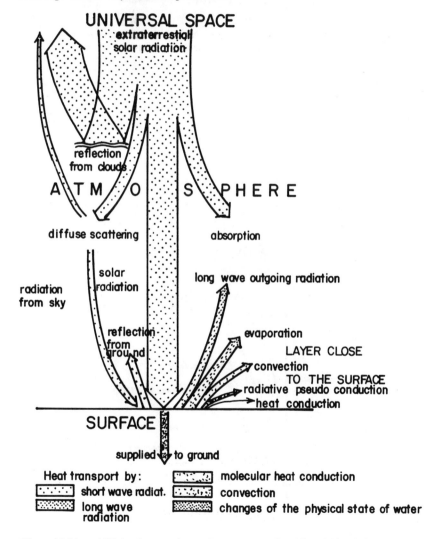

Figure 12-11. Mid-day heat exchange for a summer day (the width of the arrows corresponds to the relative amounts of transferred heat).

One type of heat transfer is due to short-wave radiation (wavelength below 1 μ; in Figure 12-11, widely dotted stripes). A large part (about 33%) of the incoming solar radiation is reflected by the surface of the clouds and is not utilized in air or ground heating. Other radiation is scattered by both air molecules and particulates. The energy from this radiation is diverted (deflected) from its source. This combined total of energy deflected by these processes and reflecting power of the earth is about 42%.[17] Radiation is also lost by ozone, water vapor and carbonic acid absorption.

Despite these losses, a large amount of solar radiation penetrates to the earth's surface, partly as direct sun radiation and partly as scattered radiation from the sky. This energy is used for warming the earth and air.

2) *Nocturnal Heat Exchange:* Solar radiation is comparatively lacking at night because there is no other energy source comparable to the sun. The nocturnal heat exchange, therefore, is characterized by no drastic, abrupt temperature gradients which can occur during the day. The heat exchange during the night depends on the heat radiation from the surface of the earth that has been stored during the day. Figure 12-12 illustrates the nocturnal heat exchange pattern.

As opposed to the diurnal heat exchange process, where large amounts of solar radiation penetrate the atmosphere and reach the earth, the nocturnal

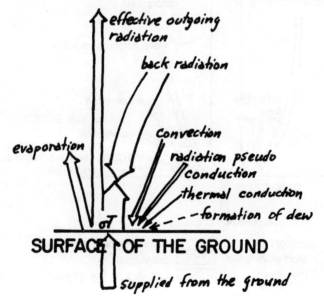

Figure 12-12. Nocturnal heat exchange. The width of arrows is proportional to heat amounts.

penetration of outgoing radiation is rather small. This is the result of water vapor and carbon dioxide absorbing radiation in certain bands of the spectrum, and primarily those of long wavelengths.[13] The maximum intensity of the earth's radiation is about 10 meters, which is a long wavelength. Therefore, the atmosphere readily admits solar radiation (0.5 μ wavelength) but limits the earth's radiation outflow. Only about 12% of the earth's nocturnal radiation is lost to space. The remainder is absorbed by the various layers of air in proportion to their water vapor and carbon dioxide content.

The strongest outgoing radiation period occurs with the nocturnal process during winter.[14] Table 12-5 shows this phenomenon in terms of the intensity of outgoing radiation per unit time. Note that the duration of the radiation period is considerably longer during the winter than in the summer. The period of outgoing radiation begins approximately two hours before sunset and ends about two hours after sunrise.

Table 12-5. Earth's Surface Temperature Related to Outgoing Radiation[14]

Surface temperature of the ground $^\circ$C	- 40	- 30	- 20	- 10	0	10
Outgoing Radiation in cal/cm^2, min	0.244	0.288	0.339	0.395	0.459	0.530
Surface temperature of the ground, $^\circ$C	20	30	40	50	60	70
Outgoing Radiation in cal/cm^2, min	0.609	0.696	0.792	0.899	1.015	1.143

Figure 12-13 shows outgoing radiation values of a cloudless sky relative to cloudiness factors. The curve shows the increasing rapidity with which radiation diminishes as cloudiness grows. Initially, the curve coincides closely with the straight line representing high clouds, for very slight nocturnal cloudiness, mostly cirrus in type. A medium amount of cloud corresponds to a middle-height cloud, with an entirely cloudy sky representing quite a thick cover. In fog conditions, the radiation levels would be 7-8% lower yet.[14]

Conduction

Molecular conduction is also known as "physical" or "true" heat conduction because the warmer molecules give off heat energy to the colder molecules. Conduction accounts for nearly all heat transmission within the earth. Ground-temperature relationships will, therefore, be examined.

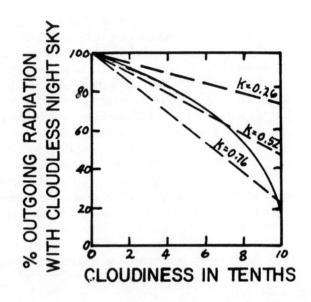

Figure 12-13. Dependence of the effective outgoing radiation on cloudiness.[14]

Figures 12-14 to 12-17 illustrate how heat movement takes place in the ground tables showing the heat cycle by lines of equal temperature. The temperature relationships during the day (incoming radiation) and the night (outgoing radiation) can be related to the ground temperatures. These figures show that for the microclimate near the ground, the ground acts as a regulating reservoir of heat. At times of heat surplus—at mid-day or in the summer—it absorbs great amounts of heat, reducing high temperatures while simultaneously storing calories. At night or in the winter, stored heat is given off preventing the temperature from falling too low.

The greater the thermal conductivity of the ground, the more effective is its role as a heat reservoir. Microclimates over soils of good conductivity exhibit year-round pleasant temperatures while soils of low conductivity have extreme microclimates associated with too-cold-by-night, too-hot-by-day conditions.

Eddy Diffusion

A heat exchange similar to that of molecular conduction takes place between the earth's surface and the adjacent air layer. Air, a good insulator, is a poorer conductor of heat than the ground, but does have good thermal diffusivity properties.

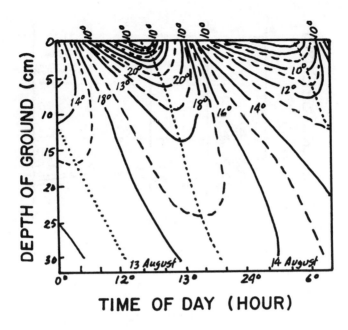

Figure 12-14. The penetration of the daily temperature cycle into the ground by heat conduction.[14]

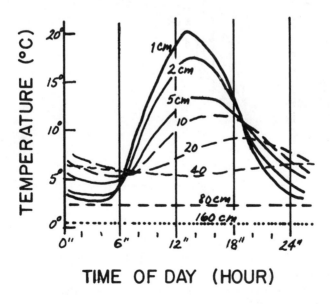

Figure 12-15. Daily temperature course in sandy soil in May.[14]

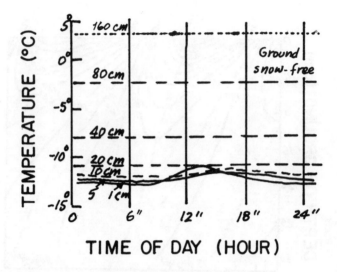

TIME OF DAY (HOUR)

Figure 12-16. Daily temperature course in sand soil in January.[14]

TEMPERATURE (°C)

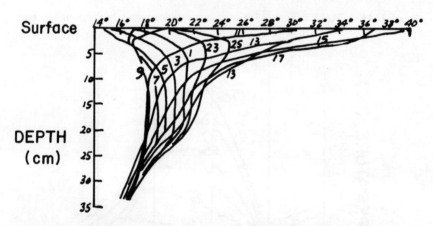

Figure 12-17. Tautochrones of the ground temperature on a radiation day in summer.[14]

There are two kinds of air circulation: 1) laminar, and 2) turbulent. In open spaces, which favors mixing, the air is almost constantly in a turbulent state. Lighter winds are not turbulent, but can gust on occasion.

Turbulence causes a continuous mixing of air masses affecting all of their properties. The parcel of air that rises at random from the earth's surface carries with it some heat, a large amount of water vapor and pollutants. Heat, water vapor, and pollutants are transported by eddy diffusion.

Humidity

Through evaporation at the ground surface the atmosphere absorbs water vapor. Transport upward of the water vapor occurs through eddy diffusion. Since the effects of eddy diffusion decrease with heights above the ground, the water vapor content of the air decreases steadily with height. Figure 12-18 illustrates this phenomenon and the relative humidity. The relative humidity is the ratio of the vapor pressure to the maximum possible at a given temperature, expressed as a percentage.

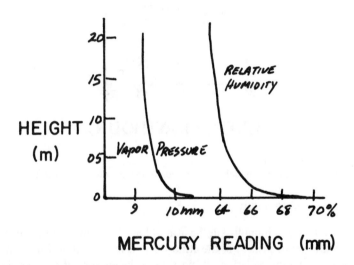

Figure 12-18. Daily mean of the relative humidity and vapor pressure in relation to altitude.[14]

There are two types of vapor pressure: 1) the wet type, produced by evaporation due to strong incoming radiation and 2) the dry type, referring to a distribution of humidity with respect to height, dry air being near the ground and that above being moist. In the case of relative humidity, normally the dry type exists during the day and the wet type at night. Figure 12-19 shows the distribution of humidity in the air layer next to the ground. Note that violent fluctuations in the relative humidity from those curves have been recorded because of the influence of temperature, which experiences unrest in the microclimate near the ground.

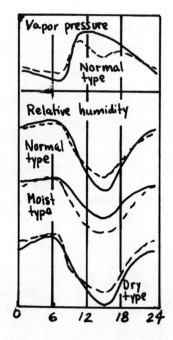

TIME OF DAY (HOUR)

Figure 12-19.   Types of humidity distribution in the layer of air next to the ground.[14]

Precipitation and Humidity

Precipitation and humidity in the atmosphere affect both temperature and winds. Water vapor present in the atmosphere is important in balancing incoming and outgoing radiation. A measure of the amount of moisture in the air is the dew point temperature. At this temperature saturation is reached, if the air is cooled without addition or loss of moisture. At the earth's surface at night, the ground releases stored radiation and the resulting temperature drop causes the airborne water vapor to condense, forming dew or frost depending upon the temperature. Fog is formed by a similar phenomenon but is the condensation of water vapor on airborne particulate matter. The formation of clouds, fog, hail, rain and frost are important measurable meteorological parameters affecting air pollution levels. Rain and snow carry large amounts of both particulate and gaseous pollutants out of the earth's atmosphere and into the soils and water. Trees and grasses act as filters to collect particulates and gases. These are natural means of "cleansing" the atmosphere of pollutants.

Topography

The influence of topography on winds and dispersion can be related directly to the time of day. During the day, slopes facing in different directions and at different angles receive lesser amounts of heat radiation than the ground level, resulting in a rising air mass. At night, the cold air moves downhill independent of slope orientation. Consequently, the cold air slips beneath warmer air. The process continues until a state of equilibrium is attained. Continued over a prolonged period of time without incoming heat radiation, such as the evening hours, this could result in an inversion or a trapping of pollutants. Such phenomena are common to hollows and valleys. In the morning, after sunrise, unless temperature and wind conditions are appropriate to enhance an upward flow of the lower air masses, the trapped pollutants could result in poor air quality.

Geophysical land forms (mountains, valleys, oceans and continents) over which air masses travel have a great influence upon the weather and meteorological conditions. Air flows are channeled, diverted and intercepted by these land forms.

Transport and Diffusion of Pollutants

The transport of pollutants and their precursors over long distances is dominated by meteorological conditions.[3,18] Studies[19-22] of transport have shown upwind sources to affect air quality over 1000 km downwind.

Regional transport and diffusion is similar to urban areas, i.e., transport by the mean wind, horizontal plume spread by diffusion, and vertical plume spread by turbulent eddy diffusion (defined by the atmospheric temperature profile. See section on meteorology, pg. 332). Two atmospheric stabilities are important in long-range pollutant transport.

1) Stability of lower layer (as defined by Pasquill Scheme A-F). Class A represents the most unstable case with large heat flux and light wind. Class F applies to very stable night time conditions with clear skies and very light wind.

2) Overall stability of boundary layer as given by the layer's lapse rate, wind and surface roughness (which is determined to an extent by topography). Very high pollution levels in an air mass are dependent on the following aspects of the transport and diffusion process:[3]

    a) slow moving or stationary air mass;

    b) reduced wind speed in the troposphere; and

    c) a mixing layer limit to vertical dispersion.

Figures 12-20 and 12-21 illustrate Pasquill's plume dispersion characteristics as a function of downwind distance and stability class. Pollutants during

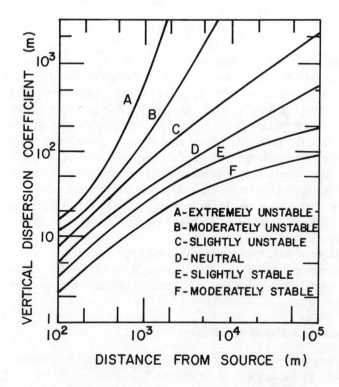

Figure 12-20.   Vertical diffusion, σ_z, vs downwind distance from source for Pasquill's stability classes.

transport and diffusion may undergo continual physical, chemical, or nuclear transformation. The transformation rates depend on temperature, pressure, water vapor, insolation, atmospheric turbulence and the pollutants involved.
    Pollutants of major interest include:

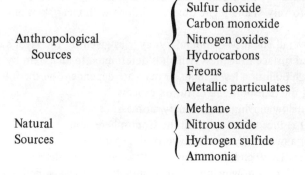

Anthropological Sources
{
Sulfur dioxide
Carbon monoxide
Nitrogen oxides
Hydrocarbons
Freons
Metallic particulates
}

Natural Sources
{
Methane
Nitrous oxide
Hydrogen sulfide
Ammonia
}

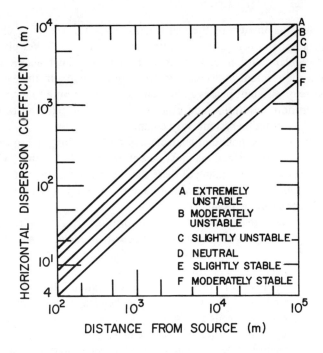

Figure 12-21. Lateral diffusion, σ_z, vs downwind distance from source for Pasquill's stability classes.

The transformation results in products including:

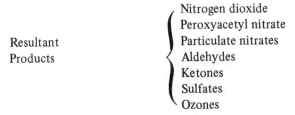

$$
\text{Resultant Products} \left\{
\begin{array}{l}
\text{Nitrogen dioxide} \\
\text{Peroxyacetyl nitrate} \\
\text{Particulate nitrates} \\
\text{Aldehydes} \\
\text{Ketones} \\
\text{Sulfates} \\
\text{Ozones}
\end{array}
\right.
$$

Regional transport involves a longer time period and thus a longer reaction time creating products that may not be entirely understood.

Lower Atmospheric Chemical Transformations

Thermal and photochemical reactions are the prime modes of chemical transformation of the lower atmosphere. Thermal reactions may occur in the liquid or gaseous phase causing an energy (heat)-related reaction. These reactions may take place between a gas and solid (different phases) or between similar phases (gas-gas).

Controlling parameters of thermal reaction rates for like phase chemicals are temperature and reactants' concentrations. A reaction takes place only after a molecular collision, and within that collision enough energy is possessed, and the reactant molecules are properly oriented. In the event the reactants are of different phases additional factors contribute to the reaction rate. Adsorption, desorption and other surface phenomena tend to make dissimilar phase reaction systems very complex.

To predict or assess pollutant characteristics over the long transport time periods these complex reactions must be better defined. Further, with the advent of a better understanding of these thermal reactions, the potential of being able to add catalysts to the atmosphere to speed up or change pollutant reactions becomes more attainable.

A photochemical reaction involves the absorption of light to effect a reaction. Light absorption is dependent on the absorbent molecule and the radiation wavelength. Thus a particular molecule is photochemically changed when it absorbs light and a reaction takes place.

The governing factor for this reaction rate is radiation intensity, effecting electromagnetic energy. The primary photochemical reaction process is relatively simple and proceeds as follows:

The energy levels of an atom or molecule increases as it absorbs light energy. This energy influences molecular activity by increasing rotational, vibrational and electronic quantum levels. However, because photochemical changes require more energy than that contained in vibrational and rotational transitions, they are dependent on electronic quantum transitions.

Several actions may occur after an atom or molecule absorbs light. These include:

1) The absorbed energy is reradiated and the molecule or atom goes to a lower quantum state.

2) Decomposition.

3) A reaction occurs with other constituents transferring the absorbed energy or a portion of it.

4) A transfer of all or a portion of the absorbed energy to other entities of nonreactive collisions.

5) Any combination of the above.

The absorption of radiation may result in no chemical reaction. A molecule may be affected forming many reaction products. This capability of light to induce chemical reactions is described as the quantum efficiency.

Due to high reactivities and short lifetimes, many reactant products are difficult to detect. There are four inorganic species of the many species that are of particular importance: 1) atomic oxygen (O); 2) molecular oxygen (O_2); 3) the hydroxyl (OH); and 4) hydroperoxyl (HO_2) radicals.

In the troposphere the important reactions involving photolysis of N_2O and ozone are:

$$N_2O + h\nu \ (2900\text{-}4300\ \text{Å}) \rightarrow O(^3P) + NO$$

and

$$O_3 + h\nu \ (2900\text{-}3500\ \text{Å}) \rightarrow O_2 + O(^1D), O(^3P)$$

where $O(^1D)$ and $O(^3P)$ are the single D and the triplet P atomic oxygens, respectively.

The hydroxyl radical, OH, is photochemically important and can be formed in the troposphere in three ways with atomic oxygen:

Water vapor with atomic oxygen interaction:

$$O(^1D) + H_2O \rightarrow 2OH$$

Nitrous acid photolysis:

$$HONO + h\nu \ (2900\text{-}4300\ \text{Å}) \rightarrow OH + NO$$

Hydrocarbons with atomic oxygen interactions:

$$O(^3P) + RH \rightarrow OH + R^0$$

where RH denotes the hydrocarbon species and R^0 is the alkyl radical.

The sulfur dioxide (SO_2) transformation to sulfates is a major ambient air quality problem. These transformations may involve the following reactions:[22]

$$O(^3P) + SO_2 \rightarrow SO_3$$

$$HO_2 + SO_2 \rightarrow SO_3 + OR$$

$$RO_2 + SO_2 \rightarrow SO_3 + OR$$

$$OH + SO_2 \rightarrow HOSO_2$$

$$OR + SO_2 \rightarrow ROSO_2$$

It is generally accepted that sulfate aerosols are formed from SO_2 in the following order: 1) oxidation of sulfur dioxide to sulfur trioxide (SO_3); 2) reaction of SO_3 with water (H_2O) to form sulfuric acid (H_2SO_4); and 3) clustering of H_2SO_4 and H_2O molecules to form prenucleation embryos, following by the heteromolecular nucleation process. Other methods of SO_2 to H_2SO_4 formation such as homogeneous photooxidation, or three-body oxidation are either too slow or produce yields too low to effectively affect the atmosphere. SO_3 molecules oxidized from SO_2 will, under atmospheric conditions, react with water to form H_2SO_4 (vapor phase) molecules. The reaction rate of SO_3 with H_2O is very fast making the generation of H_2SO_4 dependent on the conversion rate of SO_2 to SO_3. Should the steady-state concentration of H_2SO_4 exceed the equilibrium patial pressure of H_2SO_4 over a stable aqueous sulfuric acid solution (determined by the absolute humidity of the atmosphere under consideration), heteromolecular nucleation results leading to the formation of new sulfuric acid aerosols.

SO_2 is removed from the atmosphere in various manners. Aside from the oxidation of SO_2 to SO_3 and the formation of sulfate aerosols, numerous complex catalytic oxidations can occur with sufficient speed to be significant. These mechanisms normally require specific metal oxides and salts be dissolved in water droplets that are maintained at a high pH with NH_3.

Ambient concentrations of nitrogen oxide and organic matter promote SO_2 oxidation[24,25] attaining a conversion rate of (SO_2 + OH) estimated at about 0.23%/hr[26] depending on the OH concentration. In such atmospheres, the higher the OH concentration the higher the SO_2 conversion rates.[27]

Nitrogen Oxides

Nitrogen oxides, NO, NO_2 and NO_3, can be extremely reactive when in combination.

The three principal reactions of nitric oxide are:

$$NO + O(+M) \rightarrow NO_2 (+M) \text{ (fast reaction)}$$

$$2NO + O_2 \rightarrow 2NO_2 \text{ (slow reaction at low concentrations)}$$

$$NO + O_3 \rightarrow NO_2 + O_2 \text{ (fast reaction—major cause of smog)}$$

NO_2 in the lower atmosphere is part of a photochemical reaction cycle. The NO_2 absorbs sunlight reacting to form NO and O after only a few minutes after the exposure. An equilibrium state is reached described by

$$NO_2 + O_2 \leftrightharpoons NO + O_3$$

provided no other air pollutants are available to increase the concentration of nitrogen dioxide.

Three compounds can affect this nitrogen cycle resulting in increased NO_2. These are hydrocarbon, water vapor, and carbon monoxide. Automobile emissions of hydrocarbons can effect free radicals through interaction with O, O_3, OH radicals and other ambient compounds. NO_2 reactions with hydrocarbons allow photolysis of NO_2 to create additional NO_2, depending upon the hydrocarbon and concentration.

Water reacts with the nitrogen oxides to form nitrous acid:

$$HO_2 + NO \rightarrow OH + NO_2$$

This OH radical will react with CO resulting in several hundred reactions.

As the photochemical conversion of NO to NO_2 is completed, the NO and O_3 reaction rate is slowed, and ozone begins to accumulate. Other oxidation products also form including nitrates, hydrogen peroxide and sulfate particulates.

Several factors contribute to poor air quality caused by nitrogen oxide reactions, including:

- the hydrocarbon reactivity
- the hydrocarbon concentration
- the nitrogen oxides concentration
- the sunlight intensity, and
- the daily ventilation.

Regional Interactions

The interaction of atmospheric compounds such as HC, NO_x, SO_x and particulates is dependent on relative concentrations and meteorological conditions. For example, clear days promote a faster oxidation of SO_2 when there are high levels of HC and NO_x present.[28] Generally, the atmospheric occurrences can be described as follows: precursors of ozone, HC and NO_x, are entrained in a high-pressure air mass as it passes over an urban area.[29] The precursors are then converted by sunlight to ozone and other by-products. An equilibrium state of ozone concentration is attained for the day.

Without the sun, photochemical reactions cannot take place. Therefore, photochemistry halts at night. Since the ozone precursors are used up to generate ozone, they are not available to reverse the reaction and consume ozone. The layer of ozone nearest the earth's surface is absorbed, but at higher levels O_3 still exists in high concentrations. As the earth warms in the morning, mixing begins. Photochemistry also begins creating more O_3 at the expense of its precursors emitted in the morning.

Generally, high O_3 levels are related to high pressure, low wind areas with turbulent disorganized flow and relatively clean skies.

Urban areas which are the precursors to ozone have been shown to affect areas between 50 and 100 miles away.[30,31] Natural emissions of terpenes and NO_x may also contribute to photochemically produced O_3 in rural areas.[3]

Fossil-fuel-burning power plants emit SO_2 and NO_2, and their ultimate oxidation effects sulfates and nitrates. Hydrocarbon emissions from these power plants are minimal, but some organic particulates are generated. SO_2

oxidation is very slow, occurring over large regions of hundreds of miles. Three mechanisms are common: 1) homogeneous, 2) heterogeneous (aqueous media), and 3) heterogeneous (nonaqueous, like soot).

Plume sulfate formation is negligible for the first 10-25 miles but then increases as oxidants consume NO in the plume.[20] Pure atmospheric SO_2 oxidation is slow but is speeded up in the presence of catalytic aerosols or ammonia.[3] The chemistry of SO_2 depends on humidity, particulates, hydrocarbons and sunlight.[32] In the northeast U.S., sulfate ground-level concentration is at a maximum in maritime tropical air. Generally, these masses have high temperature and high absolute humidity (dew point)—important factors in sulfate generation.[33]

Figure 12-22 shows annual trends in sulfate concentrations over the eastern U.S. Higher levels in the summer are evident. Figure 12-23 illustrates the long-range transport of sulfates.

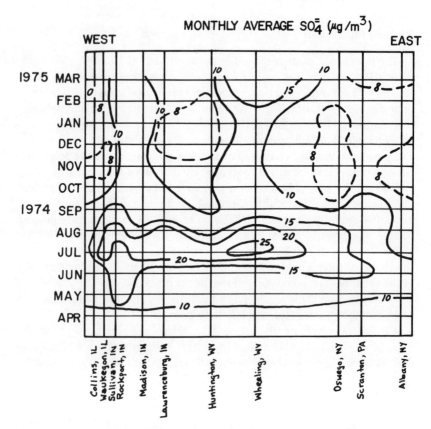

Figure 12-22. Yearly trends in sulfate over the Eastern U.S.[33]

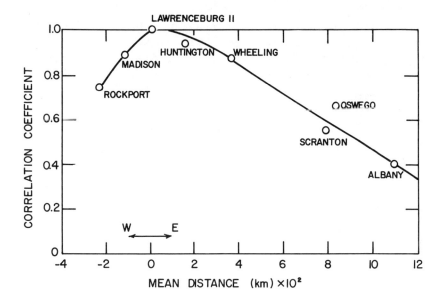

Figure 12-23. Correlation of sulfate concentration as a function of distance, from Lawrenceburg, Indiana.[33]

Air Pollution Sinks

The processes responsible for the removal of atmospheric pollutants include physical and chemical transformation, precipitation scavenging and dry deposition. These will be discussed.

Dry Deposition

The mechanisms of dry deposition include: 1) gravitational settling; 2) turbulent and Brownian diffusion; and 3) inertial, phoretic and electrical effects. These deposited materials may be rereleased to the atmosphere by desorption (in the case of gas) or by mechanical resuspension (in the case of particulates). The above mechanisms are affected by the type of pollutant, the type of deposition substrate, and by meteorological conditions.

Deposition rates are expressed in terms of deposition velocity. This parameter is a function of the substance, its physical characteristics, the atmospheric conditions and frictional drag.

Precipitation Scavenging

Precipitation scavenging is dependent on the physical state of the pollutant (*i.e.*, gas or particulate), on whether the pollutant is a gas or a particulate and whether the pollutant is within the cloud or below the cloud. The washing out of pollutants by rain, hail or other hydrometers is dependent on their properties, and their distribution in space. Collection efficiencies are influenced by many factors including:

- impact
- interception
- diffusion
- nucleation and condensation
- diffuso- and therophoresis
- electrical and turbulence effects

The first three factors are of primary importance for below-cloud scavenging, while the last three are for in-cloud effects. Precipitation scavenging of aerosols and gases are treated separately below.

The estimation of below-cloud scavenging of aerosols considers the overall scavenging efficiencies for interception and impaction, and diffusion of pollutant aerosols. The washout of a pollutant to the ground via in-cloud scavenging may involve four separate processes: 1) transport of the pollutant into the cloud system; 2) mixing of the pollutant within the cloud; 3) capture of the pollutant by the cloud droplets; and 4) delivery by falling precipitation to ground level. Precipitation scavenging of gases differs from aerosols, because gases can desorb from the hydrometeor after collection. Therefore, many factors contribute to the collection efficiency, such as mass-transfer coefficient, solubility, gas and aqueous concentrations, and physical characteristics.

Physical Transformations

Particulate physical transformations are affected by the following basic processes:

- coagulation
- condensation (or evaporation)
- nucleation

Coagulation of two aerosols may be described in terms of kinetic mechanisms. Depending on the aerosol size and ambient conditions, the coagulated aerosol will be subject to precipitation scavenging or dry deposition.

The rate of loss of particles by coagulation is a function of particle size distribution and concentration promoted by Brownian motion and turbulence.[34] Because coagulation involves aerosols distributed in size, consideration must be given to the whole size range and provision made for both gain and loss by the coagulation process.

Condensation and nucleation, depending on the particulates, may be physical or chemical processes. The importance of the nucleation, condensation and coagulation processes varies with the pollutant and the ambient conditions. Under urban conditions, the condensation process is more effective as a mechanism for aerosol size modification than is coagulation.

Regional Air Pollution Sinks

On a regional basis, chemical reactions (including radioactive decay), while existing locally, may take place due to more time for thermal oxidation and more exposure to solar radiation. Other regional sinks include precipitation and surface scavenging.

The hydrological cycle is an effective mechanism for the removal of atmospheric particulates and gases. Because of gravitational effects and greater contact area, larger particulates are removed with greater efficiency. Smaller particles, micron-sized, are not effectively removed.[35] Gaseous removal by precipitation scavenging involves the concentration of large volumes of gas into a small amount of liquid. The precipitation scavenging process involves three steps: 1) evaporation, 2) condensation, and 3) precipitation. Condensation begins with the nucleation of water vapor (and other gases) into water droplets, either by heterogeneous (taking place on foreign particles or substances) or homogeneous (taking place on molecules of the same substance) processes.[33] The resulting precipitation scavenges both within the cloud (rainout) and below the cloud (washout). Washout rates are controlled by the raindrop diameter, precipitation rate, diameter of the material to be removed and the solubility of the gas or particle.

Surface scavenging is dependent on vertical transport, height of the mixed layer, surface roughness, and surface wetness. Surface contact effects a clean atmospheric layer close to the ground.[3] Surface scavenging is particularly effective at night when the rural nocturnal inversion moves toward ground increasing ground to air contacts. The ocean is not a very effective absorber of pollutants.

Regional Modeling

Numerous complex physical and chemical processes contribute to the generation, transport, diffusion, chemistry and depletion of air pollution. Any regional air quality model must include these factors. Urban models are usually available for short distances and averaging times of a few kilometers and a few hours.[7,8] On a regional scale, however, the assumptions made for the short-term urban model are not necessarily valid. The differences between urban and regional model assumptions are outlined in Table 12-6. Basically, four assumptions are made in urban modeling which cannot be used for regional determinations. These include:

• uniform meteorological conditions
• no sink mechanisms other than chemistry

Table 12-6. Contrasts in Assumptions about Urban and Rural Air Pollution Factors[36]

Urban	Rural
Steady-state meteorology	Rarely true
Homogeneous turbulence	Never true
Uniform winds, both vertically and horizontally	Complex wind in boundary layer
No diurnal cycle in wind	For areas larger than 100 km a cycle exists
No change in thickness of mixed layer	Thickness changes daily resulting in complex-layered structure
Pollution remains in atmosphere	Scavenging by rain and surface
Perfect boundary reflection	Absorption at surface
No natural sources	They exist
No precipitation	Exists

• no diurnal variation in wind or mixing depth
• no natural sources of air pollution

These assumptions allow for the solving of mathematically unsolvable equations.

Table 12-7 illustrates the differences between some of the climatological factors in urban and rural areas. Rural areas have higher windspeeds, humidity, radiation and snowfall levels, but lower values of condensation nuclei, pollutant gases, cloudiness, fog, precipitation, temperature and calm winds.

Figure 12-24 illustrates the differences in vertical temperature profiles at different times of the day. The following list outlines troublesome areas of pollution modeling.[5]

1. Misplacement of source with respect to geography and elevation.
2. Misestimation of source strength.
3. Misestimation of wind speed at source.
4. Miscalculation of trajectory of pollutant cloud or plume, due to erroneous or uncertain wind fields.
5. Misjudgment of vertical diffusion of cloud or plume, including erroneous distribution functions, as well as parametric errors.
6. Misjudgment of the lateral diffusion of cloud or plume, including erroneous distribution functions, as well as parametric errors.
7. Misestimation of decay or loss of pollutant.
8. Misplacement of receptor or monitor.

Table 12-7. Average Changes in Climatic Elements Caused by Urbanization[37]

Element	Comparison with Rural Environment
Contaminants	
Condensation nuclei and particulates	10 times more
Gaseous admixtures	5 - 25 times more
Cloudiness	
Cover	5 - 10% more
Fog, winter	100% more
Fog, summer	30% more
Precipitation	
Totals	5 - 10% more
Days with less than 5 mm	10% more
Snowfall	5% less
Relative humidity	
Winter	2% less
Summer	8% less
Radiation	
Global	15 - 20% less
Ultraviolet, winter	30% less
Ultraviolet, summer	5% less
Sunshine duration	5 - 15% less
Temperature	
Annual mean	0.5 - 1.0°C more
Winter minima (average)	1 - 2°C more
Heating degree days	10% less
Windspeed	
Annual mean	20 - 30% less
Extreme gusts	10 - 20% less
Calms	5 - 20% more

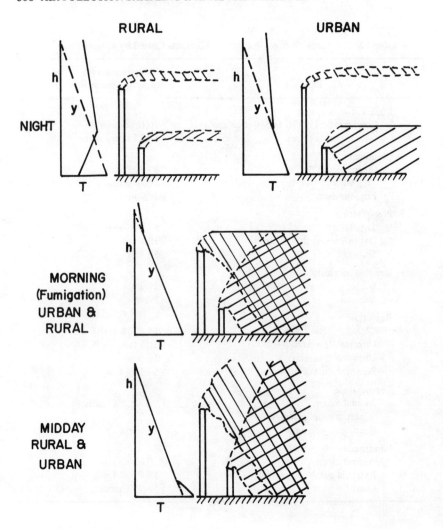

Figure 12-24. Schematic representation of the dirunal variation (clear skies) in the dispersion of elevated and low-level sources of air pollution in rural and urban settings as related to vertical temperature distribution. Y indicates the dry adiabatic lapse rate, and the solid lines that of ambient air.[37]

REFERENCES

1. Knox, J. B. "Numerical Modeling of the Transport, Diffusion and Deposition of Pollutants for Regions and Extended Scales," presented at the 66th Annual Meeting of the Air Pollution Control Association, Chicago, June, 1973.

2. Cox, R. A. *et al.* "Photochemical Oxidation of Halocarbons in the Troposphere," *Atmos. Environ.* 10:305 (1976).
3. Munn, R. E. *et al.* Dispersion and Forecasting of Air Pollution," Technical Note No. 121, WMO, Geneva (1972).
4. Clarke, J. F. "A Meteorological Analysis of Carbon Dioxide Concentrations Measured at a Rural Location," *Atmos. Environ.* 3 (1969).
5. Hilst, G. R. "Sensitivities of Air Quality Prediction to Input Errors and Uncertainties," in *Proceedings of Symposium on Multiple-Source Urban Diffusion Models,* A. C. Stern, Ed. (Washington, D. C.: U.S. Government Printing Office, 1970).
6. _____, "Routine Sulphur Dioxide Surveys around Large Modern Power Stations," *Atmos. Environ.* 10:265 (1976).
   I. Clarke, A. J., and G. Spurr. "Summary Paper."
   II. Jarman, R. T., and C. M. de Turville. "Fawley and Pembroke."
   III. Powell, A. W., and P. A. Tatchell. "Kingsnorth."
   IV. Barber, F. R., and A. Martin. "Ratcliffe-on-Soar."
   V. Dolman, D. L., and P. M. Owens. "Eggborough."
   VI. Dale, F. W. "Fiddlers Ferry."
7. Cheremisinoff, P. N., and A. C. Morresi. "Predicting Transport and Dispersion of Air Pollutants from Stacks," *Poll. Eng.* 9 (November 3, 1977).
8. Cheremisinoff, P. N., and A. C. Morresi. *Environmental Assessment Impact Statement Handbook* (Ann Arbor, Michigan: Ann Arbor Science Publishers, Inc., 1977).
9. Slade, D. H. "Meteorology and Atomic Energy," U.S. Atomic Energy Commission, Washington, D. C. (1968).
10. Plate, E. J. "Aerodynamic Characteristics of Atmospheric Boundary Layers," U.S. Atomic Energy Commission (1971).
11. Davenport, A. G. "The Relationship of Wind Structure to Wind Loading," in *Wind Effects on Buildings and Structures,* National Physical Laboratory, Symposium 16 (London: Her Majesty's Stationery Office, 1965).
12. Shellard, H. C. "The Estimation of Design Wind Speeds," in *Wind Effects on Buildings and Structures,* National Physical Laboratory, Symposium 16 (London: Her Majesty's Stationery Office, 1965).
13. Miller, A. *Meteorology* (Columbus, Ohio: Charles E. Merrill Books, 1966).
14. Geiger, R. *The Climate Near the Ground* (Cambridge, Mass.: Harvard University Press, 1950).
15. Lovelock, J. E. "Air Pollution and Climatic Change," *Atmos. Environ.* 5 (1971).
16. Beaton, J. L., A. J. Ranzieri and J. B. Skog. *Meteorology and its Influence on the Dispersion of Pollutants from Highway Line Sources,* California Division of Highways, Research Report No. CA-HWY-MR6570825(1)-72-11 (April 1972).
17. Sinclair, J. G. "Temperature of the Soil and Air in a Desert," *Monthly Weather Review,* U.S. Department of Agriculture (1922).
18. Weber, E. "Removal of Sulfur Dioxide from the Atmosphere," Proceedings of the Second Meeting of the Expert Panel on Air Pollution Modeling, NATO Committee on the Challenges of Modern Society, Paris, October 1972.

19. Westberg, H. H., *et al.* "Ozone in Rural Areas," Int. Conference on Environ. Sensing and Assessing, Sept. 1975, Las Vegas, Nevada, Inst. of Elect. and Electronic Engrs., New York, 1976.

20. "Position Paper on Regulation of Atmospheric Sulfates," U.S. Environmental Protection Agency, Office of Air Quality, Planning and Standards, Research Triangle Park (September 1975).

21. Ottar, B. "The Long Range Transport of Air Pollutants," Observation and Measurement of Air Pollution, Special Environmental Report #3.WMO No. 368, Geneva 1974.

22. *Atmospheric Chemistry; Problems and Scope,* A report of the Panel on Atmospheric Chemistry to the Committee on Atmospheric Sciences, National Research Council, National Academy of Sciences, Washington (1975).

23. Smith, F. B. "A Scheme for Estimating the Vertical Dispersion of a Plume from a Source Near Ground Level," Proceedings of the Third Meeting of the Expert Panel on Air Pollution Modeling, A Report of the Air Pollution Pilot Study, NATO Committee on the Challenges of Modern Soceity, Paris, October 1972.

24. Renzetti, N. A., and D. J. Doyle. *J. Air Poll. Control Assoc.* 2:327 (1960).

25. Cox, R. A., and S. A. Penkett. *Atmos. Environ.* 4:425 (1970).

26. Calvert, J. G. "Interactions of Air Pollutants," *NAS Conference on Health Effects of Air Pollutants,* Washington, D. C. (October 3-5, 1973).

27. Castleman, A. W. *et al.* "Kinetics of Association Reactions Pertaining to H_2SO_4 Aerosol Formation," CODATA Kinetic Symposium, Warrenton, Virginia, September 16-18, 1974.

28. Finlayson, B. J., and J. N. Pitts, Jr. "Photochemistry of the Polluted Atmosphere," *Science* 192 (April 9, 1976).

29. Chameides, W. L., and D. H. Stedman. "Ozone Formation from NO_x in 'Clean Air' " *Environ. Sci. Technol.* 10(2) (February 1976).

30. Altshuller, A. P. "Evaluation of Oxidant Results at CAMP Sites in the United States," *J. Air Poll. Control Assoc.* 25(1) (January 1975).

31. "Investigation of Rural Oxidant Levels as Related to Urban Hydrocarbon Control Strategies," prepared by the Research Triangle Institute for the EPA (March 1975).

32. Selamoglu, S. "Physical and Chemical Transformation in Air Pollution Models," Proceedings of the Second Meeting of the Expert Panel on Air Pollution Modeling, NATO Committee on the Challenges of Modern Society, Paris, July 1971.

33. Hidy, G. M. *et al.* "Design of the Sulfate Regional Experiment (SURE), Vols. I-IV," Environmental Research and Technology, Westlake Village, California (Feburary 1976).

34. Morresi, A. C. "Particle Dynamics," in *Air Pollution Control and Design Handbook* P. N. Cheremisinoff and R. A. Young, Eds. (New York: Marcel Dekker Publishers, 1977).

35. Munn, R. E., and B. Bolin. "Global Air Pollution—Meteorological Aspects," *Atmos. Environ.* 5 (1971).

36. International Conference on Environmental Sensing and Assessment, September 1975, Las Vegas, Nevada, Institute of Electical and Electronics Engineers, New York (1976).

37. "Meteorological Aspects of Air Pollution," WMO Technical Note 106, Geneva (1970).

CHAPTER 13

COMPARISON OF SOURCE EMISSION LIMITS

W. B. Rossnagel

President
Rossnagel & Associates Inc.
Cherry Hill, N. J. 08003

INTRODUCTION

This section is a guide to and comparison of source emission limits for various regional areas in the United States. It is useful in establishing emission limits from stationary point sources. Due to the brevity of the matrixes presented and changes in local regulations, the full code in any area should be consulted.

Comparison of Source Emission Limits—Northeastern States Air Pollution Control Codes

Category and Parameters	EPA[a]	N.J.[a,b]	N.Y.C.[a,c]
Incinerators			
1) Maximum Ringelmann number and time allowed	Existing #2 New #1[d]	#2 for 3 min during startup, then #1	Installation of refuse burnequipment prohibited. See Article 4, Para. 1403.2-4.03.
2) Is any visible fly ash allowed?	Yes if within Item 4[d]	No	Existing units require draft .05-.10" W.G. See special requirements Type IV incinerators.[a,c,g]
3) Odor?	— —	No	
4) Maximum particulate emission rate See Air Regulation 39	0.3 gr/scf[e] <200 lb/ hr[d]. 0.2 gr/scf[e] <200 200 lb/hr[d]. 0.8 g/ scf[e] for <50 tons/ day[e,f]	0.2 gr/scf[e] for common incinerator < 2000 lb/hr. 0.1 gr/ scf[e] for Special incinerator. >2000 lb/hr for Municipal incinerator.	
5) Special items	Test requires PHS waste & EPA Test Method 1-5.	All upgraded units two burners & scrubber required.	Efficiency test required on new models of cont. equipment.
Particulates			
6) Maximum emission rate (See EPA 40CFR 50) See Air Regulation 43	Varies with sources, i.e., Cement plant. 0.3 lb/ ton[f]. Dry catch only for test.	.02 gr/scf[e] or 99% removal, whichever is larger	Varies with source. See Para. 1403.2-9.09.
7) Special items	See 40CRF61 Hazardous Air Pollution	For fuel burning see Chapters 4 and 5.	≥#1 but <#2 Ringelmann for 2 min/hr, then always <#1.
Organics			
8) Maximum emission rate	No EPA code at present. Check for a state code.	Sub-Chap. 16 still pending. See Table 2. Now limited by Chap. 6, 7 or 9.	Based on process wt and environmental rating. See N.Y. State Part 205, and NYC Para. 1403.2-9.23.
9) SO$_x$ (SO$_2$, SO$_3$)	See Fed. Reg. 12/23/71 for boiler and sulfuric acid plant requirements. New Sources: See Air Reg. 51. Test per EPA Method 6 or 8.	Based on stack height, exhaust temp. & distance to property line. See Chap. 8. See Chap. 10 & 10 A for fuels.	Article 9, 1403.2-9.07. Article 13 for fuel standards. See also N.Y. State Part 225.
10) NO$_x$ (NO, NO$_2$) See Air Regulation 47	See Fed. Reg. 12/23/71 for boiler & nitric acid plant requirements. Test by Method 7, PDS Proc.[f]	Chap. 6 only applicable code at present time.	Article 9 Section 1403.2-9.13. See also N.Y. State Part 224 (acids), Part 227 (boilers)
Effective Date These Data	2/1/75	2/1/75	2/1/75

[a]Test requires EPA train.
[b]See N.J.D.E.P. Chapter 11 for definition of common and special incinerators.
[c]Local Law No. 49, Air Pollution Codes, by mail $2.25. See also "Specifications for Incinerator Testing." Special codes apply in NYC metropolitan region.
[d]Applies to incinerators on federal installations or for EPA approval.

N.Y.[a]	Philadelphia[a]	Pennsylvania[a]	Connecticut[a]
#1 installed after '67, #2 before.	#1 except for 30 sec/hr, 3 min/day.	<60% but >20% capacity for 3 min/hr, then <20%.	# 2 for 5 min/hr, otherwise #1.
Yes if within Item #4.	No	No	No
No	No	1200°F -.3 sec.	Annoying - No.
See Part 219 or 222, Figure 1 as applicable. Part 222 applies to NYC Nassau & Westchester City, Niagara City. Chap. IX C.	New: .08gr/scf[e]. Existing: 1 gr/scf[e]. Regulation XI effective 4/2/74.	0.1 gr/scf[e]. Only 1 emission run required on Official Test.	New: .08 gr/scf[h]. Existing: 0.4 lb/1000 lb flue gas adjusted to 50% excess air.
Can test representative models for <2000 lb/hr.	New incinerator banned at present time.	Test per method as described in Chap. 139.	Test Methods. Sec. 19-508-5 of code.[a]
See Proc. Wt A-D rating in 212 New & Mod > 105 lb/hr, <.03 gr/scf for B&C rating where Proc. Wt. N/A. <.05 gr/scf.	Regulation II, Table I.	See Chap. 123.13, Table 1 for process. Other, .04 gr/scf dry <150,000 SCFM.	Based on process wt. See Table 3.1 of code, Sec. 19-508-18ei.
See Part 227 for limits on stationary comb. sources.	<20% capacity required.	Test Method per Chapter 139.	Separate limits for process Sec. 19-508-18e2.
See Part 205, <15 lb/24 hr or <3 lb/hr, L.A. Rule 66. Exceptions <20% solvent in coatings or 30% in ink.	Reg. V fully effective 7/71, 40 lb/day maximum or reduce 85%. See Reg. V, Sec. VI for solvents.	Applicable Chap. 129 does not have emission limits.	See code Sec. 19-508-20. Reactives <3 lb/hr or <15 lb/day unless 85% reduction. Nonreactive <150 lb/hr.
<4 lb SO_2/ton acid, <.15 lb/ton acid mist for acid plants. See 212, 223 for other processes. Fuel standards, see Parts 225, 226, 230.	<.05% by vol. per Reg. III Sect. II. See also I for related ambient limits.	See Chap. 123.21 SO_x <500 ppm. See Sec. 129.12, 6.5 lb SO_2/ton 100% H_2SO_4, 0.5 lb SO_2 mist/ton 100% H_2SO_4.	Based on the process. See Sec. 19-508-19 for H_2SO_4 plant. <6.5 lb SO_2/ton acid. <9 lb/ton pulp or <500 ppm.
<3 lb NO_2/ton acid, maximum 2 hr average. See Part 224.1, 227 (boilers) and 212.	<3 lb/ton of HNO_3 maximum 2 hr average. No visual emissions allowed. See Reg. VII, Sec. III.	For HNO_3 plants lb NO_2/ton 100% HNO_3. See Chap. 129.11.	<5.5 lb NO_2/ton for HNO_3 plant. For other nonfuel burning process - <700 ppm See Sec. 19-508-22.
3/1/75	3/1/75	3/15/75	7/15/74

[e]Dry—corrected to 12% CO_2.
[f]Standards of performance for new stationary sources.
[g]Report dry particulates but include organic and inorganic solids at STP.
[h]Emission tests conducted at maximum rated burning capacity.

Comparison of Source Emission Limits–New England Area Air Pollution Control Codes

Category and Parameters	EPA[a]	Massachusetts	Maine[a]
Incinerators			
1) Maximum Ringelmann number and time allowed	Existing #2 New #1[b]	<#1	#2 except #3 for 5 min/hr or 15 min in any continuous 3-hr period. Incinerators, #1.
2) Is any visible fly ash allowed?	Yes - if within Item 4[d]	No	–
3) Odor?		Nuisance - No	–
4) Maximum Particulate Emission Rate See Air Regulation 39	0.3 gr/scf[c] <200 lb/hr.[b] 0.2 gr/scf[c] > 200 lb/hr.[b] .08 gr/scf[c] for > 50 ton/day.[b,d]	.1 gr/scf[e] .05 gr/scf lb/hr of municipal waste	0.2 gr/scf[e] for special incinerator. > 50 ton/day for 2 hr. See Sec. 601.
5) Special items	Test requires PHS waste and EPA Test Methods 1-5.	Test per EPA Method #5. Must submit pretest plan and filter tare weights.	Test per EPA Methods #1, #3, #5.
Particulates			
6) Maximum Emission Rate Note: Air requirements in EPA column are in 40CPR50 document.	Varies with sources, i.e., cement plant 0.3 lb/ton[d]. Dry catch only for test. See Air Regulation 43.	Varies with sources and process rate. Up to 2.2% S in fuel allowed 100 x 10⁶ Btu/hr if stack tests are made.	Depends on source and process rate. See Sec. 602 and Table 1. For fuel burning see Sec. 600 for table.
7) Special items	See 40CFR61 Hazardous Air Pollution	<Ringelmann #2 for 6 min/hr, otherwise <#1.	#2 except #3 for 5 min/hr or 15 min in any continuous 3 hr period.
Organics			
8) Maximum Emission Rate	No EPA code at present. Check for a state code.	Visible emissions same as incinerators. Item #1.	Ambient standards 160 µg/m³ for 3 hr period except once per year.
9) SO$_x$ (SO$_2$, SO$_3$)	See Federal Regulation (12/23/71) for boiler and sulfuric acid plant requirements. Test per Method 6 or 8[d].	New limits on Sulfur in fuel. Critical areas of concern <.05%.	For Sulfite Pulp Mills: 40 lb/air dried ton of sulfite pulp produced. Test per EPA Methods 1 and 6.
10) NO$_x$ (NO, NO$_2$)	See Federal Regulation (12/23/71) for boiler and nitric acid plant requirements. Test by Method 7, PDS Proc.[d]	NO$_x$ from fuel burning depends on area.	Ambient Standards 100 µg/m³, annual arithmetric mean of the 24-hr average NO$_2$ concentration.
Effective Date These Data	2/1/75	3/1/76	10/1/75

[a]Past property line based on objection by 15-75% of odor witnesses, depending on size of group.
[b]Applies to incinerators on federal installations or for EPA approval.
[c]Dry–corrected to 12% CO$_2$.
[d]Standards of performance for new stationary sources.

Vermont	New Hampshire	Rhode Island	Connecticut
Never $>$ #3 constructed prior to 4/30/70. $\leqslant$ #2 for 6 min/hr, then $<$ #2 Constructed after 4/30/70 $\leqslant$ #1 for 6 min/hr, then $<$ #1.	#1 for 3 min/hr, otherwise $<$ #1 after 4/15/75 all considered new.	#1 for 3 min/hr. Otherwise #1.	# 2 for 5 min/hr. Otherwise #1.
N/A	–	Yes - if within items 1 and 4	No.
Sect. 5-241[a]	–	Nuisance - No	Annoying - No.
0.1 lb/100 lb refuse burned. See Regulation 5-231 Sec. 2.	See Regulations 6 and 7. Capacity $>$ $>$ 200 lb/hr, 0.2 gr/scf. Capacity $<$ 200 lb/hr, 0.3 gr/scf.	Capacity $<$ 2000 lb/hr 16 gr/scf. Capacity $\geqslant$ 2000 lb/hr .08 gr/scf[e]. Any capacity types #4, #5, #6 .08 gr/scf[e]	New: .08 gr/scf[f]. Existing: 0.4 lb/1000 lb flue gas adjusted to 50% excess air.
EPA Type Test, see Sec. 5-404. See also special test proc.[e]	Operating instructions must be posted.	Test per EPA Method #5. Only multiple chamber units altered after 1/74.	Test Methods Sec. 19-508-5 of code.[g]
Depends on source and process rate. See Reg. 5-231, Sec. 1.4 through 6 and Table I. Asphalt Plants .07 gr/scf if built after 4/70.	Depends on source and process rate. See Reg. #4,8 (Asphalt Plants) 10 (Ferrous Foundries), 13 (Sand/gravel/Cement), 14 (Nonferrous Foundries), 15 (Pulp/Paper)	Depends on process wt. rate. See Reg. 3 Fuel Burning $>$ 250 x 10^6 Btu/hr 0.1 lb/10^6 Btu $<$ 25 x 10^6 Btu/hr 0.2 lb/ 10^6 Btu.	Based on process wt. See Table 3.1 of code, Sec. 19-508-18ei.
See Chap. 4 and 15. EPA test per Sect. 5-404.	See Reg. 8 for asphalt plants and Reg. 3 for fluorides	Test per Federal Register, 8/18/77	Separate limits for process. Sec. 19-508-18e2.
Visible emission and odor same as incinerators.	See Reg. 11. See Reg. 12 for burning tires.	40 lb/day/machine or 100 lb/day total, unless reduced 85%. See Reg. 15	See code Sec. 19-508-20. Reactives $<$ 3 lb/hr or $<$ 15 lb/day.
Percent sulfur content in fuel limitation. See Sec. 5-221.	SO_x 20 lb/ton air dried pulp. TRS 2 lb sulfur/ ton air dried pulp.	1.1 lb/10^6 Btu input. For sulfur content of fuels see Reg. 8.1.	Based on the process. See Sec. 19-508-19 for H_2SO_4 plant, $<$ 6.5 lb SO_2/ton acid. $<$ 9 lb/ton pulp or $<$ 500 ppm.
From combustion installation with capacity $>$ 250 x 10^6 Btu completed after 7/1/71 0.3 lb/10^6 Btu/hr.	See Reg. XI		$<$ 5.5 lb NO_2/ton for HNO_3 plant. For other nonfuel burning process $<$ 700 ppm. See sec. 19-508-22.
10/72	1/11/74	10/11/75	7/15/74

[e]Does not include condensibles but shows calculation for them.
[f]Emission tests conducted at maximum rated burning capacity.
[g]Test requires EPA train.

Comparison of Source Emission Limits–Capital Area Air Pollution Control Codes

Category and Parameters	EPA[a]	Delaware	District of Columbia
Incinerators			
1) Maximum Ringelmann number and time allowed.	Existing #2 New #1[b]	>#1 for 3 min/hr or 15 min/day	#2 for 4 min/hr or 24 min/ 24 hr. These emissions only allowed during start-up, soot blowing adjusting, etc.
2) Is any visible fly ash allowed?	Yes if within Item 4[b]	Yes if within Items 1 and 4.	Yes if within Item 1.
3) Odor?	–	No	No[c]
4) Maximum Particulate Emission Rate	0.3 gr/scf < 200 lb/ hr[b] 0.2 gr/scf > 200 lb/ hr[b] For > 50 ton/day .08 gr/scf[b,d,e] See Air Regulation 39.	See Table 1 for lb/hr emission rate. Example: 0.2 lb/hr for 100 lb/hr charging rate.	Existing: .08 gr/scf. New: .03 gr/scf. After 7/4/73 no incinerator of > 400 lb/hr capacity .08 gr/scf can operate.
5) Special items	Test requires PHS waste and EPA Test Methods 1-5.	No new incinerators allowed except for Type IV refuse.	After 7/4/75 new incinerators banned except where necessary.
Particulates			
6) Maximum Emission Rate (Note: Air Requirements in EPA column are in 40CFR50 document.	Varies with sources, i.e., Cement plant 0.3 lb/ton[e] Dry catch only for test. See Air Regulation 43	Varies with source. See Reg. V ≤.2 gr/scf unless specified for process on pages 20-24.	Fuel burning: .13-.02 lb/ 10^6 Btu. Sec. 8-2.708 and App. #1. Process: .03 gr/ scf Sec.8-2.710 and App. #2.
7) Special items	See 40 CRF61 Hazardous Air Pollution.	0.3 lb maximum/10^6 Btu input on all fuel-burning equipment.	Test per Federal Register dated 8/18/77
Organics			
8) Maximum Emission Rate	No EPA code at present. Check for a state code.	Included as particulates when applicable.	Solvents: 3 lb/hr or 15 lb/ 24 hr. C_2H_4: 20 lb/hr unless burned at 1300°F for 0.3 sec

Virginia	West Virignia	Maryland	Kentucky	Baltimore Area III
#1 but allows #3 for brief periods.	< #2 during 8 min startup or 6 min/hr for stoking. Then #1.	Area I, II, V, VI. Existing: #1 except #2, 4 min/hr. New: #0 except #2, 4 min/hr.	#1 for new > 50 ton/day. New units #1 for < 50 ton/ day and existing units. See Reg. AP-3.	# 0 except # 2 ≤ 4 min/hr during startup, etc.
Yes if within Items 4	No	No	Yes if within Item 1	No
Objectionable - No	Objectionable - No	Objectionable - No	–	Objectionable - No
0.14 gr/scf dry on existing units operating at design capacity. See Sec. 4.07.00. Test per ASME PTC 21 & 27. Flue fed incinerators not allowed.	Computed on incinerator capacity as found in Reg. 16-20 Series VI, page 7.	Area I,II, VI. Existing: < 200 lb/ hr 0.3 gr/scfd. > 200 lb/hr 0.2 gr/scfd. New < 1 ton/hr 0.1 gr/scfd. > 1 ton/ hr .03 gr/scfd. Area III & IV .03 gr/scfd.	New units: > 50 ton/day, .08 gr/scfd. New units: < 50 ton/day & existing units 0.2 gr/scfd. Special limits allowed for wood products. See Sec. 1 Para. 3 of AP-3.	0.03 gr/scfd. Pathological 0.1 gr/scfd.
Write to get on mailing list of Virginia Air periodical	Total of 13 regulations in effect for all air pollution sources.	Test per Federal Register Vol. 36 No.84, Part III, April 30, 1971.	Test per[a] special Kentucky Incinerator Test Procedure TS-12.	Section .060 prohibited incinerators.
Varies with source See Sec. 4.04, Table 4.4.1. Special limit cracking, asphalt & fert. see 4.03 fuel burning equipment.	Varies with source and process rate. Reg. 16-20, Series VII.	Area I,II,V,VI. Existing: .05 gr/ scfd. New: .03 gr/ scfd. Area III,IV .03 gr/scfd. All asphalt plants .03 gr/ scfd.	See Table 3.1 & 3.2 of AP-3. Cement plants < 3 lb/ton kiln feed. Indirect heat exchange (boil) < Ring. 1-Pty .1, < Ring. 2-Pty.223.	≤.03 gr/scfd. For process wt. rate see Table 2. Fuel burning see Table 1 dependent on fuel type and heat input.
Test per ASME PTC-21 or 27 or EPA train.	Note also visual requirements in Series VII.	See table based on process wt. For residual boiler .02 gr/scfd.	Test per Federal Register, 8/18/77 New Units.	Dust collector required on all fuel-burning equipment.
< 15 to < 40 lb/ day, depending on sources unless all organics reduced by 85%. See Sec. 4.05.03		Area III & IV Existing: < 200 lb/ day unless discharge reduced 85%. New: < 15 lb/day unless discharge reduced 85%.	< 15 lb/day or < 3 lb/hr unless discharge reduced 85%. See Reg. AP-5, Sec. 7.	Existing: < 200 lb/ day unless discharge reduced 95%. New: < 15 lb/day unless discharge reduced 85%.

Capital Area Air Pollution Control Codes, Continued

Category and Parameters	EPA[a]	Delaware	District of Columbia
9) SO_x (SO_2, SO_3)	Sources begun before 8/17/71 see Federal Register 8/18/77 for boiler and sulfuric acid plant requirements. New Sources: See Air Reg. 51. Test per EPA Method 6 or 8.	Varies with production rate. See Reg. VIII for fuelburning standards. See Reg. IX & X for SO_2 limits from industrial operations. Existing: <1000 ppm. New: < 500 ppm.	Process Emissions $<.05\%$ by volume. Fuel limitation Existing: $<1\%$ by weight. 1975 $<.5\%$ by weight except where control equipment is used.
10) NO_x (NO, NO_2)	See Federal Register (8/18/77) for boiler and nitric acid plant requirements. Test by Method 7, PDS Proc.[e]	See Reg. XII.	Fossil fuel fired steam generating units. $> 10^6$ Btu/hr capacity < 0.2-0.7 lb/10^6 Btu. See App. #3.
Effective Date These Data	2/1/75	1/7/72	7/7/72

[a]Test requires EPA train.
[b]Applies to incinerators on federal installations or for EPA approval.
[c]See Sec. 8-2:715. Number 1 on Barnebey-Cheney Scentometer.
[d]Dry—corrected to 12% CO_2.

Virginia	West Virginia	Maryland	Kentucky	Baltimore Area III
Basically <2000 ppm by vol. Varies by process. See Sec. 4.05.02 for curve A, D & C for SO_2 emission limits. $H_2S <$ 15 gr/100 ft^3 gas. <8000 ppm sulfur recovery.	Varies with region, type of fuel or process. See Reg. 16-20 Series X. H_2SO_4 mist limit is 35 mg/ m^3 from Series VII. HCl <210 mg/m^3. Phosphoric Acid <3 mg/m^3.	All areas. Existing: <2000 ppm SO_2. New: <500 ppm SO_2. For SO_3 & H_2SO_4, Existing: 70 mg/m^3. New: <35 mg/m^3 reported as H_2SO_4.	See AP-4 for sulfur compounds. New: <4 lb $SO_2/$ ton 98% H_2SO_4. Existing: <27 lb $SO_2/$ton 98% H_2SO_4. Other process <2000 ppm[a] Indirect heat exchangers see AP-4, Sec. 1.	Existing: <2000 ppm SO_2. New: <500 ppm SO_2. For SO_3 & H_2SO_4, Existing: 70 mg/m^3. New: <35 mg/m^3 reported as H_2SO_4. H_2S, 1300 F.s.3 sec.
<5.8 lb/ton of 100% HNO_3 from fuel combustion see Sec. IV, 4.05.05, b of Supplement and code.	HNO_3 mist or vapor <70 mg/m^3. See Reg. 16-20, Series VII.	All Areas, Existing: <5.5 lb/ton HNO_3. New: <3.0 lb/ton HNO_3. From fuel combustion[f] .2-.5 lb/10^6 Btu.	See AP-7. Existing: <5.8 lb NO_2 per ton HNO_3. New: <3 lb $NO_2/$ton HNO_3.	New: Fuel combustion[f] .2-.5 lb/10^6 Btu. $HNO_3 <3.0$ lb/ ton.
3/1/75	11/13/72	5/12/72	4/9/72	6/17/75

[e]Standards of performance for new stationary sources.
[f]For heat input of 250 x 10^6 btu/hr or more.

Comparison of Source Emission Limits–Southeastern States Air Pollution Control Codes

Category and Parameters	EPA[a]	Georgia[a]	South Carolina
Incinerators			
1) Maximum Ringelmann number and time allowed	Existing: #2 New: #1[b]	Existing: #2 or #3 for 3 min/hr. New: #1 or #2 for 3 min/0.5 hr.	Existing: #2, except 5 min/hr or 20 min/day. New: #1 except 3 min/hr or 15 min/day. See Std. #4.
2) Is any visible fly ash allowed?	Yes if within Item 4[b]	No	Yes, only if it falls on owner's property
3) Odor?	–	–	Undesirable - No
4) Maximum Particulate Emission Rate[c]	0.3 gr/scf[e] <200 lb/hr hr[b] 0.2 gr/scf >200 lb/ hr[b] For >50 ton/day, .08 gr/scf. See Air Regulation 39.	Existing: 0.2 gr/scf. New or Altered after 1/1/72: 0.1 gr/scf < 50 ton/day. For new installations Type 3-6 waste, 0.2 gr/scf.	Existing: 0.75 lb/10[6] Btu excluding auxillary fuel. New or Altered: 0.5 lb/ 10[6] Btu excluding auxillary fuel. For large municipal type see fuel-burning standards. Std. # 4, Sec. 1.
5) Special items	Test requires PHS waste & EPA Test Methods 1-5. Regulation 39.	See Chapter 391-3-1-.02, pg. 219.	
Particulates			
6) Maximum Emission Rate (Note: Air Requirements in EPA column are in 40CFR50 document.	Varies with source, i.e., Cement plant 0.3 lb/ton[d] Dry catch only for test. See Air Reg. 43.	Varies with source. See Chap. 391-3-1-.02 pages 209,213,217,218,222, 223,226 and 227.	Varies with the source and process rate. See Table A in Std. 5, pg. 30. Special limits on pulp, quarrying, cement, asphalt, cotton industry.
7) Special items	See 40CFR61 Hazardous Air Pollution	Check for special test requirements	See Reg. 4 for compliance schedule
Organics			
8) Maximum Emission Rate	No EPA code at present. Check for a state code.	No emission rates. See Chap. 391-3-1-.02 pg. 207 for Prohibited Action	
9) SO$_x$ (SO$_2$, SO$_3$)	Sources begun before 8/17/71: see Federal Register 8/18/77 for boiler and sulfuric acid plant requirements. New sources: see Air Reg. 51. Test per EPA Method 6 or 8.	H$_2$SO$_4$ Plants, Existing & New (constructed after 1/11/72) see pg. 17. For other sources: varies with stack height, area, number of units, etc. See pages 215,217,218.	H$_2$SO$_4$ Plants <10 lb SO$_2$ and 65 lb acid mist/ton acid. After 7/1/66, <4 lb SO$_2$/ ton of acid. Fuel Burning Operations: varies with fuels used. See Std. No. 2

Alabama	Tennessee[a]	North Carolina[a]	Florida
#3 for 3 min/hr. #1 otherwise	#1 for 5 min/hr or 20 min/day	Existing: #2. New: #1 for 5 min/hr	Existing: up to 7/1/75 #2. New: #1 for 3 min/hr, then no visible emissions.
–	–	–	–
No	–	No	Objectionable - No
0.20 lb/100 lb refuse for <50 ton/day. 0.10 lb/100 lb refuse for > 50 ton/day. For wood, peanut or cotton ginning waste .40 lb/100 lb material	0.2% <2000 lb/hr. 0.1% >2000 lb/hr[e,f]	0.2 lb/hr <100 lb/hr. 0.4 lb/hr <200 lb/hr. 1.0 lb/hr <500 lb/hr. 2.0 lb/hr <1000 lb/hr. 4.0 lb/hr for 2000 lb/ hr and above	Existing units: 0.1 gr/scf for >50 ton/day. New: .08 gr/ scf for >50 ton/day
Teepee units required to have underfire, forced air, variable control damper & temperature recorder	Codes shall not apply to the units with ≤2.5 ft^3 furnace volume.		Switched to EPA Test Methods in 1975
Varies with source, process rate & county classification. See Chapter 4 for special product emission limits	Existing: varies with temperature, height, velocity, etc. Emissions regulated by diffusion equation or Table 1. New: Chap. VII, Table 2.	Varies with process weight and source. See pages C-25 through C-31 for specific processes.	Varies with source. See Table I based on process weight in Chap. 17-2, pg. 5. Instack filter OK paper test.
–	Regardless of standards maximum .25 gr/scf.	See Section IV. See also Figures 1-6	Visible emission regulations vary with source.
Mobile County 3 lb/ hr or 15 lb/day[g] when in contact with flame or baked, otherwise 40 lb/day. See Chap. 6	Varies with stack temperature, height and exhaust gas volume. See Chap. VII, Sec. 2	40 lb/day or 85% reduction in discharge using control equipment. See Sec. IV, pages C-33,C-34.	–
<27.0 lb SO$_2$/ton of 100% H$_2$SO$_4$. <.5 lb acid mist/ton H$_2$SO$_4$. <.2 lb SO$_3$/ton H$_2$SO$_4$. H$_2$S <150 ppm limit. For other sources see Chapter 5	Varies with process. Chap. 1200-3-14:03. Class I: 500 ppm. Class II,III: 1000 ppm. Class IV,V,VI: 2000 ppm. Also see .03-4-5-6.	H$_2$SO$_4$ Plants: .27 lb SO$_2$ and 0.5 lb acid mist/ton of acid. Fuel Burning Equipment: Existing: < 2.3 lb SO$_2$/10^6 Btu/hr. New: (from 7/1/75) < 1.6 lb SO$_2$/10^6 Btu/hr.	H$_2$SO$_4$ Plants: 40% opacity at start, then 5% opacity. Existing: <10 lb SO$_2$/ton of 100% acid by 7/1/75. New: <4 lb SO$_2$ and < 15 lb acid mist/ton 100% acid. Steam generators: Chap. 17-2, pg. 5

Southeastern States Air Polluton Control Codes, Continued

Category and Parameters	EPA[a]	Georgia[a]	South Carolina
10) NO_x (NO, NO_2)	See Federal Register (8/18/77) for boiler and nitric acid plant requirements. Test by Method 7, PDS Proc.[d]	HNO_3 Plants before 1/1/72: <25 lb NO_2/ ton acid. After 1/1/72: 3 lb NO_2/ton acid. See 393-3-1	
Effective Date These Data	2/2/75	11/20/75	3/25/75

[a]Test requires EPA train.
[b]Applies to incinerators on federal installations or for EPA approval.
[c]Dry–corrected to 12% CO_2.
[d]Standards of performance for new stationary sources.

Alabama	Tennessee[a]	North Carolina[a]	Florida
HNO$_3$ Plants: $<$ 5.5 lb NO$_2$/ton 100% HNO$_3$. For boilers: see Chap. 8	–	HNO$_3$ Plants: $<$ 5.8 lb NO$_2$/ton acid. Boilers: see sec. IV, pg. C-34.	HNO$_3$ Plants: $<$ 3.0 lb/ ton 100% acid. Steam Generators: Chap. 17-2, pg. 6A
11/25/75	11/10/75	11/19/75	2/1/75

[e]All existing (prior to 7/1/71) units must conform to the requirements for new units after 12/31/73.
[f]Emission Standard in percent of charging rate.
[g]Mobile County only.

Comparison of Source Emission Limits—Plains States Air Pollution Control Codes

Category and Parameters	EPA[a]	Minnesota
Incinerators		
1) Maximum Ringelmann number and time allowed	Existing: #2 New: #1[d]	#1 (allowed). Maximum $> #2 < #3$ not to exceed 4 min/hr $> #1 < #2$ not to exceed 4 min/hr $> #3$ not allowed. See APC-11.
2) Is any visible fly ash allowed?	Yes if within Item 4[d]	Yes. See APC-11 limits.
3) Odor?	–	Based on odor concentration units. See APC-9.
4) Maximum Particulate Emission Rate	0.3 gr/scf < 200 lb/hr[d] 0.2 gr/scf[e] > 200 lb/hr[d] For > 50 ton/day, 0.08 gr/scf[d,f]. See Air Reg. 39	0.3 gr/scf < 200 lb/hr. 0.2 gr/scf > 2000 lb/hr, 0.1 gr/scf. See data APC-7.
5) Special items	Test requires PHS waste and EPA Test Methods 1-5.	Secondary chamber required for $1200°F$ operation
Particulates		
6) Maximum Emission Rate Note: Air Requirements in EPA column are in 40CFR50 document	Cement plants: 0.3 lb/ton $< 10\%$ opacity. See Air Reg. 43. Secondary Lead: 0.22 gr/scfd. Opacity $< 20\%$ except 2 min/60 min on > 550 lb capacity. Secondary Brass and Bronze and Iron/Steel: .022 gr/scfd. Opacity $< 10\%$ except 2 min/60 min on > 2205 lb capacity electric furnace or 550 lb/hr blast furnace. Sewage plant: 0.31 gr/scfd. Opacity $< 10\%$ except 2 min/50 min. Boilers -fossil: 1 lb/hr/10^6 Btu.	0.3 gr/scf except existing grey iron jobbing cupolas which are .4 gr/scf or 85% reduction by gas cleaning device—whichever is more stringent. See APC-5. All organics must be incinerated at $> 1200°F > 0.3$ sec. Other existing sources need $> 99\%$ efficiency or meet grain loading requirement unless outside the Metro area. Outside Metro area, efficiency $> 85\%$. New sources: $> 99.7\%$ efficiency.
7) Special items	See 40 CFR61 Hazardous Air Pollution	See APC-17 for asbestos
Organics		
8) Maximum Emission Rate	No EPA code at present. Check for a state code.	Ambient Standards: .24 ppmv hydrocarbons, .07 ppmv photochemical oxidants at property line. See APC-1.A1 and APC-9 for odor must be met.

Iowa[b]	Missouri[c]	St. Louis City[a]	Kansas
#2 (not to exceed #3 for 3 min/hr	Existing: #2. New: #1. Permitted #2 not to exceed 6 min/hr. See S-IV for tepee burners.	#1 except #2 permitted 6 min/hr.	#1 maximum
No	–	–	–
No, objectionable	No[e]	0.3 sec at $1200°F$ Animal matter.	
0.2 gr/scf $>$ 1000 lb/hr. 0.35 gr/scf $<$ 1000 lb/hr.	0.2 gr/scf over 200 lb/hr. 0.3 gr/scf for all other new incinerators.	0.2 gr/scf 200 lb/hr or more. 0.3 gr/scf all others.	0.3 gr/scf $<$ 200 lb/hr. 0.2 gr/scf $>$ 200 lb but $<$ 20,000 lb/hr 0.1 gr/scf $>$ 20,000 lb/hr.
Provisions CFR40 code Federal Register also apply for units over 50 ton/day.	–	Multiple chamber incinerator required	–
Varies with source and/or stack height. See Chap. 4 of codes for processes. For indirect fired combustion units: maximum of .6 lb/10^6 Btu input. For multiple stack effect, apply ASME APS-1. New units must meet provisions of 40 code Federal Register & in metro areas, 0.8 lb/10^6 Btu for existing units outside metro areas (metro area is $\geqslant$ 50,000 population)	Based on process wt. See Reg. S-V for process equipment. Indirect fuel burning equipment based on Btu input. See Reg. S-VI. Opacity: Existing: #2. New: #1. Permitted #3 not to exceed 6 min/hr. See Reg. S-II for approval of planned installations in "outstate" area.	Reduction of 85% particulate required for fuel burning units of $>$ 5 x 10^6 Btu/hr industry particulate based on process wt and maximum of 0.3 gr/scf. Maximum #2 Ringelmann existing units. Maximum #1 new units except #2 for maximum 6 min/hr. See code for cupolas.	Based on process wt. See 28-19-20 of Code Table F-1 for processing equipment. Indirect fired equipment emission rate based on Btu input. See 28-19-31A Table H-1 of code. Opacity: Existing units: $<$ #2 Ringelmann. New units: $<$ #1 Ringelmann.
–	Odor. See Note #2. No fugitive dust $>$ 40 μ. See Reg. S-VII.	Maximum particulate size 60 μ in all cases.	–
Also a permit to construct will not be issued if the installation prevents the attainment or maintenance of the National Ambient Air Quality Standards. Also see Subsect. 4.3C Fugitive Dust.	–	–	Only ethylene in excess of 50 lb/day covered which must be incinerated at $1300°F$ for 0.3 sec. Otherwise only limited by 28-19-13.

Plains States Air Pollution Control Codes, Continued

Category and Parameters	EPA[a]	Minnesota
9) SO_x (SO_2, SO_3)	Sources begun before 8/17/71: see Federal Register (8/18/77) for boiler and sulfuric acid plant requirements. New Sources: See Air Reg. 51. Test per EPA Method 6 or 8	New acid plant: $<$4 lb SO_2/ton acid and .15 lb/ton acid mist for acid plants. See APC-15. Fuel Standards: see APC-4. New rules now being adopted for refineries and direct combustion sources. Lone existing sulfuric acid plant to meet APC-15 requirements.
10) NO_x (NO, NO_2)	See Federal Register (12/23/71) for boiler and nitric acid plant requirements. Test by Method 7, PDS Proc.[f]	See APC-4 for boiler requirements. Priority I Air Quality Reg. only. Acid Plants: New: 3.0 lb/ton(maximum 2 hr average) No visible NO_x. See APC-16.
Effective Date These Data	2/1/75	4/13/75

[a]Also St. Louis County (Yellow Book) and St. Louis metropolitan area (Blue Book) have codes which should be checked.
[b]Code emissions based primarily on EPA requirements.
[c]Dry—corrected to 12% CO_2.

Iowa[b]	Missouri[c]	St. Louis City[a]	Kansas
<30 lb SO_2 maximum 2 hr average from existing sulfuric acid plant/ ton acid. $<.5$ lb/ton of acid mist from existing plant. Other processes 500 ppm. SO_2 maximum emission. For solid fuels: <6 lb/10^6 Btu/hr effective 7/31/75. 5 lb/10^6 Btu input effective 7/31/78	Existing: (except power plants) <2000 ppm SO_2. New: <500 ppm SO_2 H_2SO_4 or SO_3. Existing: 70 mgm/m^3. New: 35 mgm/m^3. See Reg. S-X for full details. Power plants & all other sources: Fence line SO_2 standard of $<.25$ ppm/hr or $<$ .07 ppm/24 hr.	See Ord. 50163 on sulfur content of coal and oil. 2000 ppm SO_2 existing units. <500 ppm SO_2 new units. H_2SO_4 and SO_3 from processes: Existing: <70 mgm/m^3 New: <35 mgm/m^3. See Sec. 21. See Sec. 12 for cupolas	SO_x from nonferrous smelters based on total sulfur to smelter. See 28-19-22A of code. H_2S maximum 10 ppm emission rate. See 28-19-22 (B) for existing.
Same as EPA provisions of 40 code of Federal Regulations, Part 60 (1972) shall apply.	–	–	–
5/2/75	12/23/75	2/1/75	3/1/75

[d]Applies to incinerators on federal installations or for EPA approval.
[e]Odor prohibited if discernible by Scentometer with 7 to 1 dilution ratio. This applies to all processes, not only incinerators.
[f]Standards of performance for new stationary sources.

Comparison of Source Emission Limits—Delta Region Air Pollution Control Codes

Category and Parameters	EPA[a]	Mississippi	Louisiana
Incinerators			
1) Maximum Ringelmann number and time allowed	Existing: #2 New: #1[c]	> #2 for 15 min/hr for startup & not to exceed 3 startups/day. Then ≤ #2	≥ #1 for ≤ 4 min/hr during charging, then ≤ #1
2) Is any visible fly ash allowed?	Yes if within Item 4[c]	No if public nuisance	No if public nuisance
3) Odor?	–	No if public nuisance	–
4) Maximum Particulate Emission Rate	0.3 gr/scf[e] < 200 lb/hr[c] 0.2 gr/scf > 200 lb/hr[c] For > 50 ton/day, 0.08 gr/scf[c,f].	0.2 gr/scf design capacity. 0.1 gr/scf in critical areas.	0.2 gr/scf
5) Special items	Test requires PHS waste & EPA Test Methods 1-5	New equipment must be multichamber or equivalent[a]	Must be multichambered or equivalent. All must have second burner with ≥ 1500°F temperature.[g]
Particulates			
6) Maximum Emission Rate (Note: Air Requirements in EPA column are in 40CFR50 document.)	Varies with sources, i.e., cement plant: 0.3 lb/ton[f]. Dry catch only for test. See Air Regulation 43.	General manufacturing based on process weight rate. See Regulation APC-S-1, Table I.	General process based on process weight rate. See Table 3, page 56.
7) Special items	See 40 CFR61 Hazardous Air Pollution	Separate limits for other processes. See Regulation APC-S-1, Section 3.	Item 6 does not apply to wood pulp industry, primary aluminum industry and burning of fuel for indirect heat[g]
Organics			
8) Maximum Emission Rate	No EPA code at present. Check for state code.	No limits if within property boundary of source.	Waste Gas Incineration ≥ 1300°F for ≥ 0.3 sec. Organic Solvents ≤ 3 lb/hr or ≤ 15 lb/day

Texas	Arkansas	Missouri[b]	Oklahoma
Maximum for 5 min/avg: ≤30% if built before 1/31/72; 20% if built after 1/31/72. For variance see Rule 103.1.	Existing: #2; greater allowed but not to exceed 5 min/hr. New: #1 not to exceed #3 for 5 min/ hr (maximum 3 times/ 24 hr).	Existing: #2. New: #1 permitted #2 not to exceed 6 min/hr. See S-IV for tepee burners.	#1. Exception, #3 for 5 min/hr or 20 min/24 hr period.
See Regulation I.	–	–	No. See Regulation 5, Figure 1.
–	Section 10 only applicable code at this time.	No[d]	–
Depends on flow rate and stack height. See Rule 105.	0.2 gr/scf for > 200 lb/ hr. 0.3 gr/scf for < 200 lb/hr.	0.2 gr/scf for > 200 lb/ hr. 0.3 gr/scf for all other new incinerators.	Based on charge rate. See Regulation 5, Figure 1. Solid fuels only considered part of charge.
Multichamber incinerators only unless single chamber unit shows equivalent performance.	–	–	Must be multichamber incinerator auxiliary burner of 800°F in primary combustion chamber. Second burner eliminates smoke.
Item 4 applies. For agricultural process see Rule 105-107 in Regulation I. See also General Rules (Rule 1) See also Reg. VIII.	Based on process weight weight. See Sec. 7 ofc code. Opacity not to exceed Ringelmann #1. Note: All Arkansas codes not being revised. Expected release of new version in summer 1975.	Based on process weight. See Reg. S-7 for process equipment. S-VI indirect fuel burning equipment based in Btu input. Opacity: Existing: #2. New: #2 permitted #3 not to exceed 6 min/hr.	Based on process weight rate. Existing installations must comply by 3/1/77. Temperature emissions maximum 150% of limits for < 6 min/hr. See Reg. 8, Table 1 for emission rates.
For solid fuel steam generator limit is 0.3 lb/ 10⁶ Btu input. See Rule 105.2 for ground level concentration by stat. source.	–	Odor. See Note. No fugitive dust > 40 μ. See Regulation S-VII.	Fuel burning: Existing: < 0.6 lb/10⁶ Btu input. Must meet new standards by 3/1/77. See Reg. 6, Table 1 and Figure 1 for emission standards.
Waste Gas Incineration ≥ 1300°F. See Reg. V for specific details.	Not defined. Generally limited by Section 6[d] of code.	–	Waste Gas Incineration ≥ 1300°F for ≥ 0.3 sec. heat drying, 3 lb/hr or < 15 lb/ day. See Reg. 15 for others.

Delta Region Air Pollution Control Codes, Continued

Category and Parameters	EPA[a]	Mississippi	Louisiana
9) SO_x (SO_2, SO_3)	Sources begun before 8/17/71: see Federal Register 8/18/77 for boiler and sulfuric acid plant requirements. New Sources: See Air Regulation 51. Test per EPA Method 6 or 8	Fuel burning equipment: Existing: 4.8 lb $SO_2/10^6$ Btu. Modified: 2.4 lb $SO_2/10^6$ Btu. Processes: Existing: 2000 ppm. New: <0.01 gr/scf for H_2S. 0.5 lb/ton H_2SO_4. See Reg. APC-S-1 for sulfur plants.	<6.5 lb/ton 100% H_2SO_4 or <250-500 ppmv of SO_2 depending on feed gas. <0.01 lb SO_2/lb sulfur or <1300 ppmv. Existing: from any source <2000 ppmv[h].
10) NO_x (NO, NO_2) (See Air Regulation 47)	See Federal Register (8/18/77) for boiler and nitric acid plant requirements. Test by Method 7 PDS Proc.[f]	No limits if within property boundary of source.	<6.5 lb NO_2/ton 100% HNO_3 or <500 ppm by volume[i].
Effective Date These Data	2/1/75	11/1/75	8/1/74

[a]CO_2 produced by auxillary fuel excluded from calculation to 12% CO_2.

[b]Regulations apply throughout state except for political subdivisions having exemption under Sec.203.150, RSMo or where more restrictive regulations apply.

[c]Applies to incinerators on federal installations or for EPA approval.

[d]See Regulation S-IX on Barnaby-Cheney Scentometer.

Texas	Arkansas	Missouri[b]	Oklahoma
SO$_2$ emission based on stack height, flow rate and specific process. See Reg. II for H$_2$S and and H$_2$SO$_4$. See also Reg. VIII.	SO$_2$ ambient level is: 1) $<$0.20 ppm; 2) sulfuric acid mist or SO$_3$ ambient limit is $<$ 30 μg/m^3. Both limits avg. for 30 min at property line.	Existing: (except power plants) $<$2000 ppm SO$_2$. New: $<$500 ppm SO$_2$, H$_2$SO$_4$ or SO$_3$. Existing: 70 μg/m^3. New: 35 μg/ m^3. See Reg. S-X for full details. For Power Plants and all other sources: Fence line SO$_2$ standard of $<$0.25 ppm/hr or $<$ 0.07 ppm/24 hr.	Existing: based on ambient standards. See Sec. 16.21. New: based on source. See Reg. 16, Sec. 16.31-16.35. H$_2$S, see Sec. 16.4 and 16.5.
$<$500 ppm by volume NO$_2$ for HNO$_3$ plant. For Steam Generators see Reg. VII, Rule 701. See also Reg. VIII.	Section 10 only applicable code at this time.	–	$<$ 3 lb/ton 100% HNO$_3$. Fuel burning solid, $\geqslant$ 50 x 10^6 Btu/hr $<$ 0.7 lb/10^6 Btu. Liquid fuel $\geqslant$ 50 x 10^6 Btu/hr $<$ 0.3 lb/10^6 Btu. Gas $\geqslant$ 50 x 10^6 Btu/ hr $<$ 0.2 lb/10^6 Btu.
3/5/76	3/1/75	12/23/75	9/1/75

^{e}Dry—corrected to 12% CO$_2$.
^{f}Standards of performance for new stationary sources.
^{g}Test requires EPA Methods 1-5 or equivalent methods approved by the LACC.
^{h}Test requirements: See Table 4, page 57 of the Louisiana Air Control Commission Regulations.
^{i}Test requires EPA Methods 1-4 and 7.

Comparison of Source Emission Limits–Midwestern States Air Pollution Control Codes

Category and Parameters	EPA[a]	Chicago	Illinois
Incinerators			
1) Maximum Ringelmann number and time allowed	Existing: #2 New: #1[b]	#3 for 8 min/hr, 3 times in 24 hr. Otherwise #1.5	#3 for 8 min/hr, 3 times in 24 hr. Otherwise #1.5
2) Is any visible fly ash allowed?	Yes it within Item 4[b]	No	No. See Rule 202.
3) Odor?	–	1400°F	–
4) Maximum Particulate Emission Rate	0.3 gr/scf[c] < 200 lb/hr[b] 0.2 gr/scf > 200 lb/hr[b] For > 50 ton/day, 0.08 gr/scf[b,d] See Air Regulation 39.0	Existing: 0.08 gr/scf burning > 2000 lb/hr. 0.05 gr/ scf burning > 30 ton/day. Others 0.05 gr/scf. All other New: 0.1 gr/scf. See Rule 203	Existing: 0.08 gr/scf burning > 2000 lb/hr. 0.05 gr/scf burning > 30 ton/day. Others 0.05 gr/scf. All other New: 0.1 gr/scf. See Rule 203, para. e.[e]
5) Special Items	Test requires PHS waste and EPA Test Methods 1-5.	Test per ASME PTC-27. Maximum 500 ppm. Annual inspection of incinerators required.	Test per ASME PTC-27 and ATP-1A.00. Maximum 500 ppm. See Rule 206.
Particulates			
6) Maximum Emission Rate (See EPA 40CFR50) See Air Reg. 43	Varies with sources, i.e., Cement plants: 0.3 lb/ ton[d] Dry catch only for test	Varies with source and process rate. See Rule 203.	Varies with source and process rate. See Rule 203, para. a-d.
7) Special Items	See 40CFR61 Hazardous Air Pollution.	Test per ASME PTC-27.	Test per ASME PTC-27 and ATP-1
Organics			
8) Maximum Emission Rate	No EPA code at present. Check for a state code.	8 lb/hr unless total emission reduced by 85%. See Rule 205.	8 lb/hr unless total emission reduced by 85%. See Rule 75-3 for photochemical.
9) SO_x (SO_2, SO_3)	See Federal Register 8/18/77 for boiler and sulfuric acid plant requirements. New Sources: See Air Reg. 51. Test per EPA Method 6 or 8.	Maximum 2000 ppm. New Acid Plants: < 4 lb/ ton acid produced. Acid Mist: < 0.15 lb/ton acid used or produced. See Rule 204.	Maximum 2000 ppm. New Acid Plants: 4 lb/ton acid. Acid mist: 0.15 lb/ton acid. See R74-13 for safety flares. See Rule 204. See other regulations.
10) NO_x (NO, NO_2)	See Federal Register (8/18/77) for boiler and nitric acid plant requirements. Test by Method 7 PDS Proc.[d]	Acid Plant: 3 lb/ton 100% acid. 5.5 lb/ton weak acid. See Rule 207.	Acid plant: 3 lb/ton 100% acid. 5.5 lb/ton weak acid. See Rule 207.
Effective Date These Data	2/1/75	3/1/75	3/1/75

[a]Test requires EPA train.
[b]Applies to incinerators on federal installations or for EPA approval.
[c]Dry–corrected to 12% CO_2.

Wisconsin	Indiana	Ohio	Michigan
Existing: #2 New: #1	#3 for 5 min/hr 6 times in 24 hr. Otherwise #2.	#3 for 3 min/hr. Otherwise #1.	#2 for 3 min/hr any 24 hr period. Otherwise #1.
–	Yes if within Item #4.	See EP-11-07 <20% opacity	–
1300°F, 3 sec.	–	Objectionable - No.	–
Category I: <500 lb/hr 0.3 lb/1000 lb exhaust gas. >4000 lb/hr 0.15 lb/1000 lb exhaust gas. >500 lb/hr, 0.20 lb/1000 lb exhaust gas. See Categories II, III, IV.	Capacity >1000 lb/hr 0.4 lb/1000 lb refuse.[c] All others: 0.7 lb/1000 lb refuse.[c] Must complete Refuse Disposal and Incinerator Information Form.	See EP-11-09. Capacity <100 lb/hr 0.2 lb/100 lb refuse. Capacity ≥100 lb/hr, 0.1 lb/100 lb refuse.[f] Test per EPA Methods 1-5.	<100 lb/hr 0.65[c]lb/1000 lb flue gas. >100 lb/hr 0.30[c] lb/1000 flue gas.
Must maintain 1300°F in secondary chamber.	Test per Federal Regulations.	Must complete "Intent to Test Notification"	Annual report and fee for contaminants listed in Table V.
Varies with source and process rate. See Sec. NR 154.11 Categories I-IV.	Varies with source and process rate. See Reg. APC-5. Combination of fuel Reg. APC-4R. See APC-6 for foundries.	See EP-11-11. Varies with source, process rate and priority I, II or III Region. See EP-11-10 for limits on fuel burning equipment.	See Table I for basic applications. Table II for limits based on process weight.
3 lb/hr or 15 lb/day unless total emission reduced by 85%. See NR154.13.	3 lb/hr or 15 lb/day unless total emission reduced by 85%. See Reg. APC-15.	See AP-5-07. 15 lb/day and 3 lb/hr limit unless >85% reduction. Higher limit on photochemical discharge.	
Acid Plant: <0.4 lb/ton acid produced. See NR154.12, NR154.16, NR154.18 for other sources.	Varies with stack height, area, number of units, etc. See Reg. APC-13.	See EP-11-12. SO_2 < 2000 ppmv existing, < 500 ppmv. New Sources: Acid plant: <6.5 lb/ton SO_2. Acid Mist:0.5 lb/ton. SO_3 0.2 lb/ton.	See R336.49 for power pHs. See also table III and IV.
Acid Plant: 3 lb NO_2/ton weak acid produced. See Sec. NR154.15	From fuel combustion 0.2 to 0.7 lb/10^6 Btu. See Reg. APC-17.	See AP-7-06. <0.2 lb/10^6 Btu input from gas fired, 0.30 from oil, 0.9 from coal. Nitric Acid plants: 5.5 lb/ton.	
3/31/75	1/15/75	2/1/75	4/1/75

[d]Standards of performance for new stationary sources.
[e]All other (<2000 lb/hr) existing and all "100% wood burning" incinerators 0.2 gr/scf.
[f]Emission Tests to be conducted at maximum rated burning capacity.

Comparison of Source Emission Limits—Mountain Region Air Pollution Control Codes

Category and Parameters	EPA[a]	North Dakota	Montana
Incinerators			
1) Maximum Ringelmann number and time allowed	Existing: #2 New: #1[b]	Existing: #2 New: #1. Exception: #3 for 4 min/hr.	#1 wood burn #1/2 incinerators, all others #2, except #3 for 4 min/hr.
2) Is any visible fly ash allowed?	Yes if within Item 4[b]	Yes	–
3) Odor	–	–	–
4) Maximum Particulate Emission Rate See Air Regulation 39	0.3 gr/scf[e] < 200 lb/hr[b]. 0.2 gr/scf > 200 lb/hr[b] 0.08 gr/scf for > 50 tons/day[b,d]	Table 5 sets limits. All new incinerators for waste types 2-6 must meet $1500°F$ and 0.3 sec dwell time in secondary combination chamber.	Existing: as per EPA 200 lb/hr limits. New: 0.1 gr/scf. Wood burners: Existing: 0.2 gr/scf. New: 0.1 gr/scf.
5) Special items	Test requires PHS waste and EPA Test Method 1-5.	Must be multiple chamber incinerator.	Must be multiple chamber incinerator; animal matter 0.3 sec @ $1200°F$.
Particulates			
6) Maximum Emission Rate (See EPA 40CRF50) See Air Regulation 43	Varies with sources, i.e., cement plants: 0.3 lb/ton[d] Dry catch only for test.	Based on process weight; see Table 3 or at least 99.7% efficiency at maximum capacity	Based on process weight; see table in Sec. 16-2.14(1)-S1430.
7) Special items	See 40 CFR61 Hazardous Air Pollution	Fuel burning: Existing: < 0.8 lb/10^6 Btu. New Installation: see Table 4.	Fuel burning: < 0.6 lb/10^6 Btu input. See Sec. 16-2.14 (1)-S1450.
Organics			
8) Maximum Emission Rate	No EPA code at present. Check for a state code.	Based on Ambient Air Quality Standards. See Table 1.	No state code at present.
9) SO_x (SO_2, SO_3)	See Federal Register 8/18/77 for boiler and sulfuric acid plant requirements. New Sources: See Air Reg. 51. Test per EPA Method 6 or 8.	Fuel burning: 3.0 lb SO_2/10^6 Btu. SO_2 test per Shell Development Method. SO_3 and acid mist test per Monsanto Method.	1 lb/10^6 Btu liquid or solid fuel. 50 gr/100 ft^3 gaseous fuel as H_2S at STP. Exception: Sec. 16-2.14 (1)-S1470.
10) NO_x (NO, NO_2)	See Federal Register 8/18/77 for boiler and nitric acid plant requirements. Test by Method 7, PDS Proc.[d]	No source codes. Based on Ambient Air Quality Standards. See Table 1.	No state code at present.
Effective Date These Data	2/1/75	3/9/75	7/1/75

[a]CO_2 produced by auxiliary fuel excluded from calculation to 12%.
[b]Applies to incinerators on federal installations or for EPA approval.
[c]See Regulation 2 of Colorado Air Quality Regulations.

Nebraska	South Dakota	Wyoming	Colorado
Existing: #1	#1	20% opacity. Wood burners: 20% opacity ≤6 min/hr.	20% opacity. Exceptions: see Reg. I, Para. 2 Colorado Air Quality Regulations.
–	Yes, but maximum size of 100 μ.	–	–
–	–	≤7 x dilution at property line.	No^c
0.2 gr/scf <2000 lb/ hr. 0.1 gr/scf > 2000 lb/hr	0.08 gr/scf >50 ton/ day charge. If >50 ton/ day, see para. 34:10: 06:07	0.2 lb/100 lb as per EPA 5 test method. Wood burner limits may be exceeded ≤60 min/8 hr.	Existing: 0.15 gr/scf inside designated APC areas. Outside: 0.15 gr/scf. New: 0.1 gr/scf.
–	Fuel burning <0.3 lb/ 10^6 Btu heat input.	–	Test per ASME PTC-27.
Based on process weight. See Table 2-1.	0.1 lb/10^6 Btu fuel burning.	Based on process weight. Existing: Table II, Chap. I Wyoming Air Quality Standards. New: Table I.	Based on process weight. See Reg. 1, Sec. IIc, Fig. 2 of Colorado Air Quality Control Reg. for limits.
Fuel burning based on heat input. See table, Rule 6.	Maximum opacity 40% for ≤2 min/hr fuel burning.	Fuel burning: Existing: see Fig. 1. New: <0.1 lb/ 10^6 Btu, maximum 2 hr average.	Fuel burning based on heat input. See Reg. 1, Sec. IIA.
No state code at present.	Hydrocarbons < <160 μg/m^3 3 hr once/ year ambient.	Ambient standards hydrocarbons: <160 μg/m^3 maximum 3 hr once/yr.	No state code at present.
Fuel burning: 2.5 lb/10^6 Btu input, maximum 2 hr avg. Annual and daily maximum emissions based on 1971 standards. H$_2$S <0.01 ppm for 30 min.	Fuel burning 0.8 lb/10^6 Btu liquid fuel. 1.2 lb/ 10^6 Btu solid fuel. SO$_2$ test per ASTM Method D2234-72 or D1552-64.	Ambient standards ≤acid mist 0.5 lb/ton maximum 2 hr avg. SO$_x$ see Sec. 4, Tables 4a and 4b Wyoming Air Quality Standards and Reg. H$_2$S see Sec. 7.	500 ppm SO$_x$. Exception: does not apply if emissions <5 ton SO$_2$/day.
<5.5 lb/ton of 100% HNO$_3$ or 400 ppm NO$_2$.	10% opacity, source 100 μg/m^3 annual, 50 μg/m^3 24 hr ambient.	No state code at present.	3 lb/ton of acid produced, maximum 2 hr avg. Opacity ≤10%.
6/17/75	11/6/75	2/26/75	12/18/75

^{d}Standards of performance for new stationary sources.
^{e}Dry—corrected to 12% CO$_2$.

APPENDIX

STANDARDS OF PERFORMANCE FOR NEW STATIONARY SOURCES REVISION TO REFERENCE METHODS 1 - 8*

Title 40–Protection of Environment

CHAPTER I–ENVIRONMENTAL PROTECTION AGENCY

[FRL 754-5]

PART 60–STANDARDS OF PERFORMANCE FOR NEW STATIONARY SOURCES

Revision to Reference Methods 1 - 8

AGENCY: Environmental Protection Agency.

ACTION: Final Rule.

SUMMARY: This rule revises Reference Methods 1 through 8, the detailed requirements used to measure emissions from affected facilities to determine whether they are in compliance with a standards of performance. The methods were originally promulgated December 23, 1971, and since that time several revisions became apparent which would clarify, correct and improve the methods. These revisions make the methods easier to use, and improve their accuracy and reliability.

EFFECTIVE DATE: September 19, 1977.

ADDRESSES: Copies of the comment letters are available for public inspection and copying at the U.S. Environmental Protection Agency, Public Information Reference Unit (EPA Library), Room 2922, 401 M Street, S.W., Washington, D.C. 20460. A summary of the comments and EPA's responses may be obtained upon written request from the EPA Public Information Center (PM-215), 401 M Street, S.W., Washington, D.C. 20460 (specify "Public Comment Summary: Revisions to Reference Methods 1 - 8 in Appendix A of Standards of Performance for New Stationary Sources").

---

*Reprinted from the *Federal Register,* Thursday, August 18, 1977, Part II.

FOR FURTHER INFORMATION CONTACT: Don R. Goodwin, Emission Standards and Engineering Division, Environmental Protection Agency, Research Triangle Park, North Carolina 27711, telephone No. 919-541-5271.

SUPPLEMENTARY INFORMATION: The amendments were proposed on June 8, 1976 (40 FR 23060). A total of 55 comment letters were received during the comment period — 34 from industry, 15 from governmental agencies, and 6 from other interested parties. They contained numerous suggestions which were incorporated in the final revisions.

Changes common to all eight of the reference methods are: (1) the clarification of procedures and equipment specifications resulting from the comments, (2) the addition of guidelines for alternative procedures and equipment to make prior approval of the administrator unnecessary and (3) the addition of an introduction to each reference method discussing the general use of the method and delineating the procedure for using alternative methods and equipment.

Specific changes to the methods are:

METHOD 1

1. The provision for the use of more than two traverse diameters, when specified by the Administrator, has been deleted. If one traverse diameter is in a plane containing the greatest expected concentration variation, the intended purpose of the deleted paragraph will be fulfilled.

2. Based on recent data from Fluidyne (Particulate Sampling Strategies for Large Power Plants Including Nonuniform Flow. EPA-600/2-76-170, June 1976) and Entropy Environmentalists (Determination of the Optimum Number of Traverse Points: An Analysis of Method 1 Criteria (draft), Contract No. 68-01-3172), the number of traverse points for velocity measurements has been reduced and the 2:1 length to width ratio requirement for cross-sectional layout of rectangular ducts has been replaced by a "balanced matrix" scheme.

3. Guidelines for sampling in stacks containing cyclonic flow and stacks smaller than about 0.31 meter in diameter or 0.071 m^2 in cross-sectional area will be published at a later date.

4. Clarification has been made as to when a check for cyclonic flow is necessary; also, the suggested procedure for determination of unacceptable flow conditions has been revised.

METHOD 2

1. The calibration of certain pitot tubes has been made optional. Appropriate construction and application guidelines have been included.

2. A detailed calibration procedure for temperature gauges has been included.

3. A leak check procedure for pitot lines has been included.

METHOD 3

1. The applicability of the method has been confined to fossil-fuel combustion processes and to other processes where it has been determined that components other than O_2, CO_2, CO and N_2 are not present in concentrations sufficient to affect the final results.

2. Based on recent research information (Particulate Sampling Strategies for Large Power Plants Including Nonuniform Flow, EPA-600/2-76-170, June 1976), the requirement for proportional sampling has been dropped and replaced with the requirement for constant rate sampling. Proportional and constant rate sampling have been found to give essentially the same result.

3. The "three consecutive" requirement has been replaced by "any three" for the determination of molecular weight, CO_2 and O_2.

4. The equation for excess air has been revised to account for the presence of CO.

5. A clearer distinction has been made between molecular weight determination and emission rate correction factor determination.

6. Single point, integrated sampling has been included.

METHOD 4

1. The sampling time of 1 hour has been changed to a total sampling time which will span the length of time the pollutant emission rate is being determined or such time as specified in an applicable subpart of the standards.

2. The requirement for proportional sampling has been dropped and replaced with the requirement for constant rate sampling.

3. The leak check before the test run has been made optional; the leak check after the run remains mandatory.

METHOD 5

1. The following alternatives have been included in the method:
   a. The use of metal probe liners.
   b. The use of other materials of construction for filter holders and probe liner parts.
   c. The use of polyethylene wash bottle and sample storage containers.
   d. The use of desiccants other than silica gel or calcium sulfate, when appropriate.
   e. The use of stopcock grease other than silicon grease, when appropriate.
   f. The drying of filters and probe-filter catches at elevated temperatures, when appropriate.
   g. The combining of the filter and probe washes into one container.

2. The leak check prior to a test run has been made optional. The post-test leak check remains mandatory. A method for correcting sample volume for excessive leakage rates has been included.

3. Detailed leak check and calibration procedures for the metering system have been included.

METHOD 6

1. Possible interfering agents of the method have been delineated.

2. The options of: (a) using Method 8 impinger system, or (b) determining SO_2 simultaneously with particulate matter, have been included in the method.

3. Based on recent research data, the requirement for proportional sampling has been dropped and replaced with the requirement for constant rate sampling.

4. Tests have shown that isopropanol obtained from commercial sources occasionally has peroxide impurities that will cause erroneously low SO_2 measurements. Therefore, a test for detecting peroxides in isopropanol has been included in the method.

5. The leak check before the test run has been made optional; the leak check after the run remains mandatory.

6. A detailed calibration procedure for the metering system has been included in the method.

METHOD 7

1. For variable wave length spectrophotometers, a scanning procedure for determining the point of maximum absorbance has been incorporated as an option.

METHOD 8

1. Known interfering compounds have been listed to avoid misapplication of the method.

2. The determination of filterable particulate matter (including acid mist) simultaneously with SO_3 and SO_2 has been allowed where applicable.

3. Since occasionally some commercially available quantities of isopropanol have peroxide impurities that will cause erroneously high sulfuric acid mist measurements, a test for peroxides in isopropanol has been included in the method.

4. The gravimetric technique for moisture content (rather than volumetric) has been specified because a mixture of isopropyl alcohol and water will have a volume less than the sum of the volumes of its content.

5. A closer correspondence has beem made between similar parts of Methods 8 and 5.

MISCELLANEOUS

Several commenters questioned the meaning of the term "subject to the approval of the Administrator" in relation to using alternate test methods and procedures. As defined in § 60.2 of subpart A, the "Administrator" includes any authorized representative of the Administrator of the Environmental Protection Agency. Authorized representatives are EPA officials in EPA Regional Offices or State, local, and regional governmental officials who have been delegated the responsibility of enforcing regulations under 40 CRF 60. These officials in consultation with other staff members familiar with technical aspects of source testing will render decisions regarding acceptable alternate test procedures.

In accordance with section 117 of the Act, publication of these methods was preceded by consultation with appropriate advisory committees, independent experts, and Federal departments and agencies.

(Secs. 111, 114 and 301(a) of the Clean Air Act, sec. 4(a) of Pub. L. No. 91-604, 84 Stat. 1683; sec. 4(a) of Pub. L. No. 91-604, 84 Stat. 1687; sec. 2 of Pub. L. No. 90-148, 81 Stat. 504 [42 U.S.C. 1857c-6, 1857c-9, 1857g(a)].)

NOTE—The Environmental Protection Agency has determined that this document does not contain a major proposal requiring preparation of an Economic Impact Analysis under Executive Orders 11821 and 11949 and OMB Circular A-107.

Dated: August 10, 1977.

Douglas M. Costle,
Administrator.

Part 60 of Chapter I of Title 40 of the Code of Federal Regulations is amended by revising Methods 1 through 8 of Appendix A—Reference Methods as follows:

APPENDIX A—REFERENCE METHODS

The reference methods in this appendix are referred to in § 60.8 (Performance Tests) and § 60.11 (Compliance With Standards and Maintenance Requirements) of 40 CFR Part 60, Subpart A (General Provisions). Specific uses of these reference methods are described in the standards of performance contained in the subparts, beginning with Subpart D.

Within each standard of performance, a section titled "Test Methods and Procedures" is provided to (1) identify the test methods applicable to the facility subject to the respective standard and (2) identify any special instructions or conditions to be followed when applying a method to the respective facility. Such instructions (for example, establish sampling rates, volumes, or temperatures) are to be used either in addition to, or as a substitute for procedures in a reference method. Similarly, for sources subject to emission monitoring requirements, specific instructions pertaining to any use of a reference method are provided in the subpart or in Appendix B.

Inclusion of methods in this appendix is not intended as an endorsement or denial of their applicability to sources that are not subject to standards of performance. The methods are potentially applicable to other sources; however, applicability should be confirmed by careful and appropriate evaluation of the conditions prevalent at such sources.

The approach followed in the formulation of the reference methods involves specifications for equipment, procedures, and performance. In concept, a performance specification approach would be preferable in all methods because this allows the greatest flexibility to the user. In practice, however, this approach is impractical in most cases because performance specifications cannot be established. Most of the methods described herein, therefore, involve specific equipment specifications and procedures, and only a few methods in this appendix rely on performance criteria.

Minor changes in the reference methods should not necessarily affect the validity of the results and it is recognized that alternative and equivalent methods exist. Section 60.8 provides authority for the Administrator to specify or approve (1) equivalent methods, (2) alternative methods, and (3) minor changes in the methodology of the reference methods. It should be clearly understood that unless otherwise identified all such methods and changes must have prior approval of the Administrator. An owner employing such methods or deviations from the reference methods without obtaining prior approval does so at the risk of subsequent disapproval and retesting with approved methods.

Within the reference methods, certain specific equipment or procedures are recognized as being acceptable or potentially acceptable and are specifically identified in the methods. The items identified as acceptable options may be used without approval but must be identified in the test report. The potentially approvable options are cited as "subject" to the approval of the Administrator" or as "or equivalent." Such potentially approvable techniques or alternatives may be used at the discretion of the owner without prior approval. However, detailed descriptions for applying these potentially approvable techniques or alternatives are not provided in the reference methods. Also, the potentially approvable options are not necessarily acceptable in all applications. Therefore, an owner electing to use such potentially approvable techniques or alternatives is responsible for: (1) assuring that the techniques or alternatives are in fact applicable and are properly executed; (2) including a written description of the alternative method in the test report, (the written method must be clear and must be capable of being performed without additional instruction, and the degree of detail should be similar to the detail contained in the reference methods); and (3) providing any rationale or supporting data necessary to show the validity of the alternative in the particular application. Failure to meet these requirements can result in the Administrator's disapproval of the alternative.

METHOD 1
SAMPLE AND VELOCITY TRAVERSES FOR STATIONARY SOURCES

1. *Principle and Applicability*

1.1   Principle. To aid in the representative measurement of pollutant emissions and/ or total volumetric flow rate from a stationary source, a measurement site where the effluent stream is flowing in a known direction is selected, and the cross-section of the stack is divided into a number of equal areas. A traverse point is then located within each of these equal areas.

1.2   Applicability. This method is applicable to flowing gas streams in ducts, stacks, and flues. The method cannot be used when: (1) flow is cyclonic or swirling (see Section 2.4), (2) a stack is smaller than about 0.30 meter (12 in.) in diameter, or 0.071 m^2 (113 in.^{2}) in cross-sectional area, or (3) the measurement site is less than two stack or duct diameters downstream or less than a half diameter upstream from a flow disturbance. The requirements of this method must be considered before construction of a new facility from which emissions will be measured; failure to do so may require subsequent alterations to the stack or deviation from the standard procedure. Cases involving variants are subject to approval by the Administrator, U.S. Environmental Protection Agency.

2. *Procedure*

2.1   Selection of Measurement Site. Sampling or velocity measurement is performed at a site located at least eight stack or duct diameters downstream and two diameters upstream from any flow disturbance such as a bend, expansion, or contraction in the stack, or from a visible flame. If necessary, an alternative location may be selected, at a position at least two stack or duct diameters downstream and a half diameter upstream from any flow disturbance. For a rectangular cross section, an equivalent diameter (D_e) shall be calculated from the following equation, to determine the upstream and downstream distances:

$$D_e = \frac{2LW}{L + W}$$

where L = length and W = width.

2.2   Determining the Number of Traverse Points.

2.2.1   Particulate Traverses. When the eight- and two-diameter criterion can be met, the minimum number of traverse points shall be: (1) twelve, for circular or rectangular stacks with diameters (or equivalent diameters) greater than 0.61 meter (24 in.); (2) eight, for circular stacks with diameters between 0.30 and 0.61 meter (12=24 in.); (3) nine, for rectangular stacks with equivalent diameters between 0.30 and 0.61 meter (12-24 in.).

When the eight- and two-diameter criterion cannot be met, the minimum number of traverse points is determined from Figure 1-1. Before referring to the figure, however, determine the distances from the chosen measurement site to the nearest upstream and downstream disturbances, and divide each distance by the stack diameter or equivalent diameter, to determine the distance in terms of the number of duct diameters. Then, determine from Figure 1-1 the minimum number of traverse points that corresponds: (1) to the number of duct diameters upstream; and (2) to the number of diameters downstream. Select the higher of the two minimum numbers of traverse points, or a greater value, so that for circular stacks the number is a multiple of 4, and for rectangular stacks the number is one of those shown in Table 1-1.

2.2.2   Velocity (Non-Particulate) Traverses. When velocity or volumetric flow rate is to be determined (but not particulate matter), the same procedure as that for particulate traverses (Section 2.2.1) is followed, except that Figure 1-2 may be used instead of Figure 1-1.

2.3   Cross-Sectional Layout and Location of Traverse Points.

2.3.1   Circular Stacks. Locate the traverse points on two perpendicular diameters according to Table 1-2 and the example shown in Figure 1-3. Any equation (for examples, see Citations 2 and 3 in the Bibliography) that gives the same values as those in Table 1-2 may be used in lieu of Table 1-2.

For particulate traverses, one of the diameters must be in a plane containing the greatest expected concentration variation, e.g., after bends, one diameter shall be in the

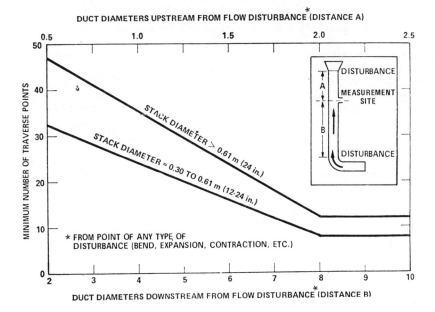

Figure 1-1. Minimum number of traverse points for particulate traverses.

Table 1-1. Cross-sectional layout for rectangular stacks

Number of Traverse Points	Matrix Layout
9	3 x 3
12	4 x 3
16	4 x 4
20	5 x 4
25	5 x 5
30	6 x 5
36	6 x 6
42	7 x 6
49	7 x 7

plane of the bend. This requirement becomes less critical as the distance from the disturbance increases; therefore, other diameter locations may be used, subject to approval of the Administrator.

In addition, for stacks having diameters greater than 0.61 m (24 in.) no traverse points shall be located within 2.5 centimeters (1.00 in.) of the stack walls; and for stack diameters equal to or less than 0.61 m (24 in.), no traverse points shall be located within 1.3 cm (0.50 in.) of the stack walls. To meet these criteria, observe the procedures given below.

2.3.1.1 Stacks With Diameters Greater Than 0.61 m (24 in.). When any of the traverse points as located in Section 2.3.1 fall within 2.5 cm (1.00 in.) of the stack walls,

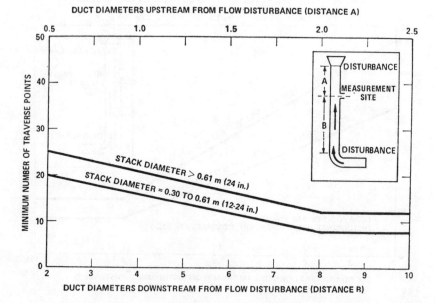

Figure 1-2. Minimum number of traverse points for velocity (nonparticulate) traverses.

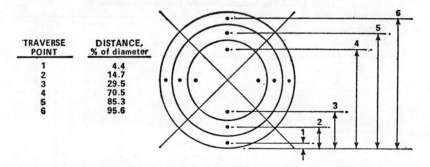

TRAVERSE POINT	DISTANCE, % of diameter
1	4.4
2	14.7
3	29.5
4	70.5
5	85.3
6	95.6

Figure 1-3.   Example showing circular stack cross section divided into 12 equal areas, with location of traverse points indicated.

relocate them away from the stack walls to: (1) a distance of 2.5 cm (1.00 in.); or (2) a distance equal to the nozzle inside diameter, whichever is larger. These relocated traverse points (on each end of a diameter) shall be the "adjusted" traverse points.

Whenever two successive traverse points are combined to form a single adjusted traverse point, treat the adjusted point as two separate traverse points, both in the sampling (or velocity measurement) procedure, and in recording the data.

2.3.1.2   Stacks With Diameters Equal to or Less Than 0.61 m (24 in.). Follow the procedure in Section 2.3.1.1, noting only that any "adjusted" points should be

Table 1-2. Location of Traverse Points in Circular Stacks
(Percent of Stack diameter from inside wall to traverse point)

Traverse point number on a diameter	Number of traverse points on a diameter											
	2	4	6	8	10	12	14	16	18	20	22	24
1	14.6	6.7	4.4	3.2	2.6	2.1	1.8	1.6	1.4	1.3	1.1	1.1
2	85.4	25.0	14.6	10.5	8.2	6.7	5.7	4.9	4.4	3.9	3.5	3.2
3		75.0	29.6	19.4	14.6	11.8	9.9	8.5	7.5	6.7	6.0	5.5
4		93.3	70.4	32.3	22.6	17.7	14.6	12.5	10.9	9.7	8.7	7.9
5			85.4	67.7	34.2	25.0	20.1	16.9	14.6	12.9	11.6	10.5
6			95.6	80.6	65.8	35.6	26.9	22.0	18.8	16.5	14.6	13.2
7				89.5	77.4	64.4	36.6	28.3	23.6	20.4	18.0	16.1
8				96.8	85.4	75.0	63.4	37.5	29.6	25.0	21.8	19.4
9					91.8	82.3	73.1	62.5	38.2	30.6	26.2	23.0
10					97.4	88.2	79.9	71.7	61.8	38.8	31.5	27.2
11						93.3	85.4	78.0	70.4	61.2	39.3	32.3
12						97.9	90.1	83.1	76.4	69.4	60.7	39.8
13							94.3	87.5	81.2	75.0	68.5	60.2
14							98.2	91.5	85.4	79.6	73.8	67.7
15								95.1	89.1	83.5	78.2	72.8
16								98.4	92.5	87.1	82.0	77.0
17									95.6	90.3	85.4	80.6
18									98.6	93.3	88.4	83.9
19										96.1	91.3	86.8
20										98.7	94.0	89.5
21											96.5	92.1
22											98.9	94.5
23												96.8
24												98.9

relocated away from the stack walls to: (1) a distance of 1.3 cm (0.50 in.); or (2) a distance equal to the nozzle inside diameter, whichever is larger.

2.3.2 Rectangular Stacks. Determine the number of traverse points as explained in Section 2.1 and 2.2 of this method. From Table 1-1, determine the grid configuration. Divide the stack cross-section into as many equal rectangular elemental areas as traverse points, and then locate a traverse point at the centroid of each equal area according to the example in Figure 1-4.

The situation of traverse points being too close to the stack walls is not expected to arise with rectangular stacks. If this problem should ever arise, the Administrator must be contacted for resolution of the matter.

2.4 Verification of Absence of Cyclonic Flow. In most stationary sources, the direction of stack gas flow is essentially parallel to the stack walls. However, cyclonic floc may exist (1) after such devices ad cyclones and inertial demisters following venturi scrubbers, or (2) in stacks having tangential inlets or other duct configurations which tend to induce swirling; in these instances, the presence or absence of cyclonic flow at the sampling location must be determined. The following techniques are acceptable for this determination.

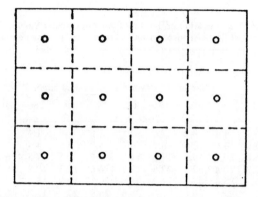

Figure 1-4.   Example showing rectangular stack cross section divided into 12 equal areas, with a traverse point at centroid of each area.

Level and zero the manometer. Connect a Type S pitot tube to the manometer. Position the Types S pitot tube at each traverse point, in succession, so that the planes of the face openings of the pitot tube are perpendicular to the stack cross-sectional plane: when the Type S pitot tube is in this position, it is as "$0°$ reference." Note the differential pressure (Δp) reading at each traverse point. If a null (zero) pitot reading is obtained at $0°$ reference at a given traverse point, an acceptable flow condition exists at that point. If the pitot reading is not zero at $0°$ reference, rotate the pitot tube (up to $\pm 90°$ yaw angle), until a null reading is obtained. Carefully determine and record the value of the rotation angle (α) to the nearest degree. After the null technique has been applied at each traverse point, calculate the average of the absolute values of α; assign α values of $0°$ to those points for which no rotation was required, and include these in the overall average. If the average value of α is greater than $10°$, the overall flow condition in the stack is unacceptable and alternative methodology, subject to the approval of the Administrator, must be used to perform accurate sample and velocity traverses.

3. *Bibliography*

1. Determining Dust Concentration in a Gas Stream. ASME. Performance Test Code No. 27. New York. 1957.

2. Devorkin, Howard, et al. Air Pollution Source Testing Manual. Air Pollution Control District. Los Angeles, CA. November 1963.

3. Methods for Determination of Velocity, Volume, Dust and Mist Content of Gases. Western Precipitation Division of Joy Manufacturing Co. Los Angeles, CA. Bulletin WP-50. 1968.

4. Standard Method for Sampling Stacks for Particulate Matter. In: 1971 Book of ASTM Standards, Part 23, ASTM Designation D-2928-71. Philadelphia, PA 1971.

5. Hanson, H. A., et al. Particulate Sampling Strategies for Large Power Plants Including Nonuniform Flow. USEPA, ORD, ESRL, Research Triangle Park, N.C. EPA-600/2-76-170. June 1976.

6. Entropy Environmentalists, Inc. Determination of the Optimum Number of Sampling Points: An Analysis of Method 1 Criteria. Environmental Protection Agency. Research Triangle Park, N.C. EPA Contract No. 68-01-3172, Task 7.

METHOD 2
DETERMINATION OF STACK GAS VELOCITY AND
VOLUMETRIC FLOW RATE (TYPE S PITOT TUBE)

1. *Principle and Applicability*

1.1 Principle. The average gas velocity in a stack is determined from the gas density and from measurement of the average velocity head with a Type S (Stausscheibe or reverse type) pitot tube.

1.2 Applicability. This method is applicable for measurement of the average velocity of a gas stream and for quantifying gas flow.

This procedure is not applicable at measurement sites which fail to meet the criteria of Method 1, Section 2.1. Also, the method cannot be used for direct measurement in cyclonic or swirling gas streams; Section 2.4 of Method 1 shows how to determine cyclonic or swirling flow conditions. When unacceptable conditions exist, alternative procedures, subject to the approval of the Administration. U.S. Environmental Protection Agency, must be employed to make accurate flow rate determinations; examples of such alternative procedures are: (1) to install straightening vanes; (2) to calculate the total volumetric flow rate stoichiometrically, or (3) to move to another measurement site at which the flow is acceptable.

2. *Apparatus*

Specifications for the apparatus are given below. Any other apparatus that has been demonstrated (subject to approval of the Administrator) to be capable of meeting the specifications will be considered acceptable.

2.1 Type S Pitot Tube. The Type S pitot tube (Figure 2-1) shall be made of metal tubing (e.g., stainless steel). It is recommended that the external tubing diameter (dimension D_t, Figure 2-2b) be between 0.48 and 0.95 centimeters (3/16 and 3/8 inch).

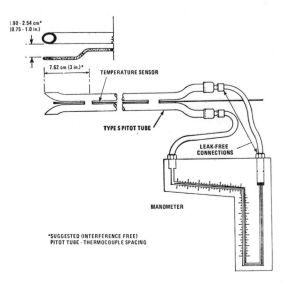

Figure 2-1. Type S pitot tube manometer assembly.

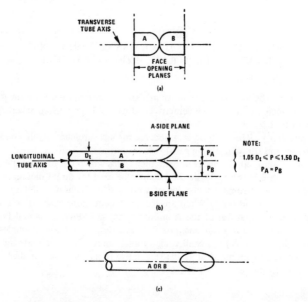

Figure 2-2. Properly constructed Type S pitot tube, shown in: (a) end view; face opening planes perpendicular to transverse axis; (b) top view; face opening planes parallel to longitudinal axis; (c) side view; both legs of equal length and centerlines coincident, when viewed from both sides. Baseline coefficient values of 0.84 may be assigned to pitot tubes constructed this way.

There shall be an equal distance from the base of each leg of the pitot tube to its face-opening plane (dimensions P_A and P_B, Figure 2-2b); it is recommended that this distance be between 1.05 and 1.50 times the external tubing diameter. The face openings of the pitot tube shall, preferably, be aligned as shown in Figure 2-2; however, slight misalignments of the openings are permissible (see Figure 2-3).

The Type S pitot tube shall have a known coefficient, determined as outlined in Section 4. An identification number shall be assigned to the pitot tube; this number shall be permanently marked or engraved on the body of the tube.

A standard pitot tube may be used instead of a Type S, provided that it meets the specifications of Sections 2.7 and 4.2; note, however, that the static and impact pressure holes of standard pitot tubes are susceptible to plugging in particulate-laden gas streams. Therefore, whenever a standard pitot tube is used to perform a traverse, adequate proof must be furnished that the openings of the pitot tube have not plugged up during the traverse period; this can be done by taking a velocity head (Δp) reading at the final traverse point, cleaning out the impact and static holes of the standard pitot tube by "back-purging" with pressurized air, and then taking another Δp reading. If the Δp readings made before and after the air purge are the same (± 5 percent), the traverse is acceptable. Otherwise, reject the run. Note that if Δp at the final traverse point is unsuitably low, another point may be selected. If "back-purging" at regular intervals is part of the procedure, then comparative Δp readings shall be taken, as above, for the last two back purges at which suitably high Δp readings are observed.

2.2 Differential Pressure Gauge. An inclined manometer or equivalent device is used. Most sampling trains are equipped with a 10-in. (water column) inclined-vertical

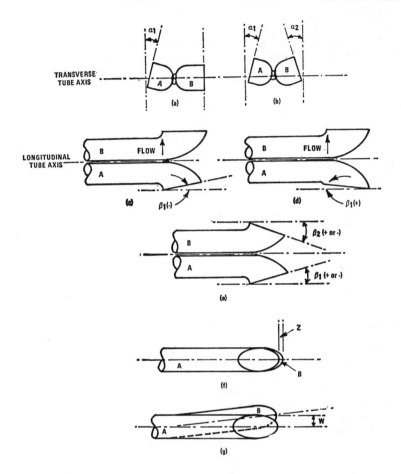

Figure 2-3. Types of face-opening misalignment that can result from field use or improper construction of Type S pitot tubes. These will not affect the baseline value of $\overline{C}p(s)$ so long as a_1 and $a_2 < 10°$, β_1 and $\beta_2 < 5°$, $z < 0.32$ cm (1/8 in.) and w < 0.08 cm (1/32 in.) (citation 11 in Section 6).

manometer, having 0.01-in. H_2O divisions on the 0- to 1-in. inclined scale, and 0.1-in. H_2O divisions on the 0- to 1-in. vertical scale. This type of manometer (or other gauge of equivalent sensitivity) is satisfactory for the measurement of Δp values as low as 1.3 mm (0.05 in.) H_2O. However, a differential pressure gauge of greater sensitivity shall be used (subject to the approval of the Administrator), if any of the following is found to be true: (1) the arithmetic average of all Δp readings at the traverse points in the stack is less than 1.3 mm (0.05 in.) H_2O; (2) for traverses of 12 or more points, more than 10 percent of the individual Δp readings are below 1.3 mm (0.05 in.) H_2O; (3) for traverses of fewer than 12 points, more than one Δp reading is below 1.3 mm (0.05 in.) H_2O. Citation 18 in Section 6 describes commercially available instrumentation for the measurement of low-range gas velocities.

As an alternative to criteria (1) through (3) above, the following calculation may be performed to determine the necessity of using a more sensitive differential pressure gauge:

$$T = \frac{\sum\limits_{i=1}^{n} \sqrt{\Delta p_i + K}}{\sum\limits_{i=1}^{n} \sqrt{\Delta p_i}}$$

where:

Δp_i = Individual velocity head reading at a traverse point, mm H_2O (in. H_2O).

n = Total number of traverse points.

K = 0.13 mm H_2O when metric units are used and 0.005 in H_2O when English units are used.

If T is greater than 1.05, the velocity head data are unacceptable and a more sensitive differential pressure gauge must be used.

NOTE—If differential pressure gauges other than inclined manometers are used (e.g., magnehelic gauges), their calibration must be checked after each test series. To check the calibration of a differential pressure gauge, compare Δp readings of the gauge with those of a gauge-oil manometer at a minimum of three points, approximately representing the range of Δp values in the stack. If, at each point, the values of Δp as read by the differential pressure gauge and gauge-oil manometer agree to within 5 percent, the differential pressure gauge shall be considered to be in proper calibration. Otherwise, the test series shall either be voided, or procedures to adjust the measured Δp values and final results shall be used, subject to the approval of the Administrator.

2.3   Temperature Gauge. A thermocouple, liquid-filled bulb thermometer, bimetallic thermometer, mercury-in-glass thermometer, or other gauge capable of measuirng temperature to within 1.5 percent of the minimum absolute stack temperature shall be used. The temperature gauge shall be attached to the pitot tube such that the sensor tip does not touch any metal; the gauge shall be in an interference-free arrangement with respect to the pitot tube face openings (see Figure 2-1 and also Figure 2-7 in Section 4). Alternate positions may be used if the pitot tube-temperature gauge system is calibrated according to the procedure of Section 4. Provided that a difference of not more than 1 percent in the average velocity measurement is introduced, the temperature gauge need not be attached to the pitot tube; this alternative is subject to the approval of the Administrator.

2.4   Pressure Probe and Gauge. A piezometer tube and mercury- or water-filled U-tube manometer capable of measuring stack pressure to within 2.5 mm (0.1 in.) Hg is used. The static tap of a standard type pitot tube or one leg of a Type S pitot tube with the face opening planes positioned parallel to the gas flow may also be used as the pressure probe.

2.5   Barometer. A mercury, aneroid, or other barometer capable of measuring atmospheric pressure to within 2.5 mm Hg (0.1 in. Hg) may be used. In many cases, the barometric reading may be obtained from a nearby national weather service station, in which case the station value (which is the absolute barometric pressure) shall be requested and an adjustment for elevation differences between the weather station and the sampling point shall be applied at a rate of minus 2.5 mm (0.1 in.) Hg per 30-meter (100 foot) elevation increase, or vice-versa, for elevation decrease.

2.6   Gas Density Determination Equipment. Method 3 equipment, if needed (see Section 3.6), to determine the stack gas dry molecular weight, and Reference Method 4 or Method 5 equipment for moisture content determination; other methods may be used subject to approval of the Administrator.

2.7   Calibration Pitot Tube. When calibration of the Type S pitot tube is necessary (see Section 4), a standard pitot tube is used as a reference. The standard pitot tube shall, preferably, have a known coefficient obtained either (1) directly from the National Bureau of Standards, Route 270, Quince Orchard Road, Gaithersburg, Maryland, or (2) by calibration against another standard pitot tube with an NBS-traceable coefficient. Alternatively, a standard pitot tube designed according to the criteria given in 2.7.1 through 2.7.5 below and illustrated in Figure 2-4 (see also Citations 7, 8, and 17 in Section 6) may be used. Pitot tubes designed according to these specifications will have baseline coefficients of about 0.99 ± 0.01.

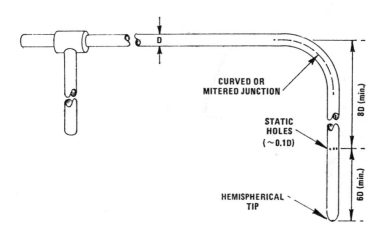

Figure 2-4. Standard pitot tube design specifications.

2.7.1   Hemispherical (shown in Figure 2-4), ellipsoidal, or conical tip.

2.7.2   A minimum of six diameters straight run (based upon D, the external diameter of the tube) between the tip and the static pressure holes.

2.7.3   A minimum of eight diameters straight run between the static pressure holes and the centerline of the external tube, following the 90 degree bend.

2.7.4   Static pressure holes of equal size (approximately 0.1 D), equally spaced in a piezometer ring configuration.

2.7.5   Ninety degree bend, with curved or mitered junction.

412 AIR POLLUTION SAMPLING & ANALYSIS DESKBOOK

2.8   Differential Pressure Gauge for Type S Pitot Tube Calibration. An inclined manometer or equivalent is used. If the single-velocity calibration technique is employed (see Section 4.1.2.3), the calibration differential pressure gauge shall be readable to the nearest 0.13 mm H_2O (0.005 in. H_2O). For multivelocity calibrations, the gauge shall be readable to the nearest 0.13 mm H_2O (0.005 in. H_2O) for Δp values between 1.3 and 25 mm H_2O (0.05 and 1.0 in. H_2O), and to the nearest 1.3 mm H_2O (0.05 in. H_2O) for Δp values above 25 mm H_2O (1.0 in. H_2O). A special, more sensitive gauge will be required to read Δp values below 1.3 mm H_2O [0.05 in. H_2O] (see Citation 18 in Section 6).

3. *Procedure*

3.1   Set up the apparatus as shown in Figure 2-1. Capillary tubing or surge tanks installed between the manometer and pitot tube may be used to dampen Δp fluctuations. It is recommended, but not required, that a pretest leak-check be conducted, as follows: (1) blow through the pitot impact opening until at least 7.6 cm (3 in.) H_2O velocity pressure registers on the manometer; then, close off the impact opening. The pressure shall remain stable for at least 15 seconds; (2) do the same for the static pressure side, except using suction to obtain the minimum of 7.6 cm (3 in.) H_2O. Other leak-check procedures, subject to the approval of the Administrator, may be used.

3.2   Level and zero the manometer. Because the manometer level and zero may drift due to vibrations and temperature changes, make periodic checks during the traverse. Record all necessary data as shown in the example data sheet (Fgirue 2-5).

3.3   Measure the velocity head and temperature at the traverse points specified by Method 1. Ensure that the proper differential pressure gauge is being used for the range of Δp values encountered (see Section 2.2). If it is necessary to change to a more sensitive gauge, do so, and remeasure the Δp and temperature readings at each traverse point. Conduct a post-test leak-check (mandatory) as described in Section 3.1 above, to validate the traverse run.

3.4   Measure the static pressure in the stack. One reading is usually adequate.

3.5   Determine the atmospheric pressure.

3.6   Determine the stack gas dry molecular weight. For combustion processes or processes that emit essentially CO_2, O_2, CO, and N_2, use Method 3. For processes emitting essentially air, an analysis need not be conducted; use a dry molecular weight of 29.0. For other processes, other methods, subject to the approval of the Administrator, must be used.

3.7   Obtain the moisture content from Reference Method 4 (or equivalent) or from Method 5.

3.8   Determine the cross-sectional area of the stack or duct at the sampling location. Whenever possible, physically measure the stack dimensions rather than using blueprints.

4. *Calibration*

4.1   Type S Pitot Tube. Before its initial use, carefully examine the Type S pitot tube in top, side, and end views to verify that the face openings of the tube are aligned within the specifications illustrated in Figure 2-2 or 2-3. The pitot tube shall not be used if it fails to meet these alignment specifications.

After verifying the face opening alignment, measure and record the following dimensions of the pitot tube: (a) the external tubing diameter (dimension D_t, Figure 2-2b); and (b) the base-to-opening plane distances (dimensions P_A and P_B, Figure 2-2b). If D_t is between 0.48 and 0.95 cm (3/16 and 3/8 in.) and if P_A and P_B are equal and between 1.05 and 1.50 R_t, there are two possible options: (1) the pitot tube may be calibrated according to the procedure outlined in Sections 4.1.2 through 4.1.5 below, or

PLANT _____

DATE _____ RUN NO. _____

STACK DIAMETER OR DIMENSIONS, m(in.) _____

BAROMETRIC PRESSURE, mm Hg (in. Hg)_____

CROSS SECTIONAL AREA, m²(ft²) _____

OPERATORS _____

PITOT TUBE I.D. NO. _____

   AVG. COEFFICIENT, C_p = _____

   LAST DATE CALIBRATED_____

SCHEMATIC OF STACK CROSS SECTION

Traverse Pt. No.	Vel. Hd., Δp mm (in.) H₂0	Stack Temperature t_s, °C (°F)	Stack Temperature T_s, °K (°R)	P_g mm Hg (in.Hg)	$\sqrt{\Delta p}$
	Average				

Figure 2-5. Velocity traverse data.

(2) a baseline (isolated tube) coefficient value of 0.84 may be assigned to the pitot tube. Note, however, that if the pitot tube is part of an assembly, calibration may still be required, despite knowledge of the baseline coefficient value (see Section 4.1.1).

If D_t, P_A, and P_B are outside the specified limits, the pitot tube must be calibrated as outlined in 4.1.2 through 4.1.5 below.

4.1.1   Type S Pitot Tube Assemblies. During sample and velocity traverses, the isolated Type S pitot tube is not always used; in many instances, the pitot tube is used in combination with other source-sampling components (thermocouple, sampling probe, nozzle) as part of an "assembly." The presence of other sampling components can sometimes affect the baseline value of the Type S pitot tube coefficient (Citation 9 in Section 6); therefore an assigned (or otherwise known) baseline coefficient value may or may not be valid for a given assembly. The baseline and assembly coefficient values will be identical only when the relative placement of the components in the assembly is such that aerodynamic interference effects are eliminated. Figures 2-6 through 2-8 illustrate interference-free component arrangements for Type S pitot tubes having external tubing diameters between 0.48 and 0.95 cm (3/16 and 3/6 in.). Type S pitot tube assemblies that fail to meet any or all of the specifications of Figures 2-6 through 2-8 shall be calibrated according to the procedure outlined in Sections 4.1.2 through 4.1.5 below, and prior to calibration, the values of the inter-component spacings (pitot-nozzle, pitot-thermocouple, pitot-probe sheath) shall be measured and recorded.

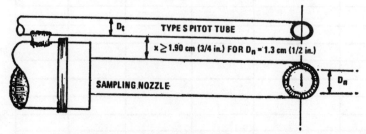

A. BOTTOM VIEW; SHOWING MINIMUM PITOT-NOZZLE SEPARATION.

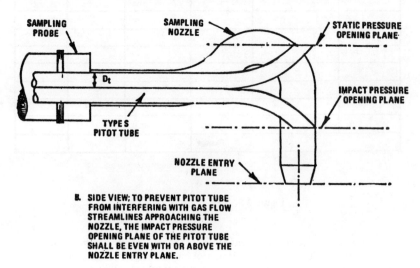

B. SIDE VIEW; TO PREVENT PITOT TUBE FROM INTERFERING WITH GAS FLOW STREAMLINES APPROACHING THE NOZZLE, THE IMPACT PRESSURE OPENING PLANE OF THE PITOT TUBE SHALL BE EVEN WITH OR ABOVE THE NOZZLE ENTRY PLANE.

Figure 2-6.   Proper pitot tube-sampling nozzle configuration to prevent aerodynamic interference; buttonhook-type nozzle; centers of nozzle and pitot opening aligned; D_t between 0.48 and 0.95 cm (3/16 and 3/8 in.).

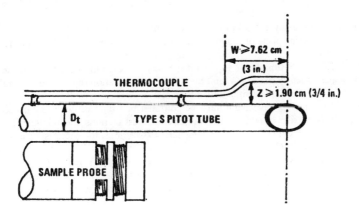

or

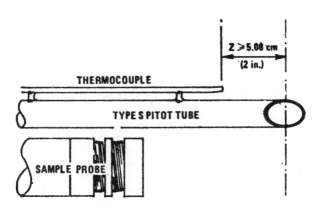

Figure 2.7. Proper thermocouple placement to prevent interference; D_t between 0.48 and 0.95 cm (3/16 and 3/8 in.).

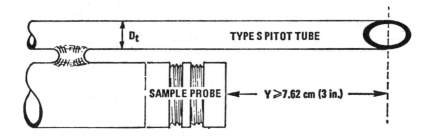

Figure 2-8. Minimum pitot-sample probe separation needed to prevent interference; D_t between 0.48 and 0.95 cm (3/16 and 3/8 in.).

NOTE–Do not use any Type S pitot tube assembly which is constructed such that the impact pressure opening plane of the pitot tube is below the entry plane of the nozzle (See Figure 2-6b).

4.1.2 Calibration Setup. If the type S pitot tube is to be calibrated, one leg of the tube shall be permanently marked A, and the other, B. Calibration shall be done in a flow system having the following essential design features:

4.1.2.1 The flowing gas stream must be confined to a duct of definite cross-sectional area, either circular or rectangular. For circular cross-sections, the minimum duct diameter shall be 30.5 cm (12 in.); for rectangular cross-sections, the width (shorter side) shall be at least 25.4 cm (10 in.).

4.1.2.2 The cross-sectional area of the calibration duct must be constant over a distance of 10 or more duct diameters. For a rectangular cross-section, use an equivalent diameter, calculated from the following equation, to determine the number of duct diameters:

$$D_t = \frac{2LW}{(L + W)}$$

Equation 2-1

where:

D_t = Equivalent diameter
L = Length
W = Width

To ensure the presence of stable, fully developed flow patterns at the calibration site, or "test section," the site must be located at least eight diameters downstream and two diameters upstream from the nearest disturbances.

NOTE–The eight- and two-diameter criteria are not absolute; other test section locations may be used (subject to approval of the Administrator), provided that the flow at the test site is stable and demonstrably parallel to the duct axis.

4.1.2.3 The flow system shall have the capacity to generate a test-section velocity around 915 m/min (3,000 ft/min). This velocity must be constant with time to guarantee steady flow during calibration. Note that Type S pitot tube coefficients obtained by single-velocity calibration at 915 m/min (3,000 ft/min) will generally be valid to within ± 3 percent for the measurement of velocities above 305 m/min (1,000 ft/min) and to within ± 5 to 6 percent for the measurement of velocities between 180 and 305 m/min (600 and 1,000 ft/min). If a more precise correlation between C_p and velocity is desired, the flow system shall have the capacity to generate at least four distinct, time-invariant test-section velocities covering the velocity range from 180 to 1,525 m/min (600 to 5,000 ft/min), and calibration data shall be taken at regular velocity intervals over this range (see Citations 9 and 14 in Section 6 for details).

4.1.2.4 Two entry ports, one each for the standard and Type S pitot tubes, shall be cut in the test section; the standard pitot entry port shall be located slightly downstream of the Type S port, so that the standard and Type S impact openings will lie in the same cross-sectional plane during calibration. To facilitate alignment of the pitot tubes during calibration, it is advisable that the test section be constructed of plexiglas or some other transparent material.

4.1.3 Calibration Procedure. Note that this procedure is a general one and must not be used without first referring to the special considerations presented in Section 4.1.5. Note also that this procedure applies only to single-velocity calibration. To obtain calibration data for the A and B sides of the Type S pitot tube, proceed as follows:

4.1.3.1   Make sure that the manometer is properly filled and that the oil is free from contamination and is of the proper density. Inspect and leak-check all pitot lines; repair or replace if necessary.

4.1.3.2   Level and zero the manometer. Turn on the fan and allow the flow to stabilize. Seal the Type S entry port.

4.1.3.3   Ensure that the manometer is level and zeroed. Position the standard pitot tube at the calibration point (determined as outlined in Section 4.1.5.1), and align the tube so that its tip is pointed directly into the flow. Particular care should be taken in aligning the tube to avoid yaw and pitch angles. Make sure that the entry port surrounding the tube is properly sealed.

4.1.3.4   Read Δp_{std} and record its value in a data table similar to the one shown in Figure 2-9. Remove the standard pitot tube from the duct and disconnect it from the manometer. Seal the standard entry port.

PITOT TUBE IDENTIFICATION NUMBER: _____ **DATE:** _____

CALIBRATED BY: _____

RUN NO.	"A" SIDE CALIBRATION			
	Δp_{std} cm H_2O (in. H_2O)	$\Delta p(s)$ cm H_2O (in. H_2O)	$C_p(s)$	DEVIATION $C_p(s) - \bar{C}_p(A)$
1				
2				
3				
	$\bar{C}_p$ (SIDE A)			

RUN NO.	"B" SIDE CALIBRATION			
	Δp_{std} cm H_2O (in. H_2O)	$\Delta p(s)$ cm H_2O (in. H_2O)	$C_p(s)$	DEVIATION $C_p(s) - \bar{C}_p(B)$
1				
2				
3				
	$\bar{C}_p$ (SIDE B)			

$$\text{AVERAGE DEVIATION} = \sigma (A \text{ OR } B) = \frac{\sum\limits_{1}^{3} \left| C_p(s) - \bar{C}_p(A \text{ OR } B) \right|}{3} \blacktriangleleft \text{MUST BE} \leqslant 0.01$$

$$\left| \bar{C}_p \text{ (SIDE A)} - \bar{C}_p \text{ (SIDE B)} \right| \blacktriangleleft \text{MUST BE} \leqslant 0.01$$

Figure 2-9. Pitot tube calibration data.

4.1.3.5   Connect the Type S pitot tube to the manometer. Open the Type S entry port. Check the manometer level and zero. Insert and align the Type S pitot tube so that its A side impact opening is at the same point as the flow. Make sure that the entry port surrounding the tube is properly sealed.

4.1.3.6   Read Δp_s and enter its value in the data table. Remove the Type S pitot tube from the duct and disconnect it from the manometer.

4.1.3.7   Repeat steps 4.1.3.5 through 4.1.3.6 above until three pairs of Δp readings have been obtained.

4.1.3.8   Repeat steps 4.1.3.3 through 4.1.3.7 above for the B side of the Type S pitot tube.

4.1.3.9   Perform calculations, as described in Section 4.1.4 below.

4.1.4   Calculations.

4.1.4.1   For each of the six pairs of Δp readings (i.e., three from side A and three from side B) obtained in Section 4.1.3 above, calculate the value of the Type S pitot tube coefficient as follows:

$$C_{p(s)} = C_{p(std)}\sqrt{\frac{\Delta p_{std}}{\Delta p_s}}$$

Equation 2-2

where:

$C_{p(s)}$ =   Type S pitot tube coefficient
$C_{p(std)}$ =   Standard pitot tube coefficient; use 0.99 if the coefficient is unknown and the tube is designed according to the criteria of Sections 2.7.1 to 2.7.5 of this method.
Δp_{std} =   Velocity head measured by the standard pitot tube, cm H_2O (in. H_2O)
Δp_s =   Velocity head measured by the Type S pitot tube, cm H_2O (in. H_2O)

4.1.4.2   Calculate $\overline{C}_p$ (side A), the mean A-side coefficient, and $\overline{C}_p$ (side B), the mean B-side coefficient; calculate the difference between those two average values.

4.1.4.3   Calculate the deviation of each of the three A-side values of $C_{p(s)}$ from $\overline{C}_p$ (side A), and the deviation of each B-side value of $C_{p(s)}$ from $\overline{C}_p$ (side B). Use the following equation:

$$\text{Deviation} = C_{p(s)} - \overline{C}_p \text{ (A or B)}$$

Equation 2-3

4.1.4.4   Calculate σ, the average deviation from the mean, for both the A and B sides of the pitot tube. Use the following equation:

$$\sigma \text{ (side A or B)} = \frac{\sum\limits_{1}^{3} |(C_{p(s)} - \overline{C}_p \text{ (A or B)}|}{3}$$

Equation 2-4

4.1.4.5   Use the Type S pitot tube only if the values of σ (side A) and σ (side B) are less than or equal to 0.01 and if the absolute value of the difference between $\overline{C}_p$ (A) and $\overline{C}_p$ (B) is 0.01 or less.

4.1.5   Special considerations.

4.1.5.1   Selection of calibration point.

4.1.5.1.1   When an isolated Type S pitot tube is calibrated, select a calibration point at or near the center of the duct, and follow the procedures outlined in Sections 4.1.3 and 4.1.4 above. The Type S pitot coefficients so obtained, i.e., $\overline{C}_p$ (side A) and $\overline{C}_p$

(side B), will be valid, so long as either: (1) the isolated pitot tube is used; or (2) the pitot tube is used with other components (nozzle, thermocouple, sample probe) in an arrangement that is free from aerodynamic interference effects (see Figures 2-6 through 2-8).

4.1.4.1.2 For Type S pitot tube-thermocouple combinations (without sample probe), select a calibration point at or near the center of the duct, and follow the procedures outlined in sections 4.1.3 and 4.1.4 above. The coefficients so obtained will be valid so long as the pitot tube-thermocouple combination is used by itself or with other components in an interference-free arrangement (Figures 2-6 through 2-8).

4.1.5.1.3 For assemblies with sample probes, the calibration point should be located at or near the center of the duct; however, insertion of a probe sheath into a small duct may cause significant cross-sectional area blockage and yield incorrect coefficient values (Citation 9 in Section 6). Therefore, to minimize the blockage effect, the calibration point may be a few inches off-center if necessary. The actual blockage effect will be negligible when the theoretical blockage, as determined by a projected area model of the probe sheath, is 2 percent or less of the duct cross-sectional area for assemblies without external sheaths (Figure 2-10a) and 3 percent or less for assemblies with external sheaths (Figure 2-10b).

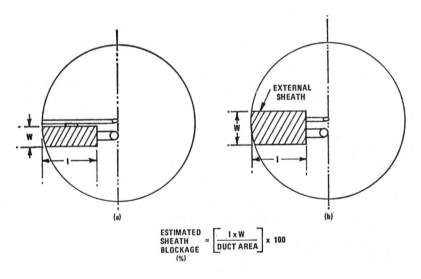

$$\text{ESTIMATED SHEATH BLOCKAGE (\%)} = \left[\frac{l \times W}{\text{DUCT AREA}}\right] \times 100$$

Figure 2-10. Projected-area models for typical pitot tube assemblies.

4.1.5.2 For those probe assemblies in which pitot tube-nozzle interference is a factor (i.e., those in which the pitot-nozzle separation distance fails to meet the specification illustrated in Figure 2-6a), the value of $C_{p(s)}$ depends upon the amount of free-space between the tube and nozzle, and therefore is a function of nozzle size. In these instances, separate calibrations shall be performed with each of the commonly used nozzle sizes in place. Note that the single-velocity calibration technique is acceptable for this purpose, even though the larger nozzle sizes (> 0.635 cm or ¼ in.) are not ordinarily used for isokinetic sampling at velocities around 915 m/min (3,000 ft/min), which is the calibration velocity; note also that it is not necessary to draw an isokinetic sample during calibration (see Citation 19 in Section 6).

4.1.5.3   For a probe assembly constructed such that its pitot tube is always used in the same orientation, only one side of the pitot tube need be calibrated (the side which will face the flow). The pitot tube must still meet the alignment specifications of Figure 2-2 or 2-3, however, and must have an average deviation (σ) value of 0.01 or less (see Section 4.1.4.4).

4.1.6   Field Use and Recalibration.

4.1.6.1   Field Use.

4.1.6.1.1   When a Type S pitot tube (isolated tube or assembly) is used in the field, the appropriate coefficient value (whether assigned or obtained by calibration) shall be used to perform velocity calculations. For calibrated Type S pitot tubes, the A side coefficient shall be used when the A side of the tube faces the flow, and the B side coefficient shall be used when the B side faces the flow; alternatively, the arithmetic average of the A and B side coefficient values may be used, irrespective of which side faces the flow.

4.1.6.1.2   When a probe assembly is used to sample a small duct (12 to 36 in. diameter), the probe sheath sometimes blocks a significant part of the duct cross-section, causing a reduction in the effective value of $\overline{C}_{p(s)}$. Consult Citation 9 in Section 6 for details. Conventional pitot-sampling probe assemblies are not recommended for use in ducts having inside diameters smaller than 12 inches (Citation 16 in Section 6).

4.1.6.2   Recalibration.

4.1.6.2.1   Isolated Pitot Tubes. After each field use, the pitot tube shall be carefully reexamined in top, side, and end views. If the pitot face openings are still aligned within the specifications illustrated in Figure 2-2 or 2-3, it can be assumed that the baseline coefficient of the pitot tube has not changed. If, however, the tube has been damaged to the extent that it no longer meets the specifications of Figure 2-2 or 2-3, the damage shall either be repaired to restore proper alignment of the face openings or the tube shall be discarded.

4.1.6.2.2   Pitot Tube Assemblies. After each field use, check the face opening alignment of the pitot tube, as in Section 4.1.6.2.1; also, remeasure the intercomponent spacings of the assembly. If the intercomponent spacings have not changed and the face opening alignment is acceptable, it can be assumed that the coefficient of the assembly has not changed. If the face opening alignment is no longer within the specifications of Figures 2-2 or 2-3, either repair the damage or replace the pitot tube (calibrating the new assembly, if necessary). If the intercomponent spacings have changed, restore the original spacings or recalibrate the assembly.

4.2   Standard pitot tube (if applicable). If a standard pitot tube is used for the velocity traverse, the tube shall be constructed according to the criteria of Section 2.7 and shall be assigned a baseline coefficient value of 0.99. If the standard pitot tube is used as part of an assembly, the tube shall be in an interference-free arrangement (subject to the approval of the Administrator).

4.3   Temperature Gauges. After each field use, calibrate dial thermometers, liquid-filled bulb thermometers, thermocouple-potentiometer systems, and other gauges at a temperature within 10 percent of the average absolute stack temperature. For temperatures up to 405°C (761°F), use an ASTM mercury-in-glass reference thermometer, or equivalent, as a reference; alternatively, either a reference thermocouple and potentiometer (calibrated by NBS) or thermometric fixed points, e.g., ice bath and boiling water (corrected for barometric pressure) may be used. For temperatures above 405°C (761°F) use an NBS-calibrated reference thermocouple-potentiometer system or an alternate reference, subject to the approval of the Administrator.

If, during calibration, the absolute temperatures measured with the gauge being calibrated and the reference gauge agree within 1.5 percent, the temperature data taken in the field shall be considered valid. Otherwise, the pollutant emission test shall either be considered invalid or adjustments (if appropriate) of the test results shall be made, subject to the approval of the Administrator.

4.4   Barometer. Calibrate the barometer used against a mercury barometer.

5. *Calculations*

Carry out calculations, retaining at least one extra decimal figure beyond that of the acquired data. Round off figures after final calculation.

5.1   Nomenclature.

A =   Cross-sectional area of stack, m^2 (ft^2).

B_{ws} =   Water vapor in the gas stream (from Method 5 or Reference Method 4), proportion by volume.

C_p =   Pitot tube coefficient, dimensionless.

K_p =   Pitot tube constant,

$$34.97 \ \frac{m}{sec} \left[\frac{\text{(g/g-mole) (mm Hg)}}{\text{(}^\circ\text{K) (mm H}_2\text{O)}} \right]^{1/2}$$

for the metric system and

$$85.49 \ \frac{ft}{sec} \left[\frac{\text{(lb/lb-mole) (in. Hg)}}{\text{(}^\circ\text{R) (in. H}_2\text{O)}} \right]^{1/2}$$

for the English system.

M_d =   Molecular weight of stack gas, dry basis (see Section 3.6) g/g-mole (lb/lb-mole).

M_s =   Molecular weight of stack gas, wet basis, g/g-mole (lb/lb-mole).

    = $M_d (1 - B_{ws}) + 18.0 \, B_{ws}$ Equation 2-5

P_{bar} =   Barometric pressure at measurement site, mm Hg (in. Hg).

P_g =   Stack static pressure, mm Hg (in. Hg).

P_s =   Absolute stack gas pressure, mm Hg (in. Hg).

    = $P_{bar} + P_g$ Equation 2-6

P_{std} =   Standard absolute pressure, 760 mm Hg (29.92 in. Hg).

Q_{sd} =   Dry volumetric stack gas flow rate corrected to standard conditions, dscm/hr (dscf/hr).

t_g =   Stack temperature, $^\circ$C ($^\circ$F).

T_s =   Absolute stack temperature, $^\circ$K ($^\circ$R).

    =   273 + t_g for metric   Equation 2-7

    =   460 + t_g for English   Equation 2-8

T_{std} =   Standard absolute temperature, 293°K (528°R)

v_s =   Average stack gas velocity, m/sec (ft/sec).

Δp =   Velocity head of stack gas, mm H$_2$O (in. H$_2$O).

3600   =   Conversion factor, sec/hr.

18.0   =   Molecular weight of water, g/g-mole (lb/lb-mole)

5.2   Average stack gas velocity.

$$v_s = K_p C_p \left(\sqrt{\Delta p} \right)_{avg} \sqrt{\frac{T_g \, (avg)}{P_s M_s}}$$

5.3   Average stack gas dry volumetric flow rate.

$$Q_{sd} = 3,600 \ (1 - B_{ws}) v_s A \ \left(\frac{T_{std}}{T_g \ (avg)} \right) \left(\frac{P_g}{P_{std}} \right)$$

Equation 2-10

6. Bibliography

1. Mark, L. S. Mechanical Engineers' Handbook. New York, McGraw-Hill Book Co., Inc. 1951.

2. Perry, J. H. Chemical Engineers' Handbook. New York. McGraw-Hill Book Co., Inc. 1960.

3. Shigehara, R. T., W. F. Todd, and W. S. Smith. Significance of Errors in Stack Sampling Measurements. U.S. Environmental Protection Agency, Research Triangle Park, N.C. (Presented at the Annual Meeting of the Air Pollution Control Association, St. Louis, Mo., June 14-19, 1970).

4. Standard Method for Sampling Stacks for Particulate Matter. In: 1971 Book of ASTM Standards, Part 23. Philadelphia, Pa. 1971. ASTM Designation D-2928-71.

5. Vennard, J. K. Elementary Fluid Mechanics. New York. John Wiley and Sons, Inc. 1947.

6. Fluid Meters—Their Theory and Application. American Society of Mechanical Engineers, New York, N.Y. 1959.

7. ASHRAE Handbook of Fundamentals. 1972. p. 208.

8. Annual Book of ASTM Standards, Part 26. 1974. p.648.

9. Vollaro, R. F. Guidelines for Type S Pitot Tube Calibration. U.S. Environmental Protection Agency, Research Triangle Park, N.C. (Presented at 1st Annual Meeting, Source Evaluation Society, Dayton, Ohio, September 18, 1975).

10. Vollaro, R. F. A Type S Pitot Tube Calibration Study. U.S. Environmental Protection Agency, Emission Measurement Branch, Research Triangle Park, N.C. July 1974.

11. Vollaro, R. F. The Effects of Impact Opening Misalignment on the Value of the Type S Pitot Tube Coefficient. U.S. Environmental Protection Agency, Emission Measurement Branch, Research Triangle Park, N.C. October 1976.

12. Vollaro, R. F. Establishment of a Baseline Coefficient Value for Properly Constructed Type S Pitot Tubes. U.S. Environmental Protection Agency, Emission Measurement Branch, Research Triangle Park, N.C. November 1976.

13. Vollaro, R. F. An Evaluation of Single-Velocity Calibration Techniques as a Means of Determining Type S Pitot Tube Coefficients. U.S. Environmental Protection Agency, Emission Measurement Branch, Research Triangle Park, N.C. August 1975.

14. Vollaro, R. F. The Use of Type S Pitot Tubes for the Measurement of Low Velocities. U.S. Environmental Protection Agency, Emission Measurement Branch, Research Triangle Park, N.C. November 1976.

15. Smith, Marvin L. Velocity Calibration of EPA Type Source Sampling Probe. United Technologies Corporation, Pratt and Whitney Aircraft Division, East Hartford, Conn. 1975.

16. Vollaro, R. F. Recommended Procedure for Sample Traverses in Ducts Smaller than 12 Inches in Diameter. U.S. Environmental Protection Agency, Emission Measurement Branch, Research Triangle Park, N.C. November 1976.

17. Ower, E., and R. C. Pankhurst. The Measurement of Air Flow, 4th Ed., London, Pergamon Press. 1966.

18. Vollaro, R. F. A Survey of Commercially Available Instrumentation for the Measurement of Low-Range Gas Velocities. U.S. Environmental Protection Agency, Emission Measurement Branch, Research Triangle Park, N.C. November 1976 (Unpublished Paper).

19. Gnyp, A. W., C. C. St. Pierre, D. S. Smith, D. Mozzon, and J. Steiner. An Experimental Investigation of the Effect of Pitot Tube-Sampling Probe Configurations on the Magnitude of the S Type Pitot Tube Coefficient for Commercially Available Source Sampling Probes. Prepared by the University of Windsor for the Ministry of the Environment, Toronto, Canada. February 1975.

METHOD 3
GAS ANALYSIS FOR CARBON DIOXIDE, OXYGEN, EXCESS AIR, AND DRY MOLECULAR WEIGHT

1. *Principle and Applicability*

1.1   Principle. A gas sample is extracted from a stack, by one of the following methods: (1) single-point, grab sampling; (2) single-point, integrated sampling; or (3) multi-point, integrated sampling. The gas sample is analyzed for percent carbon dioxide (CO_2), percent oxygen (O_2), and, if necessary, percent carbon monoxide (CO). If a dry molecular weight determination is to be made, either an Orsat or a Fyrite[1] analyzer may be used for the analysis; for excess air or emission rate correction factor determination, an Orsat analyzer must be used.

1.2   Applicability. This method is applicable for determining CO_2 and O_2 concentrations, excess air, and dry molecular weight of a sample from a gas stream of a fossil-fuel combustion process. The method may also be applicable to other processes where it has been determined that compounds other than CO_2, O_2, CO, and nitrogen (N_2) are not present in concentrations sufficient to affect the results.

Other methods, as well as modifications to the procedure described herein, are also applicable for some or all of the above determinations. Examples of specific methods and modifications include: (1) a multi-point sampling method using an Orsat analyzer to analyze individual grab samples obtained at each point; (2) a method using CO_2 or O_2 and stoichiometric calculations to determine dry molecular weight and excess air; (3) assigning a value of 30.0 for dry molecular weight, in lieu of actual measurements, for processes burning natural gas, coal, or oil. These methods and modifications may be used, but are subject to the approval of the Administrator.

2. *Apparatus*

As an alternative to the sampling apparatus and systems described herein, other sampling systems (e.g., liquid displacement) may be used provided such systems are capable of obtaining a representative sample and maintaining a constant sampling rate, and are otherwise capable of yielding acceptable results. Use of such systems is subject to the approval of the Administrator.

2.1   Grab Sampling (Figure 3-1).

2.1.1   Probe. The probe should be made of stainless steel or borosilicate glass tubing and should be equipped with an in-stack or out-stack filter to remove particulate matter (a plug of glass wool is satisfactory for this purpose). Any other material inert to O_2, CO_2, CO and N_2 and resistant to temperature at sampling conditions may be used for the probe; examples of such material are aluminum, copper, quartz glass and Teflon.

---

[1] Mention of trade names or specific products does not constitute endorsement by the Environmental Protection Agency.

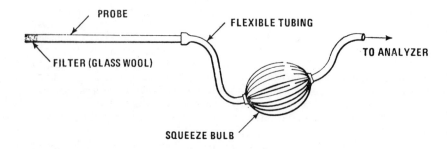

Figure 3.1. Grab-sampling train.

2.1.2 Pump. A one-way squeeze bulb, or equivalent, is used to transport the gas sample to the analyzer.

2.2   Integrated Sampling (Figure 3-2).

2.2.1 Probe. A probe such as that described in Section 2.1.1 is suitable.

2.2.2 Condenser. An air-cooled or water-cooled condenser, or other condenser that will not remove O_2, CO_2, CO and N_2 may be used to remove excess moisture which would interfere with the operation of the pump and flow meter.

2.2.3 Valve. A needle valve is used to adjust sample gas flow rate.

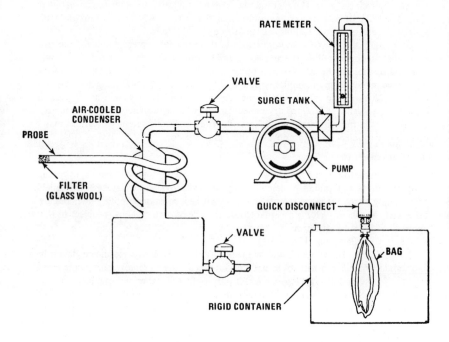

Figure 3.2. Integrated gas-sampling train.

2.2.4   Pump. A leak-free, diaphragm-type pump, or equivalent, is used to transport sample gas to the flexible bag. Install a small surge tank between the pump and rate meter to eliminate the pulsation effect of the diaphragm pump on the rotameter.

2.2.5   Rate Meter. The rotameter, or equivalent rate meter, used should be capable of measuring flow rate to within ± 2 percent of the selected flow rate. A flow rate range of 500 to 1000 cm^3/min is suggested.

2.2.6   Flexible Bag. Any leak-free plastic (e.g., Tedlar, Mylar, Teflon) or plastic-coated aluminum (e.g., aluminized Mylar) bag, or equivalent, having a capacity consistent with the selected flow rate and time length of the test run, may be used. A capacity in the range of 55 to 90 liters is suggested.

To leak-check the bag, connect it to a water manometer and pressurize the bag to 5 to 10 cm H_2O (2 to 4 in. H_2O). Allow to stand for 10 minutes. Any displacement in the water manometer indicates a leak. An alternative leak-check method is to pressurize the bag to 5 to 10 cm H_2O (2 to 4 in. H_2O) and allow to stand overnight. A deflated bag indicates a leak.

2.2.7   Pressure Gauge. A water-filled U-tube manometer, or equivalent, of about 28 cm (12 in.) is used for the flexible bag leak-check.

2.2.8   Vacuum Gauge. A mercury manometer, or equivalent, of at least 760 mm Hg (20 in. Hg) is used for the sampling train leak-check.

2.3   Analysis. For Orsat and Fyrite analyzer maintenance and operation procedures, follow the instructions recommended by the manufacturer, unless otherwise specified herein.

2.3.1   Dry Molecular Weight Determination. An Orsat analyzer or Fyrite type combustion gas analyzer may be used.

2.3.2   Emission Rate Correction Factor or Excess Air Determination. An Orsat analyzer must be used. For low CO_2 (less than 4.0 percent) or high O_2 (greater than 15.0 percent) concentrations, the measuring burette of the Orsat must have at least 0.1 percent subdivisions.

3.   *Dry Molecular Weight Determination*

Any of the three sampling and analytical procedures described below may be used for determining the dry molecular weight.

3.1   Single-Point, Grab Sampling and Analytical Procedure.

3.1.1   The sampling point in the duct shall either be at the centroid of the cross section or at a point no closer to the walls than 1.00 m (3.3 ft), unless otherwise specified by the Administrator.

3.1.2   Set up the equipment as shown in Figure 3-1, making sure all connections ahead of the analyzer are tight and leak-free. If an Orsat analyzer is used, it is recommended that the analyzer be leak-checked by following the procedure in Section 5; however, the leak-check is optional.

3.1.3   Place the probe in the stack, with the tip of the probe positioned at the sampling point; purge the sampling line. Draw a sample into the analyzer and immediately analyze it for percent CO_2 and percent O_2. Determine the percentage of the gas that is N_2 and CO by subtracting the sum of the percent CO_2 and percent O_2 from 100 percent. Calculate the dry molecular weight as indicated in Section 6.3.

3.1.4   Repeat the sampling, analysis, and calculation procedures, until the dry molecular weights of any three grab samples differ from their mean by no more than 0.3 g/g-mole (0.3 lb/lb-mole). Average these three molecular weights, and report the results to the nearest 0.1 g/g-mole (lb/lb-mole).

3.2.   Single-Point, Integrated Sampling and Analytical Procedure.

3.2.1   The sampling point in the duct shall be located as specified in Section 3.1.1.

3.2.2   Leak-check (optional) the flexible bag as in Section 2.2.6. Set up the equipment as shown in Figure 3-2. Just prior to sampling, leak-check (optional) the train by placing a vacuum gauge at the condenser inlet, pulling a vacuum of at least 250 mm Hg (10 in. Hg), plugging the outlet at the quick disconnect, and then turning off the pump. The vacuum should remain stable for at least 0.5 minute. Evacuate the flexible bag. Connect the probe and place it in the stack, with the tip of the probe positioned at the sampling point; purge the sampling line. Next, connect the bag and make sure that all connections are tight and leak free.

3.2.3   Sample at a constant rate. The sampling run should be simultaneous with, and for the same total length of time as, the pollutant emission rate determination. Collection of at least 30 liters (1.00 ft^3) of sample gas is recommended; however, smaller volumes may be collected, if desired.

3.2.4   Obtain one integrated flue gas sample during each pollutant emission rate determination. Within 8 hours after the sample is taken, analyze it for percent CO_2 and percent O_2 using either an Orsat analyzer or a Fyrite-type combustion gas analyzer. If an Orsat analyzer is used, it is recommended that the Orsat leak-check described in Section 5 be performed before this determination; however, the check is optional. Determine the percentage of the gas that is N_2 and CO by subtracting the sum of the percent CO and percent O_2 from 100 percent. Calculate the dry molecular weight as indicated in Section 6.3.

3.2.5   Repeat the analysis and calculation procedures until the individual dry molecular weights for any three analyses differ from their mean by no more than 0.3 g/g-mole (0.3 lb/lb-mole). Average these three molecular weights, and report the results to the nearest 0.1 g/g-mole (0.1 lb/lb-mole).

3.3   Multi-Point, Integrated Sampling and Analytical Procedure.

3.3.1   Unless otherwise specified by the Administrator, a minimum of eight traverse points shall be used for circular stacks having diameters less than 0.61 m (24 in.), a minimum of nine shall be used for rectangular stacks having equivalent diameters less than 0.61 m (24 in.) and a minimum of twelve traverse points shall be used for all other cases. The traverse points shall be located according to Method 1. The use of fewer points is subject to the approval of the Administrator.

3.3.2   Follow the procedures outlined in Section 3.2.2 through 3.2.5, except for the following: traverse all sampling points and sample at each point for an equal length of time. Record sampling data as shown in Figure 3-3.

4.   *Emission Rate Correction Factor or Excess Air Determination*

NOTE–A Fyrite-type combustion gas analyzer is not acceptable for excess air or emission rate correction factor determination, unless approved by the Administrator. If both percent CO_2 and percent O_2 are measured, the analytical results of any of the three procedures given below may also be used for calculating the dry molecular weight.

Each of the three procedures below shall be used only when specified in an applicable subpart of the standards. The use of these procedures for other purposes must have specific prior approval of the Administrator.

4.1   Single-Point, Grab Sampling and Analytical Procedure.

4.1.1   The sampling point in the duct shall either be at the centroid of the cross-section or at a point no closer to the walls than 1.00 m (3.3 ft), unless otherwise specified by the Administrator.

4.1.2   Set up the equipment as shown in Figure 3-1, making sure all connections ahead of the analyzer are tight and leak-free. Leak-check the Orsat analyzer according to the procedure described in Section 5. This leak-check is mandatory.

TIME	TRAVERSE PT.	Q 1pm	% DEV.[a]
	AVERAGE		

$$^{a}\% \, DEV = \left(\frac{Q - Q \, avg}{Q \, avg} \right) 100 \quad (MUST \, BE \leqslant 10\%)$$

Figure 3.3. Sampling rate data.

4.1.3   Place the probe in the stack, with the tip of the probe positioned at the sampling point; purge the sampling line. Draw a sample into the analyzer. For emission rate correction factor determination, immediately analyze the sample, as outlined in Sections 4.1.4 and 4.1.5, for percent CO_2 or percent O_2. If excess air is desired, proceed as follows: (1) immediately analyze the sample, as in Sections 4.1.4 and 4.1.5, for percent CO_2, O_2 and CO; (2) determine the percentage of the gas that is N_2 by subtracting the sum of the percent CO_2, percent O_2 and percent CO from 100 percent; and (3) calculate percent excess air as outlined in Section 6.2.

4.1.4   To ensure complete absorption of the CO_2, O_2, or if applicable, CO, make repeated passes through each absorbing solution until two consecutive readings are the same. Several passes (three or four) should be made between readings. (If constant readings cannot be obtained after three consecutive readings, replace the absorbing solution.)

4.1.5   After the analysis is completed, leak-check (mandatory) the Orsat analyzer once again, as described in Section 5. For the results of the analysis to be valid, the Orsat analyzer must pass this leak test before and after the analysis. NOTE—Since this single-point, grab sampling and analytical procedure is normally conducted in conjunction with a single-point, grab sampling and analytical procedure for a pollutant, only one analysis is ordinarily conducted. Therefore, great care must be taken to obtain a valid sample and analysis. Although in most cases only CO_2 or O_2 is required, it is recommended that both CO_2 and O_2 be measured, and that Citation 5 in the Bibliography be used to validate the analytical data.

4.2   Single-Point, Integrated Sampling and Analytical Procedure.

4.2.1   The sampling point in the duct shall be located as specified in Section 4.1.1.

4.2.2   Leak-check (mandatory) the flexible bag as in Section 2.2.6. Set up the equipment as shown in Figure 3-2. Just prior to sampling, leak-check (mandatory) the train by placing a vacuum gauge at the condenser inlet, pulling a vacuum of at least 250 mm

Hg (10 in. Hg), plugging the outlet at the quick disconnect, and then turning off the pump. The vacuum shall remain stable for at least 0.5 minute. Evacuate the flexible bag, Connect the probe and place it in the stack, with the tip of the probe positioned at the sampling point; purge the sampling line. Next, connect the bag and make sure that all connections are tight and leak free.

4.2.3 Sample at a constant rate, or as specified by the Administrator. The sampling run must be simultaneous with, and for the same total length of time as, the pollutant emission rate determination. Collect at least 30 liters (1.0 ft^3) of sample gas. Smaller volumes may be collected, subject to approval of the Administrator.

4.2.4 Obtain one integrated flue gas sample during each pollutant emission rate determination. For emission rate correction factor determination, analyze the sample within 4 hours after it is taken for percent CO_2 or percent O_2 (as outlined in Sections 4.2.5 through 4.2.7). The Orsat analyzer must be leak-checked (see Section 5) before the analysis. If excess air is desired, proceed as follows: (1) within 4 hours after the sample is taken, analyze it (as in Sections 4.2.5 through 4.2.7) for percent CO_2, O_2 and CO; (2) determine the percentage of the gas that is N_2 by subtracting the sum of the percent CO_2, percent O_2 and percent CO from 100 percent; (3) calculate percent excess air, as outlined in Section 6.2.

4.2.5 To ensure complete absorption of the CO_2, O_2, or if applicable, CO, make repeated passes through each absorbing solution until two consecutive readings are the same. Several passes (three or four) should be made between readings. (If constant readings cannot be obtained after three consecutive readings, replace the absorbing solution.)

4.2.6 Repeat the analysis until the following criteria are met:

4.2.6.1 For percent CO_2, repeat the analytical procedure until the results of any three analyses differ by no more than (a) 0.3 percent by volume when CO_2 is greater than 4.0 percent or (b) 0.2 percent by volume when CO_2 is less than or equal to 4.0 percent. Average the three acceptable values of percent CO_2 and report the results to the nearest 0.1 percent.

4.2.6.2 For percent O_2, repeat the analytical procedure until the results of any three analyses differ by no more than (1) 0.3 percent by volume when O_2 is less than 15.0 percent or (b) 0.2 percent by volume when O_2 is greater than 15.0 percent. Average the three acceptable values of percent O_2 and report the results to the nearest 0.1 percent.

4.2.6.3 For percent CO, repeat the analytical procedure until the results of any three anlayses differ by no more than 0.3 percent. Average the three acceptable values of percent CO and report the results to the nearest 0.1 percent.

4.2.7 After the analysis is completed, leak-check (mandatory) the Orsat analyzer once again, as described in Section 5. For the results of the analysis to be valid, the Orsat analyzer must pass this leak test before and after the analysis. Note: Although in most instances only CO_2 or O_2 is required, it is recommended that both CO_2 and O_2 be measured, and that Citation 5 in the Bibliography be used to validate the analytical data.

4.3 Multi-Point, Integrated Sampling and Analytical Procedure.

4.3.1 Both the minimum number of sampling points and the sampling point location shall be as specified in Section 3.3.1 of this method. The use of fewer points than specified is subject to the approval of the Administrator.

4.3.2 Follow the procedures outlined in Sections 4.2.2 through 4.2.7, except for the following: Traverse all sampling points and sample at each point for an equal length of time. Record sampling data as shown in Figure 3-3.

5. *Leak-Check Procedure for Orsat Analyzers*

Moving an Orsat analyzer frequently causes it to leak. Therefore, an Orsat analyzer should be thoroughly leak-checked on site before the flue gas sample is introduced into it. The procedure for leak-checking an Orsat analyzer is:

5.1.1   Bring the liquid level in each pipette up to the reference mark on the capillary tubing and then close the pipette stopcock.

5.1.2   Raise the leveling blub sufficiently to bring the confining liquid meniscus onto the graduated portion of the burette and then close the manifold stopcock.

5.1.3   Record the meniscus position.

5.1.4   Observe the meniscus in the burette and the liquid level in the pipette for movement over the next 4 minutes.

5.1.5   For the Orsat anlayzer to pass the-leak-check, two conditions must be met.

5.1.5.1   The liquid level in each pipette must not fall below the bottom of the capillary tubing during this 4-minute interval.

5.1.5.2   The meniscus in the burette must not change by more than 0.2 ml during this 4-minute interval.

5.1.6   If the analyzer fails the leak-check procedure, all rubber connections and stopcocks should be checked until the cause of the leak is identified. Leaking stopcocks must be disassembled, cleaned, and regreased. Leaking rubber connections must be replaced. After the analyzer is reassembled, the leak-check procedure must be repeated.

6. *Calculations*

6.1   Nomenclature.

M_d = Dry molecular weight, g/g-mole (lb/lb-mole)

% EA = Percent excess air.

$\%CO_2$ = Percent CO_2 by volume (dry basis).

$\%O_2$ = Percent O_2 by volume (dry basis).

$\%CO$ = Percent CO by volume (dry basis).

$\%N_2$ = Percent N_2 by volume (dry basis).

0.264 = Ratio of O_2 to N_2 in air, v/v.

0.280 = Molecular weight of N_2 or CO, divided by 100.

0.320 = Molecular weight of O_2 divided by 100.

0.440 = Molecular weight of CO_2 divided by 100.

6.2   Percent Excess Air. Calculate the percent excess air (if applicable), by substituting the appropriate values of percent O_2, CO and N_2 (obtained from Section 4.1.3 or 4.2.4) into Equation 3-1.

$$\%EA = \left[\frac{\%O_2 - 0.5\ \%CO}{0.264\ \%N_2\ (\%O_2 - 0.5\ \%CO)} \right] 100$$

Equation 3-1

NOTE—The equation above assumes that ambient air is used as the source of O_2 and that the fuel does not contain appreciable amounts of N_2 (as do coke oven or blast furnace gases). For those cases when appreciable amounts of N_2 are present (coal, oil, and natural gas do not contain appreciable amounts of N_2) or when oxygen enrichment is used, alternate methods, subject to approval of the Administrator, are required.

6.3   Dry Molecular Weight. Use Equation 3-2 to calculate the dry molecular weight of the stack gas

$$M_d = 0.440\ (\%CO_2) + 0.320\ (\%O_2) + 0.280\ (\%N_2 + \%CO)$$

Equation 3-2

NOTE–The above equation does not consider argon in air (about 0.9 percent, molecular weight of 37.7). A negative error of about 0.4 percent is introduced. The tester may opt to include argon in the analysis using procedures subject to approval of the Administrator.

7. *Bibliography*

1. Altshuller, A. P. Storage of Gases and Vapors in Plastic Bags. International Journal of Air and Water Pollution. *6*:75-81-1963.
2. Conner, William D. and J. S. Nader. Air Sampling Plastic Bags. Journal of the American Industrial Hygiene Association. *25*:291-297. 1964.
3. Burrell Manual for Gas Analysts, Seventh edition. Burrell Corporation, 2223 Fifth Avenue, Pittsburgh, Pa. 15219. 1951.
4. Mitchell, W. J. and M. R. Midgett. Field Reliability of the Orsat Analyzer. Journal of Air Pollution Control Association *26*:491-495. May 1976.
5. Shigehara, R. T., R. M. Neulieht, and W. S. Smith. Validating Orsat Analysis Data from Fossil Fuel-Fired Units. Stack Sampling News. *4*(2):21-26. August, 1976.

METHOD 4
DETERMINATION OF MOISTURE CONTENT IN STACK GASES

1. *Principle and Applicability*

1.1   Principle. A gas sample is extracted at a constant rate from the source; moisture is removed from the sample stream and determined either volumetrically or gravimetrically.

1.2   Applicability. This method is applicable for determining the moisture content of stack gas.

Two procedures are given. The first is a reference method, for accurate determinations of moisture content (such as are needed to calculate emission data). The second is an approximation method, which provides estimates of percent moisture to aid in setting isokinetic sampling rates prior to a pollutant emission measurement run. The approximation method described herein is only a suggested approach; alternative means for approximating the moisture content, e.g., drying tubes, wet bulb-dry bulb techniques, condensation techniques, stoichiometric calculations, previous experience, etc., are also acceptable.

The reference method is often conducted simultaneously with a pollutant emission measurement run; when it is, calculation of percent isokinetic, pollutant emission rate, etc., for the run shall be based upon the results of the reference method of irs equivalent; these calculations shall not be based upon the results of the approximation method, unless the approximation method is shown, to the satisfaction of the Administrator, U.S. Environmental Protection Agency, to be capable of yielding results within 1 percent H_2O of the reference method.

NOTE–The reference method may yield questionable results when applied to saturated gas streams or to streams that contain water droplets. Therefore, when these conditions exist or are suspected, a second determination of the moisture content shall be made simultaneously with the reference method, as follows: Assume that the gas stream is saturated. Attach a temperature sensor [capable of measuring to $\pm 1°C$ $(2°F)$] to the reference method probe. Measure the stack gas temperature at each traverse point (see Section 2.2.1) during the reference method traverse; calculate the average stack gas temperature. Next, determine the moisture percentage, either by: (1) using a psychrometric chart and making appropriate corrections if stack pressure is different from that of the chart, or (2) using saturation vapor pressure tables. In cases where the psychrometric chart or the saturation vapor pressure tables are not applicable (based on evaluation of the process), alternate methods, subject to the approval of the Administrator, shall be used.

2. Reference Method

The procedure described in Method 5 for determining moisture content is acceptable as a reference method.

2.1   Apparatus. A schematic of the sampling train used in this reference method is shown in Figure 4-1. All components shall be maintained and calibrated according to the procedure outlined in Method 5.

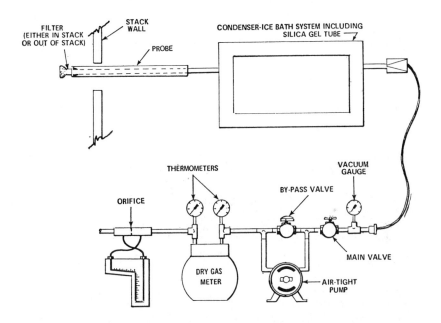

Figure 4-1. Moisture sampling train—reference method.

2.1.1 Probe. The probe is constructed of stainless steel or glass tubing, sufficiently heated to prevent water condensation, and is equipped with a filter, either in-stack (e.g., a plug of glass wool inserted into the end of the probe) or heated out-stack (e.g., as described in Method 5), to remove particulate matter.

When stack conditions permit, other metals or plastic tubing may be used for the probe, subject to the approval of the Administrator.

2.1.2 Condenser. The condenser consists of four impingers connected in series with ground glass, leak-free fittings or any similarly leak-free non-contaminating fittings. The first, third, and fourth impingers shall be of the Greenburg-Smith design, modified by replacing the tip with a 1.3 centimeter (1/2 inch) ID glass tube extending to about 1.3 cm (1/2 in.) from the bottom of the flask. The second impinger shall be of the Greenburg-Smith design with the standard tip. Modifications (e.g., using flexible connections between the impingers, using materials other than glass, or using flexible vacuum lines to connect the filter holder to the condenser) may be used, subject to the approval of the Administrator.

The first two impingers shall contain known volumes of water, the third shall be empty, and the fourth shall contain a known weight of 6- to 16-mesh indicating type silica gel, or equivalent desiccant. If the silica gel has been previously used, dry at $175°C$ ($350°F$) for 2 hours. New silica gel may be used as received. A thermometer, capable of measuring temperature to within $1°C$ ($2°F$) shall be placed at the outlet of the fourth impinger, for monitoring purposes.

Alternatively, any system may be used (subject to the approval of the Administrator) that cools the sample gas stream and allows measurement of both the water that has been condensed and the moisture leaving the condenser, each to within 1 ml or 1 g. Acceptable means are to measure the condensed water, either gravimetrically or volumetrically, and to measure the moisture leaving the condenser by: (1) monitoring the temperature and pressure at the exit of the condenser and using Dalton's law of partial pressures, or (2) passing the sample gas stream through a tared silica gel (or equivalent desiccant) trap, with exit gases kept below $20°C$ ($68°F$), and determining the weight gain.

If means other than silica gel are used to determine the amount of moisture leaving the condenser, it is recommended that silica gel (or equivalent) still be used between the condenser system and pump, to prevent moisture condensation in the pump and metering devices and to avoid the need to make corrections for moisture in the metered volume.

2.1.3 Cooling System. An ice bath container and crushed ice (or equivalent) are used to aid in condensing moisture.

2.1.4 Metering System. This system includes a vacuum gauge, leak-free pump, thermometers capable of measuring temperature to within $3°C$ ($5.4°F$), dry gas meter capable of measuring volume to within 2 percent, and related equipment as shown in Figure 4-1. Other metering systems, capable of maintaining a constant sampling rate and determining sample gas volume, may be used, subject to the approval of the Administrator.

2.1.5 Barometer. Mercury, aneroid, or other barometer capable of measuring atmospheric pressure to within 2.5 mm Hg (0.1 in. Hg) may be used. In many cases, the barometric reading may be obtained from a nearby national weather service station, in which case the station value (which is the absolute barometric pressure) shall be requested and an adjustment for elevation differences between the weather station and the sampling point shall be applied at a rate of minus 2.5 mm Hg (0.1 in. Hg) per 30 m (100 ft) elevation increase or vice versa for elevation decrease.

2.1.6 Graduated Cylinder and/or Balance. These items are used to measure condensed water and moisture caught in the silica gel to within 1 ml or 0.5 g. Graduated cylinders shall have subdivisions no greater than 2 ml. Most laboratory balances are capable of weighing to the nearest 0.5 g or less. These balances are suitable for use here.

2.2 Procedure. The following procedure is written for a condenser system (such as the impinger system described in Section 2.1.2) incorporating volumetric analysis to measure the condensed moisture, and silica gel and gravimetric analysis to measure the moisture leaving the condenser.

2.2.1 Unless otherwise specified by the Administrator, a minimum of eight traverse points shall be used for circular stacks having diameters less than 0.61 m (24 in.), a minimum of nine points shall be used for rectangular stacks having equivalent diameters less than 0.61 m (24 in.) and a minimum of twelve traverse points shall be used in all other cases. The traverse points shall be located according to Method 1. The use of fewer points is subject to the approval of the Administrator. Select a suitable probe and probe length such that all traverse points can be sampled. Consider sampling from opposite sides of the stack (four total sampling ports) for large stacks, to permit use of shorter probe lengths. Mark the probe with heat resistant tape or by some other method to denote the proper distance into the stack or duct for each sampling point. Place known volumes of water in the first two impingers. Weigh and record the weight of the silica gel to the nearest 0.5 g, and transfer the silica gel to the fourth impinger; alternatively, the silica gel may first be transferred to the impinger, and the weight of the silica gel plus impinger recorded.

2.2.2 Select a total sampling time such that a minimum total gas volume of 0.60 scm (21 scf) will be collected, at a rate no greater than 0.021 m³/min (0.75 cfm). When both moisture content and pollutant emission rate are to be determined, the moisture determination shall be simultaneous with, and for the same total length of time as, the pollutant emission rate run, unless otherwise specified in an applicable subpart of the standards.

2.2.3 Set up the sampling train as shown in Figure 4-1. Turn on the probe heater and (if applicable) the filter heating system to temperatures of about 120°C (248°F) to prevent water condensation ahead of the condenser; allow time for the temperatures to stabilize. Place crushed ice in the ice bath container. It is recommended, but not required, that a leak check be done as follows: Disconnect the probe from the first impinger or (if applicable) from the filter holder. Plug the inlet to the first impinger (or filter holder) and pull a 380 mm (15 in.) Hg vacuum; a lower vacuum may be used, provided that it is not exceeded during the test. A leakage rate in excess of 4 percent of the average sampling rate or 0.00057 m³/min (0.02 cfm), whichever is less, is unacceptable. Following the leak check, reconnect the probe to the sampling train.

2.2.4 During the sampling run, maintain a sampling rate within 10 percent of constant rate, or as specified by the Administrator. For each run, record the data required on the example data sheet shown in Figure 4-2. Be sure to record the dry gas meter reading at the beginning and end of each sampling time increment and whenever sampling is halted. Take other appropriate readings at each sample point, at least once during each time increment.

2.2.5 To begin sampling, position the probe tip at the first traverse point. Immediately start the pump and adjust the flow to the desired rate. Traverse the cross section, sampling at each traverse point for an equal length of time. Add more ice and, if necessary, salt to maintain a temperature of less than 20°C (68°F) at the silica gel outlet.

2.2.6 After collecting the sample, disconnect the probe from the filter holder (or from the first impinger) and conduct a leak check (mandatory) as described in Section 2.2.3. Record the leak rate. If the leakage rate exceeds the allowable rate, the tester shall either reject the test results or shall correct the sample volume as in Section 6.3 of Method 5. Next, measure the volume of the moisture condensed to the nearest ml. Determine the incease in weight of the silica gel (or silica gel plus impinger) to the nearest 0.5 g. Record this information (see example data sheet, Figure 4-3) and calculate the moisture percentage, as described in 2.3 below.

2.3 Calculations. Carry out the following calculations, retaining at least one extra decimal figure beyond that of the acquired data. Round off figures after final calculation.

2.3.1 Nomenclature.

B_{ws} = Proportion of water vapor, by volume, in the gas stream.

M_W = Molecular weight of water, 18.0 g/g-mole (18.0 lb/lb-mole).

P_m = Absolute pressure (for this method, same as barometric pressure) at the dry gas meter, mm Hg (in. Hg).

P_{std} = Standard absolute pressure, 760 mm Hg (29.92 in. Hg).

PLANT

OCATION

OPERATOR

DATE

RUN NO.

AMBIENT TEMPERATURE

BAROMETRIC PRESSURE

PROBE LENGTH m(ft)

SCHEMATIC OF STACK CROSS SECTION

TRAVERSE POINT NUMBER	SAMPLING TIME (θ), min.	STACK TEMPERATURE °C (°F)	PRESSURE DIFFERENTIAL ACROSS ORIFICE METER (ΔH), mm(in.) H_2O	METER READING GAS SAMPLE VOLUME m^3 (ft3)	ΔV_m m^3 (ft3)	GAS SAMPLE TEMPERATURE AT DRY GAS METER		TEMPERATURE OF GAS LEAVING CONDENSER OR LAST IMPINGER, °C (°F)
						INLET (Tm_{in}), °C (°F)	OUTLET (Tm_{out}), °C (°F)	
TOTAL						Avg.	Avg.	
AVERAGE							Avg.	Avg.

Figure 4-2. Field moisture determination—reference method.

	IMPINGER VOLUME, ml	SILICA GEL WEIGHT, g
FINAL		
INITIAL		
DIFFERENCE		

Figure 4-3. Analytical data—reference method.

R = Ideal gas constant, 0.06236 (mm Hg) (m^3)/(g-mole) $(^\circ K)$ for metric units and 21.85 (in. Hg) (ft^3)/(lb-mole) $(^\circ R)$ for English units.

T_m = Absolute temperature at meter, $^\circ K$ $(^\circ R)$.

T_{std} = Standard absolute temperature, $293^\circ K$ $(528^\circ R)$.

V_m = Dry gas volume measured by dry gas meter, dcm (dcf).

ΔV_m = Incremental dry gas volume measured by dry gas meter at each traverse point, dcm (dcf).

$V_{m(std)}$ = Dry gas volume measured by the dry gas meter, corrected to standard conditions, scm (scf).

$V_{wc(std)}$ = Volume of water vapor condensed corrected to standard conditions, scm (scf).

$V_{wsg(std)}$ = Volume of water vapor collected in silica gel corrected to standard conditions, scm (scf).

V_f = Final volume of condenser water, ml.

V_i = Initial volume, if any, of condenser water, ml.

W_f = Final weight of silica gel or silica gel plus impinger, g.

W_i = Initial weight of silica gel or silica gel plus impinger, g.

Y = Dry gas meter calibration factor.

P_w = Density of water, 0.9982 g/ml (0.002201 lb/ml).

2.3.2 Volume of water vapor condensed.

$$V_{wc(std)} = \frac{(V_f - V_i) P_w R T_{std}}{P_{std} M_w}$$

$$= K_1 (V_f - V_i)$$

Equation 4-1

where:

K_1 = 0.001333 m^3/ml for metric units
= 0.04707 ft^3/ml for English units

2.3.3 Volume of water vapor collected in silica gel.

$$V_{wsg(std)} = \frac{(W_f - W_i) R T_{std}}{P_{std} M_w}$$

$$= K_2 (W_f - W_i)$$

Equation 4-2

where:

K_2 = 0.001335 m^3/g for metric units
= 0.04715 ft^3/g for English units

2.3.4   Sample gas volume.

$$V_{m(std)} = V_m Y \frac{(P_m)(T_{std})}{(P_{std})(T_m)}$$

$$= K_3 Y \frac{V_m P_m}{T_m}$$

Equation 4-3

where:

K_3 = $0.3858\,°K/mm$ Hg for metric units
    = $17.64\,°R/in.$ Hg for English units

NOTE–If the post-test leak rate (Section 2.2.6) exceeds the allowable rate, correct the value of V_m in Equation 4-3, as described in Section 6.3 of Method 5.

2.3.5   Moisture Content.

$$B_{ws} = \frac{V_{wc(std)} + V_{wsg(std)}}{V_{wc(std)} + V_{wsg(std)} + V_{m(std)}}$$

Equation 4-4

NOTE–In saturated or moisture droplet-laden gas streams, two calculations of the moisture content of the stack gas shall be made, one using a value based upon the saturated conditions (see Section 1.2), and another based upon the results of the impinger analysis. The lower of these two values of B_{ws} shall be considered correct.

2.3.6   Verification of constant sampling rate. For each time increment, determine the ΔV_m. Calculate the average. If the value for any time increment differs from the average by more than 10 percent, reject the results and repeat the run.

3. *Approximation Method*

The approximation method described below is presented only as a suggested method (see Section 1.2).

3.1   Apparatus.

3.1.1   Probe. Stainless steel or glass tubing, sufficiently heated to prevent water condensation and equipped with a filter (either in-stack or heated out-stack) to remove particulate matter. A plug of glass wool, inserted into the end of the probe, is a satisfactory filter.

3.1.2   Impingers. Two midget impingers, each with 30 ml capacity, or equivalent.

3.1.3   Ice Bath. Container and ice, to aid in condensing moisture in impingers.

3.1.4   Drying Tube. Tube packed with new or regenerated 6- to 16-mesh indicating-type silica gel (or equivalent desiccant), to dry the sample gas and to protect the meter and pump.

3.1.5   Valve. Needle valve, to regulate the sample gas flow rate.

3.1.6   Pump. Leak-free diaphragm type or equivalent, to pull the gas sample through the train.

3.1.7   Volume meter. Dry gas meter, sufficiently accurate to measure the sample volume within 2%, and calibrated over the range of flow rates and conditions actually encountered during sampling.

3.1.8   Rate Meter. Rotameter, to measure the flow range from 0 to 31 pm (0 to 0.11 cfm).

3.1.9   Graduated Cylinder. 25 ml.

3.1.10   Barometer. Mercury, aneroid, or other barometer, as described in Section 2.1.5 above.

3.1.11   Vacuum Gauge. At least 760 mm Hg (30 in. Hg) gauge, to be used for the sampling leak check.

3.2   Procedure.

3.2.1   Place exactly 5 ml distilled water in each impinger. Assemble the apparatus without the probe as shown in Figure 4-4. Leak check the train by placing a vacuum gauge at the inlet to the first impinger and drawing a vacuum of at least 250 mm Hg (10 in. Hg), plugging the outlet of the rotameter, and then turning off the pump. The vacuum shall remain constant for at least one minute. Carefully release the vacuum gauge before unplugging the rotameter end.

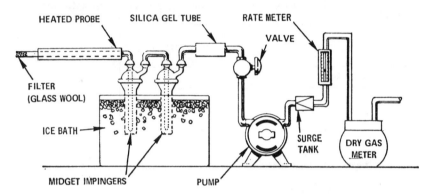

Figure 4-4. Moisture sampling train—approximation method.

3.2.2   Connect the probe, insert it into the stack, and sample at a constant rate of 2 lpm (0.071 cfm). Continue sampling until the dry gas meter registers about 30 liters (1.1 ft^3) or until visible liquid droplets are carried over from the first impinger to the second. Record temperature, pressure, and dry gas meter readings as required by Figure 4-5.

3.2.3   After collecting the sample, combine the contents of the two impingers and measure the volume to the nearest 0.5 ml.

3.3   Calculations. The calculation method presented is designed to estimate the moisture in the stack gas; therefore, other data, which are only necessary for accurate moisture determinations, are not collected. The following equations adequately estimate the moisture content, for the purpose of determining isokinetic sampling rate settings.

3.3.1   Nomenclature.

B_{wm} =   Approximate proportion, by volume, of water vapor in the gas stream leaving the second impinger, 0.025.

B_{ws} =   Water vapor in the gas stream, proportion by volume.

M_w =   Molecular weight of water, 18.0 g/g-mole (18.0 lb/lb-mole).

P_m =   Absolute pressure (for this method, same as barometric pressure) at the dry gas meter.

P_{std} =   Standard absolute pressure, 760 mm Hg (29.92 in. Hg).

R =   Ideal gas constant, 0.06236 (mm Hg) (m^3)/(g-mole) ($^\circ$K) for metric units and 21.85 (in. Hg) (ft^3)/(lb-mole) ($^\circ$R) for English units.

T_m =   Absolute temperature at meter, $^\circ$K ($^\circ$R)

LOCATION_____ COMMENTS

TEST_____

DATE_____

OPERATOR_____

BAROMETRIC PRESSURE_____

CLOCK TIME	GAS VOLUME THROUGH METER, (Vm), m^3 (ft^3)	RATE METER SETTING m^3/min. (ft^3/min.)	METER TEMPERATURE, °C (°F)

Figure 4-5. Field moisture determination–approximation method.

T_{std} = Standard absolute temperature, 293°K (528°R).
V_f = Final volume of impinger contents, ml.
V_i = Initial volume of impinger contents, ml.
V_m = Dry gas volume measured by dry gas meter, dcm (dcf).
$V_{m(std)}$ = Dry gas volume measured by dry gas meter, corrected to standard conditions, dscm (dscf).
$V_{wc(std)}$ = Volume of water vapor condensed, corrected to standard conditions, scm (scf).
ρ_w = Density of water, 0.9982 g/ml (0.002201 lb/ml).

3.3.2 Volume of water vapor collected.

$$V_{wc} = \frac{(V_f - V_i)\, \rho_w\, R\, T_{std}}{P_{std}\, M_w}$$

$$= K_1\, (V_f - V_i)$$

Equation 4-5

where:

K_1 = 0.001333 m^3/ml for metric units
= 0.04707 ft^3/ml for English units.

3.3.3 Gas volume.

$$V_{m(std)} = V_m \left(\frac{P_m}{P_{std}} \right) \left(\frac{T_{std}}{T_m} \right)$$

$$= K_2\, \frac{V_m\, P_m}{T_m}$$

Equation 4-6

where:

K_2 = 0.3858 °K/mm Hg for metric units
= 17.64 °R/in. Hg for English units.

3.3.4 Approximate moisture content.

$$B_{ws} = \frac{V_{wc}}{V_{wc} + V_{m(std)}} + B_{wm}$$

$$= \frac{V_{wc}}{V_{wc} + V_{m(std)}} + (0.025)$$

Equation 4-7

4. Calibration

4.1 For the reference method, calibrate equipment as specified in the following sections of Method 5: Section 5.3 (metering system); Section 5.5 (temperature gauges); and Section 5.7 (barometer). The recommended leak check of the metering system (Section 5.6 of Method 5) also applies to the reference method. For the approximation method, use the procedures outlined in Section 5.1.1 of Method 6 to calibrate the metering system, and the procedure of Method 5, Section 5.7 to calibrate the barometer.

5. Bibliography

1. Air Pollution Engineering Manual (Second Edition). Danielson, J. A. (ed.). U.S. Environmental Protection Agency, Office of Air Quality Planning and Standards. Research Triangle Park, N.C. Publication No. AP-40. 1973.

2. Devorkin, Howard, et al. Air Pollution Source Testing Manual. Air Pollution Control District, Los Angeles, Calif. November, 1963.

3. Methods for Determination of Velocity, Volume, Dust and Mist Content of Gases. Western Precipitation Division of Joy Manufacturing Co., Los Angeles, Calif. Bulletin WP-50. 1968.

METHOD 5
DETERMINATION OF PARTICULATE EMISSIONS FROM STATIONARY SOURCES

1. Principle and Applicability

1.1 Principle. Particulate matter is withdrawn isokinetically from the source and collected on a glass fiber filter maintained at a temperature in the range of 120 ± 14°C (248 ± 25°F) or such other temperature as specified by an applicable subpart of the standards or approved by the Administrator, U.S. Environmental Protection Agency, for a particular application. The particulate mass, which includes any material that condenses at or above the filtration temperature, is determined gravimetrically after removal of uncombined water.

1.2 Applicability. This method is applicable for the determination of particulate emissions from stationary sources.

2. Apparatus

2.1 Sampling Train. A schematic of the sampling train used in this method is shown in Figure 5-1. Complete construction details are given in APTD-0581 (Citation 2 in Section 7); commercial models for this train are also available. For changes from APTD-0581 and for allowable modifications of the train shown in Figure 5-1, see the following subsections.

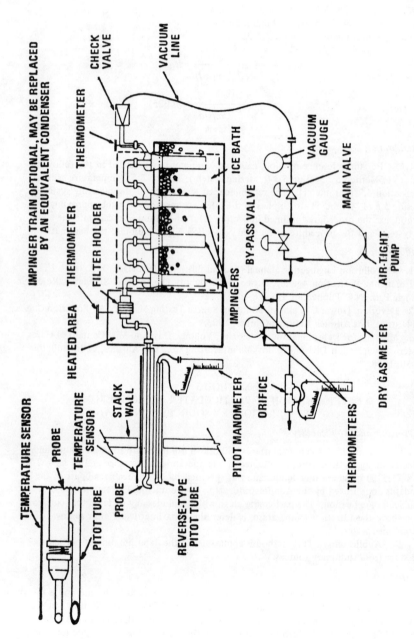

Figure 5-1. Particulate sampling train.

The operating and maintenance procedures for the sampling train are described in APTD-0576 (Citation 3 in Section 7). Since correct usage is important in obtaining valid results, all users should read APTD-0576 and adopt the operating and maintenance procedures outlined in it, unless otherwise specified herein. The sampling train consists of the following components:

2.1.1 Probe Nozzle. Stainless steel (316) or glass with sharp, tapered leading edge. The angle of taper shall be $\leqslant 30°$ and the taper shall be on the outside to preserve a constant internal diameter. The probe nozzle shall be of the button-hook or elbow design, unless otherwise specified by the Administrator. If made of stainless steel, the nozzle shall be constructed from seamless tubing; other materials of construction may be used, subject to the approval of the Administrator.

A range of nozzle sizes suitable for isokinetic sampling should be available, e.g., 0.32 to 1.27 cm (1/8 to 1/2 in.)—or larger if higher volume sampling trains are used—inside diameter (ID) nozzles in increments of 0.16 cm (1/16 in.). Each nozzle shall be calibrated according to the procedures outlined in Section 5.

2.1.2 Probe Liner. Borosilicate or quartz glass tubing with a heating system capable of maintaining a gas temperature at the exit end during sampling of $120 \pm 14°C$ ($248 \pm 25°F$), or such other temperature as specified by an applicable subpart of the standards or approved by the Administrator for a particular application. (The tester may opt to operate the equipment at a temperature lower than that specified.) Since the actual temperature at the outlet of the probe is not usually monitored during sampling, probes constructed according to APTD-0581 and utilizing the calibration curves of APTD-0576 (or calibrated according to the procedure outlined in APTD-0576) will be considered accpetable.

Either borosilicate or quartz glass probe liners may be used for stack temperatures up to about $480°C$ ($900°F$); quartz liners shall be used for temperatures between 480 and $900°C$ (900 and $1,650°F$). Both types of liners may be used at higher temperatures than specified for short periods of time, subject to the approval of the Administrator. The softening temperature for borosilicate is $820°C$ ($1,508°F$), and for quartz it is $1,500°C$ ($2,732°F$).

Whenever practical, every effort should be made to use borosilicate or quartz glass probe liners. Alternatively, metal liners (e.g., 316 stainless steel, Incoloy 825,[2] or other corrosion resistant metals) made of seamless tubing may be used, subject to the approval of the Administrator.

2.1.3 Pitot Tube. Type S, as described in Section 2.1 of Method 2, or other device approved by the Administrator. The pitot tube shall be attached to the probe (as shown in Figure 5-1) to allow constant monitoring of the stack gas velocity. The impact (high pressure) opening plane of the pitot tube shall be even with or above the nozzle entry plane (see Method 2, Figure 2-6b) during sampling. The Type S pitot tube assembly shall have a known coefficient, determined as outlined in Section 4 of Method 2.

2.1.4 Differential Pressure Gauge. Inclined manometer or equivalent devices (two), as described in Section 2.2 of Method 2. One manometer shall be used for velocity head (Δp) readings, and the other, for orifice differential pressure readings.

2.1.5 Filter Holder. Borosilicate glass, with a glass frit filter support and a silicone rubber gasket. Other materials of construction (e.g., stainless steel, Teflon, Viton) may be used, subject to approval of the Administrator. The holder design shall provide a positive seal against leakage from the outside or around the filter. The holder shall be attached immediately at the outlet of the probe (or cyclone, if used).

---

[2]Mention of trade names or specific products does not constitute endorsement by the Environmental Protection Agency.

2.1.6 Filter Heating System. Any heating system capable of maintaining a temperature around the filter holder during sampling of $120 \pm 14°C$ ($248 \pm 25°F$), or such other temperature as specified by an applicable subpart of the standards or approved by the Administrator for a particular application. Alternatively, the tester may opt to operate the equipment at a temperature lower than that specified. A temperature gauge capable of measuring temperature to within $3°C$ ($5.4°F$) shall be installed so that the temperature around the filter holder can be regulated and monitored during sampling. Heating systems other than the one shown in APTD-0581 may be used.

2.1.7 Condenser. The following system shall be used to determine the stack gas moisture content: Four impingers connected in series with leak-free ground glass fittings or any similar leak-free non-contaminating fittings. The first, third, and fourth impingers shall be of the Greenburg-Smith design, modified by replacing the tip with 1.3 cm (1/2 in.) ID glass tube extending to about 1.3 cm (1/2 in.) from the bottom of the flask. The second impinger shall be of the Greenburg-Smith design with the standard tip. Modifications (e.g., using flexible connections between the impingers, using materials other than glass, or using flexible vacuum lines to connect the filter holder to the condenser) may be used, subject to the approval of the Administrator. The first and second impingers shall contain known quantities of water (Section 4.1.3), the third shall be empty, and the fourth shall contain a known weight of silica gel, or equivalent desiccant. A thermometer, capable of measuring temperature to within $1°C$ ($2°F$) shall be placed at the outlet of the fourth impinger for monitoring purposes.

Alternatively, any system that cools the sample gas stream and allows measurement of the water condensed and moisture leaving the condenser, each to within 1 ml or 1 g may be used, subject to the approval of the Administrator. Acceptable means are to measure the condensed water either gravimetrically or volumetrically and to measure the moisture leaving the condenser by: (1) monitoring the temperature and pressure at the exit of the condenser and using Dalton's law of partial pressures; or (2) passing the sample gas stream through a tared silica gel (or equivalent desiccant) trap with exit gases kept below $20°C$ ($68°F$) and determining the weight gain.

If means other than silica gel are used to determine the amount of moisture leaving the condenser, it is recommended that silica gel (or equivalent) still be used between the condenser system and pump to prevent moisture condensation in the pump and metering devices and to avoid the need to make corrections for moisture in the metered volume.

NOTE—If a determination of the particulate matter collected in the impingers is desired in addition to moisture content, the impinger system described above shall be used, without modification. Individual States or control agencies requiring this information shall be contacted as to the sample recovery and analysis of the impinger contents.

2.1.8 Metering System. Vacuum gauge, leak-free pump, thermometers capable of measuring temperature to within $3°C$ ($5.4°F$), dry gas meter capable of measuring volume to within 2 percent, and related equipment, as shown in Figure 5-1. Other metering systems capable of maintaining sampling rates within 10 percent of isokinetic and of determining sample volumes to within 2 percent may be used, subject to the approval of the Administrator. When the metering system is used in conjunction with a pitot tube, the system shall enable checks of isokinetic rates.

Sampling trains utilizing metering systems designed for higher flow rates than that described in APTD-0581 or APTD-0576 may be used provided that the specifications of this method are met.

2.1.9 Barometer. Mercury, aneroid, or other barometer capable of measuring atmospheric pressure to within 2.5 mm Hg (0.1 in. Hg). In many cases, the barometric reading may be obtained from a nearby national weather service station, in which case the station value (which is the absolute barometric pressure) shall be requested and an adjustment for elevation differences between the weather station and sampling point shall be applied at a rate of minus 2.5 mm Hg (0.1 in. Hg) per 30 mm (100 ft) elevation increase or vice versa for elevation decrease.

    2.1.10   Gas Density Determination Equipment. Temperature sensor and pressure gauge, as described in Section 2.3 and 2.4 of Method 2, and gas analyzer, if necessary, as described in Method 3. The temperature sensor shall, preferably, be permanently attached to the pitot tube or sampling probe in a fixed configuration, such that the tip of the sensor extends beyond the leading edge of the probe sheath and does not touch any metal. Alternatively, the sensor may be attached just prior to use in the field. Note, however, that if the temperature sensor is attached in the field, the sensor must be placed in an interference-free arrangement with respect to the Type S pitot tube openings (see Method 2, Figure 2-7). As a second alternative, if a difference of not more than 1 percent in the average velocity measurement is to be introduced, the temperature gauge need not be attached to the probe or pitot tube. (The alternative is subject to the approval of the Administrator.)

    2.2   Sample Recovery. The following items are needed:

    2.2.1   Probe-Liner and Probe-Nozzle Brushes. Nylon bristle brushes with stainless steel wire handles. The probe brush shall have extensions (at least as long as the probe) of stainless steel, Nylon, Teflon, or similarly inert material. The brushes shall be properly sized and shaped to brush out the probe liner and nozzle.

    2.2.2   Wash Bottles—Two. Glass wash bottles are recommended; polyethylene wash bottles may be used at the option of the tester. It is recommended that acetone not be stored in polyethylene bottles for longer than a month.

    2.2.3   Glass Sample Storage Containers. Chemically resistant, borosilicate glass bottles, for acetone washes, 500 ml or 1000 ml. Screw cap liners shall either be rubber-backed Teflon or shall be constructed so as to be leak-free and resistant to chemical attack by acetone. (Narrow mouth glass bottles have been found to be less prone to leakage.) Alternatively, polyethylene bottles may be used.

    2.2.4   Petri Dishes. For filter samples, glass or polyethylene, unless otherwise specified by the Administrator.

    2.2.5   Graduated Cylinder and/or Balance. To measure condensed water to within 1 ml or 1 g. Graduated cylinders shall have subdivisions no greater than 2 ml. Most laboratory balances are capable of weighing to the nearest 0.5 g or less. Any of these balances is suitable for use here and in Section 2.3.4.

    2.2.6   Plastic Storage Containers. Air-tight containers to store silica gel.

    2.2.7   Funnel and Rubber Policeman. To aid in transfer of silica gel to container; not necessary if silica gel is weighed in the field.

    2.2.8   Funnel. Glass or polyethylene, to aid in sample recovery.

    2.3   Analysis. For analysis, the following equipment is needed.

    2.3.1   Glass Weighing Dishes.

    2.3.2   Desiccator.

    2.3.3   Analytical Balance. To measure to within 0.1 mg.

    2.3.4   Balance. To measure to within 0.5 g.

    2.3.5   Beakers. 250 ml.

    2.3.6   Hygrometer. To measure the relative humidity of the laboratory environment.

    2.3.7   Temperature Gauge. To measure the temperature of the laboratory environment.

3. *Reagents*

    3.1   Sampling. The reagents used in sampling are as follows:

3.1.1 Filters. Glass fiber filters, without organic binder, exhibiting at least 99.95 percent efficiency ($\leqslant 0.05$ percent pentration) on 0.3-micron dioctyl phthalate smoke particles. The filter efficiency test shall be conducted in accordance with ASTM standard method D 2986-71. Test data from the supplier's quality control program are sufficient for this purpose.

3.1.2 Silica Gel. Indicating type, 6 to 16 mesh. If previously used, dry at 175°C (350°F) for 2 hours. New silica gel may be used as received. Alternatively, other types of desiccants (equivalent or better) may be used, subject to the approval of the Administrator.

3.1.3 Water. When analysis of the material caught in the impingers is required, distilled water shall be used. Run blanks prior to field use to eliminate a high blank on test samples.

3.1.4 Crushed Ice.

3.1.5 Stopcock Grease. Acetone-insoluble, heat-stable silicone grease. This is not necessary if screw-on connectors with Teflon sleeves, or similar, are used. Alternatively, other types of stopcock grease may be used, subject to the approval of the Administrator.

3.2 Sample Recovery. Acetone—reagent grade, $\leqslant 0.001$ percent residue, in glass bottles—is required. Acetone from metal containers generally has a high residue blank and should not be used. Sometimes, suppliers transfer acetone to glass bottles from metal containers; thus, acetone blanks shall be run prior to field use and only acetone with low blank values ($\leqslant 0.001$ percent) shall be used. In no case shall a blank value of greater than 0.001 percent of the weight of acetone used be subtracted from the sample weight.

3.3 Analysis. Two reagents are required for the analysis:

3.3.1 Acetone. Same as 3.2.

3.3.2 Desiccant. Anhydrous calcium sulfate, indicating type. Alternatively, other types of desiccants may be used, subject to the approval of the Administrator.

4. *Procedure*

4.1 Sampling. The complexity of this method is such that, in order to obtain reliable results, testers should be trained and experienced with the test procedures.

4.1.1 Pretest Preparation. All the components shall be maintained and calibrated according to the procedure described in APTD-0576, unless otherwise specified herein.

Weigh several 200 to 300g portions of silica gel in air-tight containers to the nearest 0.5 g. Record the total weight of the silica gel plus container, on each container. As an alternative, the silica gel need not be preweighed, but may be weighed directly in its impinger or sampling holder just prior to train assembly.

Check filters visually against light for irregularities and flaws or pinhole leaks. Label filters of the proper diameter on the back side near the edge using numbering machine ink. As an alternative, label the shipping containers (glass or plastic petri dishes) and keep the filters in these containers at all times except during sampling and weighing.

Desiccate the filters at $20 \pm 5.6^\circ$C ($68 \pm 10^\circ$F) and ambient presure for at least 24 hours and weigh at intervals of at least 6 hours to a constant weight, i.e., <0.5 mg change from previous weighing; record results to the nearest 0.1 mg. During each weighing the filter must not be exposed to the laboratory atmosphere for a period greater than 2 minutes and a relative humidity above 50 percent. Alternatively (unless otherwise specified by the Administrator), the filters may be oven dried at 105°C (220°F) for 2 to 3 hours, desiccated for 2 hours, and weighed. Procedures other than those described, which account for relative humidity effects, may be used, subject to the approval of the Administrator.

4.1.2 Preliminary Determinations. Select the sampling site and the minimum number of sampling points according to Method 1 or as specified by the Administrator. Determine the stack pressure, temperature, and the range of velocity heads using Method 2; it is recommended that a leak-check of the pitot lines (see Method 2, Section 3.1) be performed. Determine the moisture content using Approximation Method 4 or its alternatives for the purpose of making isokinetic sampling rate settings. Determine the stack gas dry molecular weight, as described in Method 2, Section 3.6; if integrated Method 3 sampling is used for molecular weight determination, the integrated bag sample shall be taken simultaneously with, and for the same total length of time as, the particulate sample run.

Select a nozzle size based on the range of velocity heads, such that it is not necessary to change the nozzle size in order to maintain isokinetic sampling rates. During the run, do not change the nozzle size. Ensure that the proper differential pressure gauge is chosen for the range of velocity heads encountered (see Section 2.2 of Method 2).

Select a suitable probe liner and probe length such that all traverse points can be sampled. For large stacks, consider sampling from opposite sides of the stack to reduce the length of probes.

Select a total sampling time greater than or equal to the minimum total sampling time specified in the test procedures for the specific industry such that (1) the sampling time per point is not less than 2 min (or some greater time interval as specified by the Administrator), and (2) the sample volume taken (corrected to standard conditions) will exceed the required minimum total gas sample volume. The latter is based on an approximate average sampling rate.

It is recommended that the number of minutes sampled at each point be an integer or an integer plus one-half minute, in order to avoid timekeeping errors.

In some circumstances, e.g., batch cycles, it may be necessary to sample for shorter times at the traverse points and to obtain smaller gas sample volumes. In these cases, the Administrator's approval must first be obtained.

4.1.3 Preparation of Collection Train. During preparation and assembly of the sampling train, keep all openings where contamination can occur covered until just prior to assembly or until sampling is about to begin.

Place 100 ml of water in each of the first two impingers, leave the third impinger empty, and transfer approximately 200 to 300 g of preweighed silica gel from its container to the fourth impinger. More silica gel may be used, but care should be taken to ensure that it is not entrained and carried out from the impinger during sampling. Place the container in a clean place for later use in the sample recovery. Alternatively, the weight of the silica gel plus impinger may be determined to the nearest 0.5 g and recorded.

Using a tweezer or clean disposable surgical gloves, place a labeled (identified) and weighed filter in the filter holder. Be sure that the filter is properly centered and the gasket properly placed so as to prevent the sample gas stream from circumventing the filter. Check the filter for tears after assembly is completed.

When glass liners are used, install the selected nozzle using a Viton A O-ring when stack temperatures are less than 260°C (500°F) and an asbestos string gasket when temperatures are higher. See APTD-0576 for details. Other connecting systems using either 316 stainless steel or Teflon ferrules may be used. When metal liners are used, install the nozzle as above or by a leak-free direct mechanical connection. Mark the probe with heat resistant tape or by some other method to denote the proper distance into the stack or duct for each sampling points.

Set up the train as in Figure 5-1, using (if necessary) a very light coat of silicone grease on all ground glass joints, greasing only the outer portion (see APTD-0576) to avoid possibility of contamination by the silicone grease. Subject to the approval of the Administrator, a glass cyclone may be used between the probe and filter holder when the total particulate catch is expected to exceed 100 mg or when water droplets are present in the stack gas.

Place crushed ice around the impingers.

4.1.4 Leak-Check Procedures.

4.1.4.1 Pretest Leak-Check. A pretest leak-check is recommended, but not required. If the tester opts to conduct the pretest leak-check, the following procedure shall be used.

After the sampling train has been assembled, turn on and set the filter and probe heating systems at the desired operating temperatures. Allow time for the temperatures to stabilize. If a Viton A O-ring or other leak-free connection is used in assembling the probe nozzle to the probe liner, leak-check the train at the sampling site by plugging the nozzle and pulling a 380 mm Hg (15 in. Hg) vacuum.

NOTE—A lower vacuum may be used, provided that it is not exceed during the test.

If an asbestos string is used, do not connect the probe to the train during the leak-check. Instead, leak-check the train by first plugging the inlet to the filter holder (cyclone, if applicable) and pulling a 380 mm Hg (15 in. Hg) vacuum (see Note immediately above). Then connect the probe to the train and leak-check at about 25 mm Hg (1 in. Hg) vacuum; alternatively, the probe may be leak-checked with the rest of the sampling train, in one step, at 380 mm Hg (15 in. Hg) vacuum. Leakage rates in excess of 4 percent of the average sampling rate or 0.00057 m^3/min (0.02 cfm), whichever is less, are unacceptable.

The following leak-check instructions for the sampling train described in APTD-0576 and APTD-0581 may be helpful. Start the pump with bypass valve fully open and coarse adjust valve completely closed. Partially open the coarse adjust valve and slowly close the bypass valve until the desired vacuum is reached. Do not reverse direction of bypass valve; this will cause water to back up into the filter holder. If the desired vacuum is exceeded, either leak-check at this higher vacuum or end the leak check as shown below and start over.

When the leak-check is completed, first slowly remove the plug from the inlet to the probe, filter holder, or cyclone (if applicable) and immediately turn off the vacuum pump. This prevents the water in the impingers from being forced backward into the filter holder and silica gel from being entrained backward into the third impinger.

4.1.4.2 Leak-Checks During Sample Run. If, during the sampling run, a component (e.g., filter assembly or impinger) change becomes necessary, a leak-check shall be conducted immediately before the change is made. The leak-check shall be done according to the procedure outlined in Section 4.1.4.1 above, except that it shall be done at a vacuum equal to or greater than the maximum value recorded up to that point in the test. If the leakage rate is found to be no greater than 0.00057 m^3/min (0.02 cfm) or 4 percent of the average sampling rate (whichever is less), the results are acceptable, and no correction will need to be applied to the total volume of dry gas metered; if, however, a higher leakage rate is obtained, the tester shall either record the leakage rate and plan to correct the sample volume as shown in Section 6.3 of this method, or shall void the sampling run.

Immediately after component changes, leak-checks are optional; if such leak-checks are done, the procedure outlined in Section 4.1.4.1 above shall be used.

4.1.4.3 Post-Test Leak-Check. A leak-check is mandatory at the conclusion of each sampling run. The leak-check shall be done in accordance with the procedures outlined in Section 4.1.4.1, except that it shall be conducted at a vacuum equal to or greater than the maximum value reached during the sampling run. If the leakage rate is found to be no greater than 0.00057 m^3/min (0.02 cfm) or 4 percent of the average sampling rate (whichever is less), the results are acceptable, and no correction need be applied to the total volume of dry gas metered. If, however, a higher leakage rate is obtained, the tester shall either record the leakage rate and correct the sample volume as shown in Section 6.3 of this method, or shall void the sampling run.

4.1.5 Particulate Train Operation. During the sampling run, maintain an isokinetic sampling rate (within 10 percent of true isokinetic unless otherwise specified by the Administrator) and a temperature around the filter of $120 \pm 14^\circ C$ ($248 \pm 25^\circ F$), or such other temperature as specified by an applicable subpart of the standards or approved by the Administrator.

For each run, record the data required on a data sheet such as the one shown in Figure 5-2. Be sure to record the initial dry gas meter reading. Record the dry gas meter readings at the beginning and end of each sampling time increment, when changes in flow rates are made, before and after each leak check, and when sampling is halted. Take other readings required by Figure 5-2 at least once at each sample point during each time increment and additional readings when significant changes (20 percent variation in velocity head readings) necessitate additional adjustments in flow rate. Level and zero the manometer. Because the manometer level and zero may drift due to vibrations and temperature changes, make periodic checks during the traverse.

Clean the portholes prior to the test run to minimize the change of sampling deposited material. To begin sampling, remove the nozzle cap, verify that the filter and probe heating systems are up to temperature, and that the pitot tube and probe are properly positioned. Position the nozzle at the first traverse point with the tip pointing directly into the gas stream. Immediately start the pump and adjust the flow to isokinetic conditions. Nomographs are available, which aid in the rapid adjustment of the isokinetic sampling rate without excessive computations. These nomographs are designed for use when the Type S pitot tube coefficient is 0.85 ± 0.02, and the stack gas equivalent density (dry molecular weight) is equal to 29 ± 4. APTD-0576 details the procedure for using the nomographs. If C_p and M_d are outside the above stated ranges do not use the nomographs unless appropriate steps (see Citation 7 in Section 7) are taken to compensate for the deviations.

When the stack is under significant negative pressure (height of impinger stem), take care to close the coarse adjust valve before inserting the probe into the stack to prevent water from backing into the filter holder. If necessary, the pump may be turned on with the coarse adjust valve closed.

When the probe is in position, block off the openings around the probe and porthole to prevent unrepresentative dilution of the gas stream.

Traverse the stack cross-section, as required by Method 1 or as specified by the Administrator, being careful not to bump the probe nozzle into the stack walls when sampling near the walls or when removing or inserting the probe through the portholes; this minimizes the chance of extracting deposited material.

During the test run, make periodic adjustments to keep the temperature around the filter holder at the proper level; add more ice and, if necessary, salt to maintain a temperature of less than $20^\circ C$ ($68^\circ F$) at the condenser/silica gel outlet. Also, periodically check the level and zero of the manometer.

If the pressure drop across the filter becomes too high, making isokinetic sampling difficult to maintain, the filter may be replaced in the midst of a sample run. It is recommended that another complete filter assembly be used rather than attempting to change the filter itself. Before a new filter assembly is installed, conduct a leak-check (see Section 4.1.4.2). The total particulate weight shall include the summation of all filter assembly catches.

A single train shall be used for the entire sample run, except in cases where simultaneous sampling is required in two or more separate ducts or at two or more different

PLANT _____
LOCATION _____
OPERATOR _____
DATE _____
RUN NO. _____
SAMPLE BOX NO. _____
METER BOX NO. _____
METER ΔH@ _____
C FACTOR _____
PITOT TUBE COEFFICIENT, C_p _____

AMBIENT TEMPERATURE _____
BAROMETRIC PRESSURE _____
ASSUMED MOISTURE, % _____
PROBE LENGTH, m (ft) _____
NOZZLE IDENTIFICATION NO. _____
AVERAGE CALIBRATED NOZZLE DIAMETER, cm (in.) _____
PROBE HEATER SETTING _____
LEAK RATE, m^3/min, (cfm) _____
PROBE LINER MATERIAL _____
STATIC PRESSURE, mm Hg (in. Hg) _____
FILTER NO. _____

SCHEMATIC OF STACK CROSS SECTION

TRAVERSE POINT NUMBER	SAMPLING TIME (θ), min.	VACUUM mm Hg (in. Hg)	STACK TEMPERATURE (T_s) °C (°F)	VELOCITY HEAD (ΔP_s), mm(in.)H_2O	PRESSURE DIFFERENTIAL ACROSS ORIFICE METER mm H_2O (in. H_2O)	GAS SAMPLE VOLUME m^3 (ft^3)	GAS SAMPLE TEMPERATURE AT DRY GAS METER		FILTER HOLDER TEMPERATURE, °C (°F)	TEMPERATURE OF GAS LEAVING CONDENSER OR LAST IMPINGER, °C (°F)
							INLET °C (°F)	OUTLET °C (°F)		
TOTAL										
AVERAGE							Avg.	Avg.	Avg.	

Figure 5-2. Particulate field data.

locations within the same duct, or, in cases where equipment failure necessitates a change of trains. In all other situations, the use of two or more trains will be subject to the approval of the Administrator.

Note that when two or more trains are used, separate analyses of the front-half and (if applicable) impinger catches from each train shall be performed, unless identical nozzle sizes were used on all trains, in which case, the front-half catches from the individual trains may be combined (as may the impinger catches) and one analysis of front-half catch and one analysis of impinger catch may be performed. Consult with the Administrator for details concerning the calculation of results when two or more trains are used.

At the end of the sample run, turn off the coarse adjust valve, remove the probe and nozzle from the stack, turn off the pump, record the final dry gas meter reading, and conduct a post-test leak-check, as outlined in Section 4.1.4.3. Also, leak-check the pitot lines as described in Method 2, Section 3.1; the lines must pass this leak-check, in order to validate the velocity head data.

4.1.6 Calculation of Percent Isokinetic. Calculate percent isokinetic (see Calculations, Section 6) to determine whether the run was valid or another test run should be made. If there was difficulty in maintaining isokinetic rates due to source conditions, consult with the Administrator for possible variance on the isokinetic rates.

4.2 Sample Recovery. Proper cleanup procedure begins as soon as the probe is removed from the stack at the end of the sampling period. Allow the probe to cool.

When the probe can be safely handled, wipe off all external particulate matter near the tip of the probe nozzle and place a cap over it to prevent losing or gaining particulate matter. Do not cap off the probe tip tightly while the sampling train is cooling down as this would create a vacuum in the filter holder, thus drawing water from the impingers into the filter holder.

Before moving the sample train to the cleanup site, remove the probe from the sample train, wipe off the silicone grease, and cap the open outlet of the probe. Be careful not to lose any condensate that might be present. Wipe off the silicone grease from the filter inlet where the probe was fastened and cap it. Remove the umbilical cord from the last impinger and cap the impinger. If a flexible line is used between the first impinger or condenser and the filter holder, disconnect the line at the filter holder and and let any condensed water or liquid drain into the impingers or condenser. After wiping off the silicone grease, cap off the filter holder outlet and impinger inlet. Either ground-glass stoppers, plastic caps, or serum caps may be used to close these openings.

Transfer the probe and filter-impinger assembly to the cleanup area. This area should be clean and protected from the wind so that the chances of contaminating or losing the sample will be minimized.

Save a portion of the acetone used for cleanup as a blank. Take 200 ml of this acetone directly from the wash bottle being used and place it in a glass sample container labeled "acetone blank."

Inspect the train prior to and during disassembly and note any abnormal conditions. Treat the samples of follows:

Container No. 1. Carefully remove the filter from the filter holder and place it in its identified petri dish container. Use a pair of tweezers and/or clean disposable surgical gloves to handle the filter. If it is necessary to fold the filter, do so such that the particulate cake is inside the fold. Carefully transfer to the petri dish any particulate matter and/or filter fibers which adhere to the filter holder gasket, by using a dry nylon bristle brush and/or a sharp-edged blade. Seal the container.

Container No. 2. Taking care to see that dust on the outside of the probe or other exterior surfaces does not get into the sample, quantitatively recover particulate matter or any condensate from the probe nozzle, probe fitting, probe liner and front half of the filter holder by washing these components with acetone and placing the wash in a glass container. Distilled water may be used instead of acetone when approved by the Administrator and shall be used when specified by the Administrator; in these cases, save a water blank and follow the Administrator's directions on analysis. Perform the acetone rinses as follows:

Carefully remove the probe nozzle and clean the inside surface by rinsing with acetone from a wash bottle and brushing with a nylon bristle brush. Brush until the acetone rinse shows no visible particles, after which make a final rinse of the inside surface with acetone.

Brush and rinse the inside parts of the Swagelok fitting with acetone in a similar way until no visible particles remain.

Rinse the probe liner with acetone by tilting and rotating the probe while squirting acetone into its upper end so that all inside surface will be wetted with acetone. Let the acetone drain from the lower end into the sample container. A funnel (glass or polyethylene) may be used to aid in transferring liquid washes to the container. Follow the acetone rinse with a probe brush. Hold the probe in an inclined position, squirt acetone into the upper end as the probe brush is being pushed with a twisting action through the probe; hold a sample container underneath the lower end of the probe, and catch any acetone and particulate matter which is brushed from the probe. Run the brush through the probe three times or more until no visible particulate matter is carried out with the acetone or until none remains in the probe liner on visual inspection. With stainless steel or other metal probes, run the brush through in the above prescribed manner at least six times since metal probes have small crevices in which particulate matter can be entrapped. Rinse the brush with acetone, and quantitatively collect these washings in the sample container. After the brushing, make a final acetone rinse of the probe as described above.

It is recommended that two people be used to clean the probe to minimize sample losses. Between sampling runs, keep brushes clean and protected from contamination.

After ensuring that all joints have been wiped clean of silicone grease, clean the inside of the front half of the filter holder by rubbing the surfaces with a nylon bristle brush and rinsing with acetone. Rinse each surface three times or more if needed to remove visible particulate. Make a final rinse of the brush and filter holder. Carefully rinse out the glass cyclone, also (if applicable). After all acetone washings and particulate matter have been collected in the sample container, tighten the lid on the sample container so that acetone will not leak out when it is shipped to the laboratory. Mark the height of the fluid level to determine whether or not leakage occurred during transport. Label the container to clearly identify its contents.

Container No. 3. Note the color of the indicating silica gel to determine if it has been completely spent and make a notation of its condition. Transfer the silica gel from the fourth impinger to its original container and seal. A funnel may make it easier to pour the silica gel without spilling. A rubber policeman may be used as an aid in removing the silica gel from the impinger. It is not necessary to remove the small amount of dust particles that may adhere to the impinger wall and are difficult to remove. Since the gain in weight is to be used for moisture calculations, do not use any water or other liquids to transfer the silica gel. If a balance is available in the field, follow the procedure for container No. 3 in Section 4.3

Impinger Water. Treat the impingers as follows: Make a notation of any color or film in the liquid catch. Measure the liquid which is in the first three impingers to within ± 1 ml by using a graduated cylinder or by weighing it to within ± 0.5 g by using a balance (if one is available). Record the volume or weight of liquid present. This information is required to calculate the moisture content of the effluent gas.

Discard the liquid after measuring and recording the volume or weight, unless analysis of the impinger catch is required (see Note, Section 2.1.7).

If a different type of condenser is used, measure the amount of moisture condensed either volumetrically or gravimetrically.

Whenever possible, containers should be shipped in such a way that they remain upright at all times.

4.3 Analysis. Record the data required on a sheet such as the one shown in Figure 5-3. Handle each sample container as follows:

Container No. 1. Leave the contents in the shipping container or transfer the filter and any loose particulate from the sample container to a tared glass weighing dish. Desiccate for 24 hours in a desiccator containing anhydrous calcium sulfate. Weigh to a constant weight and report the results to the nearest 0.1 mg. For purposes of this Section, 4.3, the term "constant weight" means a difference of no more than 0.5 mg or 1 percent of total weight less tare weight, whichever is greater, between two consecutive weighings, with no less than 6 hours of desiccation time between weighings.

Alternatively, the sample may be oven dried at 105°C (220°F) for 2 to 3 hours, cooled in the desiccator, and weighed to a constant weight, unless otherwise specified by the Administrator. The tester may also opt to oven dry the sample at 105°C (220°F) for 2 to 3 hours, weigh the sample, and use this weight as a final weight.

Container No. 2. Note the level of liquid in the container and confirm on the analysis sheet whether or not leakage occurred during transport. If a noticeable amount of leakage has occurred, either void the sample or use methods, subject to the approval of the Administrator, to correct the final results. Measure the liquid in this container either volumetrically to ± 1 ml or gravimetrically to ± 0.5 g. Transfer the contents to a tared 250-ml beaker and evaporate to dryness at ambient temperature and pressure. Desiccate for 24 hours and weigh to a constant weight. Report the results to the nearest 0.1 mg.

Container No. 3. Weigh the spent silica gel (or silica gel plus impinger) to the nearest 0.5 g using a balance. This step may be conducted in the field.

"Acetone Blank" Container. Measure acetone in this container either volumetrically or gravimetrically. Transfer the acetone to a tared 250-ml beaker and evaporate to dryness at ambient temperature and pressure. Desiccate for 24 hours and weigh to a constant weight. Report the results to the nearest 0.1 mg.

NOTE—At the option of the tester, the contents of Container No. 2 as well as the acetone blank container may be evaporated at temperatures higher than ambient. If evaporation is done at an elevated temperature, the temperature must be below the boiling point of the solvent; also, to prevent "bumping," the evaporation process must be closely supervised, and the contents of the beaker must be swirled occasionally to maintain an even temperature. Use extreme care, as acetone is highly flammable and has a low flash point.

5. *Calibration*

Maintain a laboratory log of all calibrations.

5.1 Probe Nozzle. Probe nozzles shall be calibrated before their initial use in the field. Using a micrometer, measure the inside diameter of the nozzle to the nearest 0.025 mm (0.001 in.). Make three separate measurements using different diameters each time, and obtain the average of the measurements. The difference between the high and low numbers shall not exceed 0.1 mm (0.004 in.). When nozzles become nicked, dented, or corroded, they shall be reshaped, sharpened, and recalibrated before use. Each nozzle shall be permanently and uniquely identified.

5.2 Pitot Tube. The Type S pitot tube assembly shall be calibrated according to the procedure outlined in Section 4 of Method 2.

Plant_____

Date_____

Run No. _____

Filter No. _____

Amount liquid lost during transport _____

Acetone blank volume, ml _____

Acetone wash volume, ml_____

Acetone blank concentration, mg/mg (equation 5-4)_____

Acetone wash blank, mg (equation 5-5) _____

CONTAINER NUMBER	WEIGHT OF PARTICULATE COLLECTED, mg		
	FINAL WEIGHT	TARE WEIGHT	WEIGHT GAIN
1			
2			
TOTAL	✕	✕	
Less acetone blank			
Weight of particulate matter			

	VOLUME OF LIQUID WATER COLLECTED	
	IMPINGER VOLUME, ml.	SILICA GEL WEIGHT, g
FINAL		
INITIAL		
LIQUID COLLECTED		
TOTAL VOLUME COLLECTED		g* ml

* CONVERT WEIGHT OF WATER TO VOLUME BY DIVIDING TOTAL WEIGHT INCREASE BY DENSITY OF WATER (1g/ml).

$$\frac{\text{INCREASE, g}}{1 \text{ g/ml}} = \text{VOLUME WATER, ml}$$

Figure 5-3. Analytical data.

5.3   Metering System. Before its initial use in the field, the metering system shall be calibrated according to the procedure outlined in APTD-0576. Instead of physically adjusting the dry gas meter dial readings to correspond to the wet test meter readings, calibration factors may be used to mathematically correct the gas meter dial readings to the proper values. Before calibrating the metering system, it is suggested that a leak-check be conducted. For metering systems having diaphragm pumps, the normal leak-check procedure will not detect leakages within the pump. For these cases the following leak-check procedure is suggested: make a 10-minute calibration run at 0.00057 m^3/min (0.02 cfm); at the end of the run, take the difference of the measured wet test meter and dry gas meter volumes; divide the difference by 10 to get the leak rate. The leak rate should not exceed 0.00057 m^3/min (0.02 cfm).

After each field use, the calibration of the metering system shall be checked by performing three calibration runs at a single, intermediate orifice setting (based on the previous field test), with the vacuum set at the maximum value reached during the test series. To adjust the vacuum, insert a valve between the wet test meter and the inlet of the metering system. Calculate the average value of the calibration factor. If the calibration has changed by more than 5 percent, recalibrate the meter over the full range of orifice settings, as outlined in APTD-0576.

Alternative procedures, e.g., using the orifice meter coefficients, may be used, subject to the approval of the Administrator.

NOTE—If the dry gas meter coefficient values obtained before and after a test series differ by more than 5 percent, the test series shall either be voided, or calculations for the test series shall be performed using whichever meter coefficient value (i.e., before or after) gives the lower value of total sample volume.

5.4   Probe Heater Calibration. The probe heating system shall be calibrated before its initial use in the field according to the procedure outlined in APTD-0576. Probes constructed according to APTD-0581 need not be calibrated if the calibration curves in APTD-0576 are used.

5.5   Temperature Gauge. Use the procedure in Section 4.3 of Method 2 to calibrate in-stack temperature gauges. Dial thermometers, such as are used for the dry gas meter and condenser outlet, shall be calibrated against mercury-in-glass thermometers.

5.6   Leak Check of Metering System Shown in Figure 5-1. That portion of the sampling train from the pump to the orifice meter should be leak-checked prior to initial use and after each shipment. Leakage after the pump will result in less volume being recorded than is actually sampled. The following procedure is suggested (see Figure 5-4):

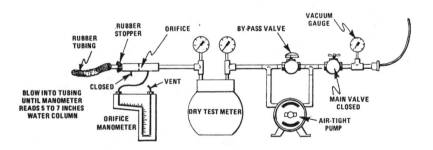

Figure 5-4. Leak check of meter box.

Close the main valve on the meter box. Insert a one-hole rubber stopper with rubber tubing attached into the orifice exhaust pipe. Disconnect and vent the low side of the orifice manometer. Close off the low side orifice tap. Pressurize the system to 13 to 18 cm (5 to 7 in.) water column by blowing into the rubber tubing. Pinch off the tubing and observe the manometer for one minute. A loss of pressure on the manometer indicates a leak in the meter box; leaks, if present, must be corrected.

    5.7   Barometer. Calibrate against a mercury barometer.

6. Calculations

Carry out calculations, retaining at least one extra decimal figure beyond that of the acquired data. Round off figures after the final calculation. Other forms of the equations may be used as long as they give equivalent results.

    6.1   Nomenclature

A_n	=	Cross-sectional area of nozzle, m^2 (ft^2).
B_{ws}	=	Water vapor in the gas stream, proportion by volume.
C_a	=	Acetone blank residue concentrations, mg/g.
c_g	=	Concentration of particulate matter in stack gas, dry basis, corrected to standard conditions, g/dscm (g/dscf).
I	=	Percent of isokinetic sampling.
L_a	=	Maximum acceptable leakage rate for either a pretest leak-check or for a leak check following a component change; equal to $0.00057 \ m^3/$ min (0.02 cfm) or 4 percent of the average sampling rate, whichever is less.
L_i	=	Individual leakage rate observed during the leak check conducted prior to the "ith" component change ($i = 1, 2, 3 \ldots n$), m^3/min (cfm).
L_p	=	Leakage rate observed during the post-test leak check, m^3/min (cfm).
m_n	=	Total amount of particulate matter collected, mg.
M_w	=	Molecular weight of water, 18.0 g/g-mole (18.0 lb/lb-mole).
m_a	=	Mass of residue of acetone after evaporation, mg.
P_{bar}	=	Barometric pressure at the sampling site, mm Hg (in. Hg).
P_s	=	Absolute stack gas pressure, mm Hg (in. Hg).
P_{std}	=	Standard absolute pressure, 760 mm Hg (29.92 in. Hg).
R	=	Ideal gas constant, 0.06236 mm Hg-$m^3/^\circ$K-g-mole (21.85 in. Hg-ft$^3/$ $^\circ$R-lb-mole).
T_m	=	Absolute average dry gas meter temperature (see Figure 5-2), $^\circ$K ($^\circ$R).
T_s	=	Absolute average stack gas temperature (see Figure 5-2) $^\circ$K ($^\circ$R).
T_{std}	=	Standard absolute temperature, 293°K (528°R).
V_a	=	Volume of acetone blank, ml.
V_{aw}	=	Volume of acetone used in wash, ml.
V_{lc}	=	Total volume of liquid collected in impingers and silica gel (see Figure 5-3), ml.
V_m	=	Volume of gas sample as measured by dry gas meter, dcm (dcf).
$V_{m(std)}$	=	Volume of gas sample measured by the dry gas meter, corrected to standard conditions, dscm (dscf).
$V_{w(std)}$	=	Volume of water vapor in the gas sample, corrected to standard conditions, scm (scf).
V_s	=	Stack gas velocity, calculated by Method 2, Equation 2-9, using data obtained from Method 5, m/sec (ft/sec).
W_a	=	Weight of residue in acetone wash, mg.

Y = Dry gas meter calibration factor.

ΔH = Average pressure differential across the orifice meter (see Figure 5-2) mm H_2O (in. H_2O).

ρ_a = Density of acetone, mg/ml (see label on bottle).

ρ_w = Density of water, 0.9982 g/ml (0.002201 lb/ml).

θ = Total sampling time, min.

θ_1 = Sampling time interval, from the beginning of a run until the first component change, min.

θ_i = Sampling time interval, between two successive component changes, beginning with the interval between the first and second changes, min.

θ_p = Sampling time interval, from the final (nth) component change until the end of the sampling run, min.

13.6 = Specific gravity of mercury.

60 = Sec/min.

100 = Conversion to percent.

6.2 Average dry gas meter temperature and average orifice pressure drop. See data sheet (Figure 5-2).

6.3 Dry Gas Volume. Correct the sample volume measured by the dry gas meter to standard conditions ($20°C$, 760 mm Hg or $68°F$, 29.92 in. Hg) by using Equation 5-1.

$$V_{m(std)} = V_m Y \left(\frac{T_{std}}{T_m} \right) \left[\frac{P_{bar} + \dfrac{\Delta H}{13.6}}{P_{std}} \right]$$

$$= K_1 V_m Y \frac{P_{bar} + (\Delta H/13.6)}{T_m}$$

Equation 5-1

where:

m = $0.3858°K$/mm Hg for metric units

= $17.64°R$/in. Hg for English units

NOTE—Equation 5-1 can be used as written unless the leakage rate observed during any of the mandatory leak checks (i.e., the post-test leak check or leak checks conducted prior to component changes) exceeds L_a. If L_p or L_i exceeds L_a, Equation 5-1 must be modified as follows:

(a) Case I. No component changes made during sampling run. In this case, replace V_m in Equation 5-1 with the expression:

$$[V_m - (L_v - L_a) \theta]$$

(b) Case II. One or more component changes made during the sampling run. In this case, replace V_m in Equation 5-1 by the expression: and substitute only

$$\left[V_m - (L_1 - L_a) \theta_1 - \sum_{i=2}^{n} (L_i - L_a) \theta_i - (L_v - L_a) \theta_v \right]$$

and substitute only for those leakage rates (l_i or L_v) which exceed L_a.

6.4 Volume of water vapor.

$$V_{w(std)} = V_{1c} \left(\frac{\rho_w}{M_w} \right) \left(\frac{RT_{std}}{P_{std}} \right) = K_2 V_{1c}$$

Equation 5-2

where:

K_2 = 0.001333 m^3/ml for metric units
= 0.04707 ft^3/ml for English units.

6.5   Moisture Content.

$$B_{ws} = \frac{V_{w(std)}}{V_{m(std)} + V_{w(std)}}$$

Equation 5-3

NOTE—In saturated or water droplet-laden gas streams, two calculations of the moisture content of the stack gas shall be made, one from the impinger analysis (Equation 5-3), and a second from the assumption of saturated conditions. The lower of the two values of B_{ws} shall be considered correct. The procedure for determining the moisture content based upon assumption of saturated conditions is given in the Note of Section 1.2 of Method 4. For the purposes of this method, the average stack gas temperature from Figure 5-2 may be used to make this determination, provided that the accuracy of the in-stack temperature sensor is ± 1°C (2°F).

6.6   Acetone Blank Concentration.

$$C_a = \frac{m_a}{V_a \rho_a}$$

Equation 5-4

6.7   Acetone Wash Blank.

$$W_a = C_a V_{aw} \rho_a$$

Equation 5-5

6.8   Total Particulate Weight. Determine the total particulate catch from the sum of the weights obtained from containers 1 and 2 less the acetone blank (see Figure 5-3). NOTE—Refer to Section 4.1.5 to assist in calculation of results involving two or more filter assemblies or two or more sampling trains.

6.9   Particulate Concentration.

$$c_g = (0.001 \, g/mg) \, (m_n/V_{m(std)})$$

Equation 5-6

6.10   Conversion Factors:

From	To	Multiply by
scf	m^3	0.02832
g/ft^3	gr/ft^3	15.43
g/ft^3	lb/ft^3	2.205 x 10^{-3}
g/ft^3	g/m^3	35.31

6.11   Isokinetic Variation.

6.11.1   Calculation From Raw Data.

$$I = \frac{100 T_s [K_2 V_{1c} + (V_m/T_m) \, (P_{bar} + \Delta H/13.6)]}{60 \, \theta v_s P_s A_n}$$

Equation 5-7

where:

K_3 = 0.003454 mm Hg-m^3/ml-$^\circ$K for metric units.

= 0.002669 in. Hg-ft^3/ml-$^\circ$R for English units.

6.11.2   Calculation From Intermediate Values.

$$I = \frac{T_s V_{m(std)} P_{std} \, 100}{T_{std} \, v_s \, \theta \, A_n P_s \, 60(1 - B_{ws})}$$

$$= K_4 \frac{T_s V_{m(std)}}{P_s V_s A_n \, \theta \, (1 - B_{ws})}$$

Equation 5-8

where:

K_4 = 4.320 for metric units

= 0.09450 for English units.

6.12   Acceptable Results. If 90 percent $\leqslant I \leqslant 110$ percent, the results are acceptable. If the results are low in comparison to the standard and I is beyond the acceptable range, or if I is less than 90 percent, the Administrator may opt to accept the results. Use Citation 4 to make judgments. Otherwise, reject the results and repeat the test.

7. *Bibliography*

1. Addendum to Specifications for Incinerator Testing at Federal Facilities. PHS, NCAPC. Dec. 6, 1967.

2. Martin, Robert M. Construction Details of Isokinetic Source-Sampling Equipment. Environmental Protection Agency. Research Triangle Park, N.C. APTD-0581. April, 1971.

3. Rom, Jerome J. Maintenance, Calibration, and Operation of Isokinetic Source Sampling Equipment. Environmental Protection Agency. Research Triangle Park, N.C. APTD-0576. March 1972.

4. Smith, W. S., R. T. Shigehara, and W. F. Todd. A Method of Interpreting Stack Sampling Data. Paper Presented at the 63d Annual Meeting of the Air Pollution Control Association, St. Louis, Mo. June 14-19, 1970.

5. Smith, W. S., et al. Stack Gas Sampling Improved and Simplified With New Equipment. APCA Paper No. 67-119. 1967.

6. Specifications for Incinerator Testing at Federal Facilities. PHS, NCAPC. 1967.

7. Shigehara, R. T. Adjustments in the EPA Nomograph for Different Pitot Tube Coefficients and Dry Molecular Weights. Stack Sampling News 2:4-11. October 1974.

8. Vollaro, R. F. A Survey of Commercially Available Instrumentation For the Measurement of Low-Range Gas Velocities. U.S. Environmental Protection Agency, Emission Measurement Branch. Research Triangle Park, N.C. November 1976 (unpublished paper).

9. Annual Book of ASTM Standards. Part 26. Gaseous Fuels; Coal and Coke; Atmospheric Analysis. American Society for Testing and Materials. Philadelphia, Pa. 1974. pp.617-622.

METHOD 6
DETERMINATION OF SULFUR DIOXIDE EMISSIONS
FROM STATIONARY SOURCES

1. *Principle and Applicability*

1.1   Principle. A gas sample is extracted from the sampling point in the stack. The sulfuric acid mist (including sulfur trioxide) and the sulfur dioxide are separated. The sulfur dioxide fraction is measured by the barium-thorin titration method.

1.2   Applicability. This method is applicable for the determination of sulfur dioxide emissions from stationary sources. The minimum detectable limit of the method has been determined to be 3.4 milligrams (mg) of SO_2/m^3 (2.12 x 10^{-7} lb/ft^3). Although no upper limit has been established, tests have shown that concentrations as high as 80,000 mg/m^3 of SO_2 can be collected efficiently in two midget impingers, each containing 15 milliliters of 3 percent hydrogen peroxide, at a rate of 1.0 lpm for 20 minutes. Based on theoretical calculations, the upper concentration limit in a 20-liter sample is about 93,300 mg/m^3.

Possible interferents are free ammonia, water-soluble cations, and fluorides. The cations and fluorides are removed by glass wool filters and an isopropanol bubbler, and hence do not affect the SO_2 analysis. When samples are being taken from a gas stream with high concentrations of very fine metallic fumes (such as in inlets to control devices), a high-efficiency glass fiber filter must be used in place of the glass wool plug (i.e., the one in the probe) to remove the cation interferents.

Free ammonia interferes by reacting with SO_2 to form particulate sulfite and by reacting with the indicator. If free ammonia is present (this can be determined by knowledge of the process and noticing white particulate matter in the probe and isopropanol bubbler), alternative methods, subject to the approval of the Administrator, U.S. Environmental Protection Agency, are required.

2. *Apparatus*

2.1   Sampling. The sampling train is shown in Figure 6-1, and component parts are discussed below. The tester has the option of substituting sampling equipment described in Method 8 in place of the midget impinger equipment of Method 6. However, the Method 8 train must be modified to include a heated filter between the probe and isopropanol impinger, and the operation of the sampling train and sample analysis must be at the flow rates and solution volumes defined in Method 8.

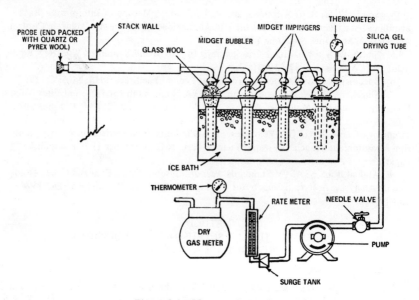

Figure 6-1. SO_2 sampling train.

The tester also has the option of determining SO_2 simultaneously with particulate matter and moisture determinations by (1) replacing the water in a Method 5 impinger system with 3 percent peroxide solution, or (2) by replacing the Method 5 water impinger system with a Method 8 isopropanol-filter-peroxide system. The analysis for SO_2 must be consistent with the procedure in Method 8.

2.1.1 Probe. Borosilicate glass, or stainless steel (other materials of construction may be used, subject to the approval of the Administrator), approximately 6-mm inside diameter, with a heating system to prevent water condensation and a filter (either in-stack or heated out-stack) to remove particulate matter, including sulfuric acid mist. A plug of glass wool is a satisfactory filter.

2.1.2 Bubbler and Impingers. One midget bubbler, with medium-coarse glass frit and borosilicate or quartz glass wool packed in top (see Figure 6-1) to prevent sulfuric acid mist carryover, and three 30-ml midget impingers. The bubbler and midget impingers must be connected in series with leak-free glass connectors. Silicone grease may be used, if necessary, to prevent leakage.

At the option of the tester, a midget impinger may be used in place of the midget bubbler.

Other collection absorbers and flow rates may be used, but are subject to the approval of the Administrator. Also, collection efficiency must be shown to be at least 99 percent for each test run and must be documented in the report. If the efficiency is found to be acceptable after a series of three tests, further documentation is not required. To conduct the efficiency test, an extra absorber must be added and analyzed separately. This extra absorber must not contain more than 1 percent of the total SO_2.

2.1.3 Glass Wool. Borosilicate or quartz.

2.1.4 Stopcock Grease. Acetone-insoluble, heat-stable silicone grease may be used, if necessary.

2.1.5 Temperature Gauge. Dial thermometer, or equivalent, to measure temperature of gas leaving impinger train to within $1°C$ ($2°F$).

2.1.6 Drying Tube. Tube packed with 6- to 16-mesh indicating type silica gel, or equivalent, to dry the gas sample and to protect the meter and pump. If the silica gel has been used previously, dry at $175°C$ ($350°F$) for 2 hours. New silica gel may be used as received. Alternatively, other types of desiccants (equivalent or better) may be used, subject to approval of the Administrator.

2.1.7 Valve. Needle valve, to regulate sample gas flow rate.

2.1.8 Pump. Leak-free diaphragm pump, or equivalent, to pull gas through the train. Install a small tank between the pump and rate meter to eliminate the pulsation effect of the diaphragm pump on the rotameter.

2.1.9 Rate Meter. Rotameter, or equivalent, capable of measuring flow rate to within 2 percent of the selected flow rate of about 1000 cc/min.

2.1.10 Volume Meter. Dry gas meter, sufficiently accurate to measure the sample volume within 2 percent, calibrated at the selected flow rate and conditions actually encountered during sampling, and equipped with a temperature gauge (dial thermometer, or equivalent) capable of measuring temperature to within $3°C$ ($5.4°F$).

2.1.11 Barometer. Mercury, aneroid, or other barometer capable of measuring atmospheric pressure to within 2.5 mm Hg (0.1 in. Hg). In many cases, the barometric reading may be obtained from a nearby national weather service station, in which case the station value (which is the absolute barometric pressure) shall be requested and an adjustment for elevation differences between the weather station and sampling point shall be applied at a rate of minus 2.5 mm Hg (0.1 in. Hg) per 30 m (100 ft) elevation increase or vice versa for elevation decrease.

2.1.12   Vacuum Gauge. At least 760 mm Hg (30 in. Hg) gauge, to be used for leak check of the sampling train.
2.2   Sample Recovery.
2.2.1   Wash bottles. Polyethylene or glass, 500 ml, two.
2.2.2   Storage Bottles. Polyethylene, 100 ml, to store impinger samples (one per sample).
2.3   Analysis.
2.3.1   Pipettes. Volumetric type, 5-ml, 20-ml (one per sample), and 25-ml sizes.
2.3.2   Volumetric Flasks. 100-ml size (one per sample) and 100-ml size.
2.3.3   Burettes. 5- and 50-ml sizes.
2.3.4   Erlenmeyer Flasks. 250-ml size (one for each sample, blank, and standard).
2.3.5   Dropping Bottle. 125-ml size, to add indicator.
2.3.6   Graduated Cylinder. 100-ml size.
2.3.7   Spectrophotometer. To measure absorbance at 352 nanometers.

3. *Reagents*

Unless otherwise indicated, all reagents must conform to the specifications established by the Committee on Analytical Reagents of the American Chemical Society. Where such specifications are not available, use the best available grade.
3.1   Sampling.
3.1.1   Water. Deionized, distilled to conform to ASTM specification D1193-74, Type 3. At the option of the analyst, the $KMnO_4$ test for oxidizable organic matter may be omitted when high concentrations of organic matter are not expected to be present.
3.1.2   Isopropanol, 80 percent. Mix 80 ml of isopropanol with 20 ml of deionized, distilled water. Check each lot of isopropanol for peroxide impurities as follows: shake 10 ml of isopropanol with 10 ml of freshly prepared 10 percent potassium iodide solution. Prepare a blank by similarly treating 10 ml of distilled water. After 1 minute, read the absorbance at 352 nanometers on a spectrophotometer. If absorbance exceeds 0.1, reject alcohol for use.
Peroxides may be removed from isopropanol by redistilling or by passage through a column of activated alumina; however, reagent grade isopropanol with suitably low peroxide levels may be obtained from commercial sources. Rejection of contaminated lots may, therefore, be a more efficient procedure.
3.1.3   Hydrogen Peroxide, 3 Percent. Dilute 30 percent hydrogen peroxide 1:9 (v/v) with deionized, distilled water (30 ml is needed per sample). Prepare fresh daily.
3.1.4   Potassium Iodide Solution, 10 Percent. Dissolve 10.0 grams KI in deionized, distilled water and dilute to 100 ml. Prepare when needed.
3.2   Sample Recovery.
3.2.1   Water. Deionized, distilled, as in 3.1.1.
3.2.2   Isopropanol, 80 Percent. Mix 80 ml of isopropanol with 20 ml of deionized, distilled water.
3.3   Analysis.
3.3.1   Water. Deionized, distilled, as in 3.1.1.
3.3.2   Isopropanol, 100 percent.
3.3.3   Thorin Indicator. 1-(o-arsonophenylazo)-2-naphthol-3,6-disulfonic acid, disodium salt, or equivalent. Dissolve 0.20 g in 100 ml of deionized, distilled water.
3.3.4   Barium Perchlorate Solution, 0.0100 N. Dissolve 1.95 g of barium perchlorate trihydrate $[Ba(ClO_4)_2 \cdot 3H_2O]$ in 200 ml distilled water and dilute to 1 liter with isopropanol. Alternatively, 1.22 g of $[BaCl_2 \cdot 2H_2O]$ may be used instead of the perchlorate. Standardize as in Section 5.5.

3.3.5 Sulfuric Acid Standard, 0.0100 N. Purchase or standardize to ± 0.002 N against 0.0100 N NaOH which has previously been standardized against potassium acid phthalate (primary standard grade).

4. *Procedure.*

4.1 Sampling.

4.1.1 Preparation of collection train. Measure 15 ml of 80 percent isopropanol into the midget bubbler and 15 ml of 3 percent hydrogen peroxide into each of the first two midget impingers. Leave the final midget impinger dry. Assemble the train as shown in Figure 6-1. Adjust probe heater to a temperature sufficient to prevent water condensation. Place crushed ice and water around the impingers.

4.1.2 Leak-check procedure. A leak-check prior to the sampling run is optional; however, a leak-check after the sampling run is mandatory. The leak-check procedure is as follows:

With the probe disconnected, place a vacuum gauge at the inlet to the bubbler and pull a vacuum of 250 mm (10 in.) Hg; plug or pinch off the outlet of the flow meter, and then turn off the pump. The vacuum shall remain stable for at least 30 seconds. Carefully release the vacuum gauge before releasing the flow meter end to prevent back flow of the impinger fluid.

Other leak-check procedures may be used, subject to the approval of the Administrator, U.S. Environmental Protection Agency. The procedure used in Method 5 is not suitable for diaphragm pumps.

4.1.3 Sample collection. Record the initial dry gas meter reading and barometric pressure. To begin sampling, position the tip of the probe at the sampling point, connect the probe to the bubbler, and start the pump. Adjust the sample flow to a constant rate of approximately 1.0 liter/min as indicated by the rotameter. Maintain this constant rate (± 10 percent) during the entire sampling run. Take readings (dry gas meter, temperatures at dry gas meter and at impinger outlet and rate meter) at least every 5 minutes. Add more ice during the run to keep the temperature of the gases leaving the last impinger at $20°C$ ($69°F$) or less. At the conclusion of each run, turn off the pump, remove probe from the stack, and record the final readings. Conduct a leak check as in Section 4.1.2 (This leak check is mandatory.) If a leak is found, void the test run. Drain the ice bath, and purge the remaining part of the train by drawing clean ambient air through the system for 15 minutes at the sampling rate.

Clean ambient air can be provided by passing air through a charcoal filter or through an extra midget impinger with 15 ml of 3 percent H_2O_2. The tester may opt to simply use ambient air, without purification.

4.2 Sample Recovery. Disconnect the impingers after purging. Discard the contents of the midget bubbler. Pour the contents of the midget impingers into a leak-free polyethylene bottle for shipment. Rinse the three midget impingers and the connecting tubes with deionized, distilled water, and add the washings to the same storage container. Mark the fluid level. Seal and identify the sample container.

4.3 Sample Analysis. Note level of liquid in container, and confirm whether any sample was lost during shipment; note this on analytical data sheet. If a noticeable amount of leakage has occurred, either void the sample or use methods, subject to the approval of the Administrator, to correct the final results.

Transfer the contents of the storage container to a 100-ml volumetric flask and dilute to exactly 100 ml with deionized, distilled water. Pipette a 20-ml aliquot of this solution into a 250-ml Erlenmeyer flask, add 80 ml of 100 percent isopropanol and two to four drops of thorin indicator, and titrate to a pink endpoint using 0.0100 N barium perchlorate. Repeat and average the titration volumes. Run a blank with each series of samples. Replicate titrations must agree within 1 percent or 0.2 ml, whichever is larger.

(NOTE—Protect the 0.0100 N barium perchlorate solution from evaporation at all times.)

5. *Calibration*

5.1   Metering System.

5.1.1   Initial Calibration. Before its initial use in the field, first leak check the metering system (drying tube, needle valve, pump, rotameter, and dry gas meter) as follows: place a vacuum gauge at the inlet to the drying tube and pull a vacuum of 250 mm (10 in.) Hg; plug or pinch off the outlet or the flow meter, and then turn off the pump. The vacuum shall remain stable for at least 30 seconds. Carefully release the vacuum gauge before releasing the flow meter end.

Next, calibrate the metering system (at the sampling flow rate specified by the method) as follows: connect an appropriately sized wet test meter (e.g., 1 liter per revolution) to the inlet of the drying tube. Make three independent calibration runs, using at least five revolutions of the dry gas meter per run. Calculate the calibration factor, Y (wet test meter calibration volume divided by the dry meter volume, both volumes adjusted to the same reference temperature and pressure), for each run, and average the results. If any Y value deviates by more than 2 percent from the average, the metering system is unacceptable for use. Otherwise, use the average as the calibration factor for subsequent test runs.

5.1.2   Post-Test Calibration Check. After each field test series, conduct a calibration check as in Section 5.1.1 above, except for the following variations: (1) the leak check is not to be conducted, (b) three, or more revolutions of the dry gas meter may be used, and (c) only two independent runs need be made. If the calibration factor does not deviate by more than 5 percent from the initial calibration factor (determined in Section 5.1.1), then the dry gas meter volumes obtained during the test series are acceptable. If the calibration factor deviates by more than 5 percent, recalibrate the metering system as in Section 5.1.1, and for the calculations, use the calibration factor (initial or recalibration) that yields the lower gas volume for each test run.

5.2   Thermometers. Calibrate against mercury-in-glass thermometers.

5.3   Rotameter. The rotameter need not be calibrated but should be cleaned and maintained according to the manufacturer's instruction.

5.4   Barometer. Calibrate against a mercury barometer.

5.5   Barium Perchlorate Solution. Standardize the barium perchlorate solution against 25 ml of standard sulfuric acid to which 100 ml of 100 percent isopropanol has been added.

6. *Calculations*

Carry out calculations, retaining at least one extra decimal figure beyond that of the acquired data. Round off figures after final calculation.

6.1   Nomenclature.

C_{SO_2} = Concentration of sulfur dioxide, dry basis corrected to standard conditions, mg/dscm (lb/dscf).

N = Normality of barium perchlorate titrant, milliequivalents/ml.

P_{bar} = Barometric pressure at the exit orifice of the dry gas meter, mm Hg (in. Hg).

P_{std} = Standard absolute pressure, 760 mm Hg (29.92 in. Hg).

T_m = Average dry gas meter absolute temperature, $^\circ$K ($^\circ$R).

T_{std} = Standard absolute temperature, 293°K (528°R).

V_a = Volume of sample aliquot titrated, ml.
V_m = Dry gas volume as measured by the dry gas meter, dcm (dcf).
$V_{m(std)}$ = Dry gas volume measured by the dry gas meter, corrected to standard conditions, dscm (dscf).
V_{soln} = Total volume of solution in which the sulfur dioxide sample is contained, 100 ml
V_t = Volume of barium perchlorate titrant used for the sample, ml (average of replicate titrations).
V_{tb} = Volume of barium perchlorate titrant used for the blank, ml.
Y = Dry gas meter calibration factor.
32.03 = Equivalent weight of sulfur dioxide.

6.2   Dry sample gas volume, corrected to standard conditions.

$$V_{m(std)} = V_m \, Y \left(\frac{T_{std}}{T_m} \right) \left(\frac{P_{bar}}{P_{std}} \right) = K_1 \, Y \, \frac{V_m P_{bar}}{T_m}$$

Equation 6-1

where:

K_1 = 0.3858 $^\circ$K/mm Hg for metric units.
    = 17.64 $^\circ$R/in. Hg for English units.

6.3   Sulfur dioxide concentration.

$$C_{SO_2} = K_2 \, \frac{(V_t - V_{tb}) N \left(\dfrac{V_{soln}}{V_a} \right)}{V_{m(std)}}$$

Equation 6-2

where:

K_3 = 32.03 mg/meq. for metric units
    = 7.061 x 10^{-5} lb/meq. for English units.

7. *Bibliography*

1. Atmospheric Emissions from Sulfuric Acid Manufacturing Processes. U.S. DHEW, PHS, Division of Air Pollution. Public Health Service Publication No. 99-AP-13. Cincinnati, Ohio. 1965.

2. Corbett, P. F. The Determination of SO_2 and SO_3 in Flue Gases. Journal of the Institute of Fuel. 24:237-243, 1961.

3. Matty, R. E. and E. K. Diehl. Measuring Flue-Gas SO_2 and SO_3. Power. 101:94-97. November 1957.

4. Patton, W. F. and J. A. Brink, Jr. New Equipment and Techniques for Sampling Chemical Process Gases. J. Air Pollution Control Association. 13:162. 1963.

5. Rom, J. J. Maintenance, Calibration and Operation of Isokinetic Source-Sampling Equipment. Office of Air Programs, Environmental Protection Agency. Research Triangle Park, N.C. APTD-0576. March 1972.

6. Hamil, H. F. and D. E. Camann. Collaborative Study of Method for the Determination of Sulfur Dioxide Emissions from Stationary Sources (Fossil-Fuel Fired Steam Generators). Environmental Protection Agency, Research Triangle Park, N.C. EPA-650/4-74-024. December 1973.

7. Annual Book of ASTM Standards. Part 31; Water, Atmospheric Analysis. American Society for Testing and Materials. Philadelphia, Pa. 1974. pp.40-42.

8. Knoll, J. E. and M. R. Midgett. The Application of EPA Method 6 to High Sulfur Dioxide Concentrations. Environmental Protection Agency, Research Triangle Park, N.C. EPA-600/4-76-038. July 1976.

METHOD 7
DETERMINATION OF NITROGEN OXIDE EMISSIONS
FROM STATIONARY SOURCES

1. *Principle and Applicability*

1.1　Principle. A grab sample is collected in an evacuated flask containing a dilute sulfuric acid-hydrogen peroxide absorbing solution, and the nitrogen oxides, except nitrous oxide, are measured colorimetrically using the phenoldisulfonic acid (PDS) procedure.

1.2　Applicability. This method is applicable to the measurement of nitrogen oxides emitted from stationary sources. The range of the method has been determined to be 2 to 400 milligrams NO_x (as NO_2) per dry standard cubic meter, without having to dilute the sample.

2. *Apparatus*

2.1　Sampling (see Figure 7-1). Other grab sampling systems or equipment, capable of measuring sample volume to within ± 2.0 percent and collecting a sufficient sample volume to allow analytical reproducibility to within ± 5 percent, will be considered acceptable alternatives, subject to approval of the Administrator, U.S. Environmental Protection Agency. The following equipment is used in sampling:

2.1.1　Probe. Borosilicate glass tubing, sufficiently heated to prevent water condensation and equipped with an in-stack or out-stack filter to remove particulate matter

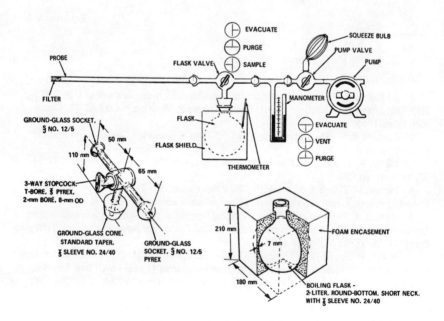

Figure 7-1. Sampling train, flask valve and flask.

(a plug of glass wool is satisfactory for this purpose). Stainless steel or Teflon[3] tubing may also be used for the probe. Heating is not necessary if the probe remains dry during the purging period.

2.1.2 Collection Flask. Two-liter borosilicate, round bottom flask, with short neck and 24/40 standard taper opening, protected against implosion or breakage.

2.1.3 Flask Valve. T-bore stopsock connected to a 24/40 standard taper joint.

2.1.4 Temperature Gauge. Dial-type thermometer, or other temperature gauge, capable of measuring $1^{\circ}C$ ($2^{\circ}F$) intervals from -5 to $50^{\circ}C$ (25 to $125^{\circ}F$).

2.1.5 Vacuum Line. Tubing capable of withstanding a vacuum of 75 mm Hg (3 in. Hg) absolute pressure, with "T" connection and T-bore stopcock.

2.1.6 Vacuum Gauge. U-tube manometer, 1 meter (36 in.) with 1-mm (0.1-in.) divisions, or other gauge capable of measuring presure to within ± 2.5 mm Hg (0.10 in. Hg).

2.1.7 Pump. Capable of evacuating the collection flask to a pressure equal to or less than 75 mm Hg (3 in. Hg) absolute.

2.1.8 Squeeze Bulb. One-way.

2.1.9 Stopcock and Ground Joint Grease. A high-vacuum, high-temperature chlorofluorocarbon grease is required. Halocarbon 25-5S has been found to be effective.

2.1.11 Barometer. Mercury, aneroid, or other barometer capable of measuring atmospheric pressure to within 2.5 mm Hg (0.1 in. Hg). In many cases, the barometric reading may be obtained from a nearby national weather service station, in which case the station value (which is the absolute barometric pressure) shall be requested and an adjustment for elevation differences between the weather station and sampling point shall be applied at a rate of minus 2.5 mm Hg (0.1 in. Hg) per 30 mm (100 ft) elevation increase, or vice versa for elevation decrease.

2.2 Sample Recovery. The following equipment is required for sample recovery:

2.2.1 Graduated Cylinder. 50-ml with 1-ml divisions.

2.2.2 Storage Containers. Leak-free polyethylene bottles.

2.2.3 Wash Bottle. Polyethylene or glass.

2.2.4 Glass Stirring Rod.

2.2.5 Test Paper for Indicating pH. To cover the pH range of 7 to 14.

2.3 Analysis. For the analysis, the following equipment is needed:

2.3.1 Volumetric Pipettes. Two 1 ml, two 2 ml, one 3 ml, one 4 ml, two 10 ml, and one 25 ml for each sample and standard.

2.3.2 Porcelain Evaporating Dishes. 175- to 150-ml capacity with lip for pouring, one for each sample and each standard. The Coors No. 45006 (shallow-form, 195 ml) has been found to be satisfactory. Alternatively, polymethyl pentene beakers (Nalge No. 1203, 150 ml), or glass beakers (150 ml) may be used. When glass beakers are used, etching of the beakers may cause solid matter to be present in the analytical step. The solids should be removed by filtration (see Section 4.3).

2.3.3 Steam Bath. Low-temperature ovens or thermostatically controlled hot plates kept below $70^{\circ}C$ ($160^{\circ}F$) are acceptable alternatives.

2.3.4 Dropping Pipette or Dropper. Three required.

2.3.5 Polyethylene Policeman. One for each sample and each standard.

2.3.6 Graduated Cylinder. 100 ml with 1-ml divisions.

2.3.7 Volumetric Flasks. 50 ml (one for each sample), 100 ml (one for each sample and each standard, and one for the working standard KNO_3 solution), and 1000 ml (one).

2.3.8 Spectrophotometer. To measure absorbance at 410 nm.

[3]Mention of trade names or specific products does not constitute endorsement by the Environmental Protection Agency.

2.3.9   Graduated Pipette. 10 ml with 0.1-ml divisions.

2.3.10   Test Paper for Indicating pH. To cover the pH range of 7 to 14.

2.3.11   Analytical Balance. To measure to within 0.1 mg.

3. Reagents

Unless otherwise indicated, it is intended that all reagents conform to the specifications established by the Committee on Analytical Reagents of the American Chemical Society, where such specifications are available; otherwise, use the best available grade.

3.1   Sampling. To prepare the absorbing solution, cautiously add 2.8 ml concentrated H_2SO_4 to 1 liter of deionized, distilled water. Mix well and add 6 ml of 3 percent hydrogen peroxide, freshly prepared from 30 percent hydrogen peroxide solution. The absorbing solution should be used within 1 week of its preparation. Do not expose to extreme heat or direct sunlight.

3.2   Sample Recovery. Two reagents are required for sample recovery:

3.2.1   Sodium Hydroxide ($1N$). Dissolve 40 g NaOH in deionized, distilled water and dilute to 1 liter.

3.2.2   Water. Deionized, distilled to conform to ASTM specification D1193-74, Type 3. At the option of the analyst, the $KMnO_4$ test for oxidizable organic matter may be omitted when high concentrations of organic matter are not expected to be present.

3.3   Analysis. For the analysis, the following reagents are required:

3.3.1   Fuming Sulfuric Acid. 15 to 18 percent by weight free sulfur trioxide. HANDLE WITH CAUTION.

3.3.2   Phenol. White solid.

3.3.3   Sulfuric Acid. Concentrated, 95 percent minimum assay. HANDLE WITH CAUTION.

3.3.4   Potassium Nitrate. Dried at 105 to 110°C (220 to 230°F) for a minimum of 2 hours just prior to preparation of standard solution.

3.3.5   Standard KNO_3 Solution. Dissolve exactly 2.198 g of dried potassium nitrate (KNO_3) in deionized, distilled water and dilute to 1 liter with deionized, distilled water in a 1000-ml volumetric flask.

3.3.6   Working Standard KNO_3 Solution. Dilute 10 ml of the standard solution to 100 ml with deionized distilled water. One milliliter of the working standard solution is equivalent to 100 μg nitrogen dioxide (NO_2).

3.3.7   Water. Deionized, distilled as in Section 3.2.2.

3.3.8   Phenoldisulfonic Acid Solution. Dissolve 25 g of pure white phenol in 150 ml concentrated sulfuric acid on a steam bath. Cool, add 75 ml fuming sulfuric acid, and heat at 100°C (212°F) for 2 hours. Store in a dark, stoppered bottle.

4. Procedure

4.1   Sampling.

4.1.1   Pipette 25 ml of absorbing solution into a sample flask, retaining a sufficient quantity for use in preparing the calibration standards. Insert the flask valve stopper into the flask with the valve in the "purge" position. Assemble the sampling train as shown in Figure 7-1 and place the probe at the sampling point. Make sure that all fittings are tight and leak-free, and that all ground glass joints have been properly greased with a high-vacuum, high-temperature chlorofluorocarbon-based stopcock grease. Turn the flask valve and the pump valve to their "evacuate" positions. Evacuate the flask to 77 mm Hg (3 in. Hg) absolute pressure, or less. Evacuation to a pressure approaching the vapor pressure of water at the existing temperature is desirable. Turn the pump valve to its "vent" position and turn off the pump. Check for leakage by observing the manometer for any pressure fluctuation. (Any variation greater than 10 mm Hg (0.4 in. Hg) over a period of 1 minute is not acceptable, and the flask is not to be used until the leakage problem in corrected. Pressure in the flask is not to exceed 75 mm Hg (3 in. Hg) absolute at the time sampling is commenced. Record the volume of the flask and valve (V_f), the flask temperature (T_i), and the barometric pressure. Turn the flask valve counterclockwise to its "purge" position and do the same with the pump valve. Purge the probe and the vacuum tube using the squeeze bulb. If condensation occurs in the probe and the flask valve area, heat the probe and purge until the condensation disappears. Next, turn the pump valve to its "vent" position. Turn the flask valve clockwise to its "evacuate" position and record the difference in the mercury levels in the manometer. The absolute internal pressure in the flask (P_i) is equal to the barometric pressure less the manometer reading. Immediately turn the flask valve to the "sample" position and permit the gas to enter the flask until pressures in the flask and sample line (i.e., duct, stack) are equal. This will usually require about 15 seconds; a longer period indicates a "plug" in the probe, which must be corrected before sampling is continued. After collecting the sample, turn the flask valve to its "purge" position and disconnect the flask from the sampling train. Shake the flask for at least 5 minutes.

4.1.2 If the gas being sampled contains insufficient oxygen for the conversion of NO to NO_2 (e.g., an applicable subpart of the standard may require taking a sample of a calibration gas mixture of NO in N_2), then oxygen shall be introduced into the flask to permit this conversion. Oxygen may be introduced into the flask by one of three methods; (1) Before evacuating the sampling flask, flush with pure cylinder oxygen, then evacuate flask to 75 mm Hg (3 in. Hg) absolute pressure or less; or (2) inject oxygen into the flask after sampling; or (3) terminate sampling with a minimum of 50 mm Hg (2 in. Hg) vacuum remaining in the flask, record this final pressure, and then vent the flask to the atmosphere until the flask pressure is almost equal to atmospheric pressure.

4.2 Sample Recovery. Let the flask set for a minimum of 16 hours and then shake the contents for 2 minutes. Connect the flask to a mercury filled U-tube manometer. Open the valve from the flask to the manometer and record the flask temperature (T_f), the barometric pressure, and the difference between the mercury levels in the manometer. The absolute internal pressure in the flask (P_f) is the barometric pressure less the manometer reading. Transfer the contents of the flask to a leak-free polyethlene bottle. Rinse the flask twice with 5-ml portions of deionized, distilled water and add the rinse water to the bottle. Adjust the pH to between 9 and 12 by adding sodium hydroxide $(1N)$, dropwise (about 25 to 35 drops). Check the pH by dipping a stirring rod into the solution and then touching the rod to the pH test paper. Remove as little material as possible during this step. Mark the height of the liquid level so that the container can be checked for leakage after transport. Label the container to clearly identify its contents. Seal the container for shipping.

4.3 Analysis. Note the level of the liquid in container and confirm whether or not any sample was lost during shipment; note this on the analytical data sheet. If a noticeable amount of leakage has occurred, either void the sample or use methods, subject to the approval of the Administrator, to correct the final results. Immediately prior to analysis, transfer the contents of the shipping container to a 50-ml volumetric flask, and rinse the container twice with 5-ml portions of deionized, distilled water. Add the rinse water to the flask and dilute to the mark with deionized, distilled water; mix thoroughly. Pipette a 25-ml aliquot into the procelain evaporating dish. Return any unused portion of the sample to the polyethylene storage bottle. Evaporate the 25-ml aliquote to dryness on a steam bath and allow to cool. Add 2 ml phenoldisulfonic acid solution to the dried residue and titurate thoroughly with a polyethylene policeman. Make sure the solution contacts all the residue. Add 1 ml deionized, distilled water and four drops of concentrated sulfuric acid. Heat the solution on a steam bath for 3 minutes with occasional stirring. Allow the solution to cool, add 20 ml deionized, distilled water, mix well by stirring, and add concentrated ammonium hydroxide, dropwise, with constant stirring, until the pH is 10 (as determined by pH paper). If the sample contains solids, these must be removed by filtration (centrifugation is an acceptable alternative, subject to the approval of the Administrator), as follows: filter through Whatman No. 41 filter paper into a 100-ml volumetric flask; rinse the evaporating dish with three 5-ml portions of deionized, distilled water; filter these three rinses. Wash the filter with at least three 15-ml portions of deionized, distilled water. Add the filter washings to the contents of the volumetric flask and dilute to the mark with deionized, distilled water. If solids are absent, the solution can be transferred directly to the 100-ml volumetric flask and diluted to the mark with deionized, distilled water. Mix the contents of the flask thoroughly, and measure the absorbance at the optimum wavelength used for the standards (Section 5.2.1), using the blank solution as a zero reference. Dilute the sample and the blank with equal volumes of deionized, distilled water if the absorbance exceeds A_4, the absorbance of the 400 μg NO_2 standard (see Section 5.2.2).

5. *Calibration*

5.1   Flask-Volume. The volume of the collection flask-flask valve combination must be known prior to sampling. Assemble the flask and flask valve and fill with water, to the stopcock. Measure the volume of water to ± 10 ml. Record this volume on the flask.

5.2   Spectrophotometer Calibration.

5.2.1   Optimum Wavelength Determination. For both fixed and variable wavelength spectrophotometers, calibrate against standard certified wavelength of 410 nm, every 6 months. Alternatively, for variable wavelength spectrophotometers, scan the spectrum between 400 and 415 nm using a 200 μg NO_2 standard solution (see Section 5.2.2). If a peak does not occur, the spectrophotometer is probably malfunctioning, and should be repaired. When a peak is obtained within the 400 to 415 nm range, the wavelength at which this peak occurs shall be the optimum wavelength for the measurement of absorbance for both the standards and samples.

5.2.2   Determination of Spectrophotometer Calibration Factor K_C. Add 0.0, 1.0, 2.0, 3.0 and 4.0 ml of the KNO_3 working standard solution (1 ml = 100 μg NO_2) to a series of five porcelain evaporating dishes. To each, add 25 ml of absorbing solution, 10 ml deionized, distilled water, and sodium hydroxide (1N), dropwise, until the pH is between 9 and 12 (about 25 to 35 drops each). Beginning with the evaporation step, follow the analysis procedure of Section 4.3, until the solution has been transferred to the 100 ml volumetric flask and diluted to the mark. Measure the absorbance of each solution, at the optimum wavelength, as determined in Section 5.2.1. This calibration procedure must be repeated on each day that samples are analyzed. Calculate the spectrophotometer calibration factor as follows:

$$K_C = 100 \frac{A_1 + 2A_2 + 3A_3 + 4A_4}{A_1{}^2 + A_2{}^2 + A_3{}^2 + A_4{}^2}$$

Equation 7-1

where:

K_C =   Calibration factor
A_1 =   Absorbance of the 100-μg NO_2 standard
A_2 =   Absorbance of the 200-μg NO_2 standard
A_3 =   Absorbance of the 300-μg NO_2 standard
A_4 =   Absorbance of the 400-μg NO_2 standard

5.3   Barometer. Calibrate against a mercury barometer.

5.4   Temperature Gauge. Calibrate dial thermometers against mercury-in-glass thermometers.

5.5   Vacuum Gauge. Calibrate mechanical gauges, if used, against a mercury manometer such as that specified in 2.1.6.

5.6   Analytical Balance. Calibrate against standard weights.

6. *Calculations*

Carry out the calculations, retaining at least one extra decimal figure beyond that of the acquired data. Round off figures after final calculations.

6.1   Nomenclature.

A = Absorbance of sample.

C = Concentration of NO_X as NO_2, dry basis, corrected to standard conditions, mg/dscm (lb/dscf).

F = Dilution factor (i.e., 25/5, 25/10, etc., required only if sample dilution was needed to reduce the absorbance into the range of calibration.

K_c = Spectrophotometer calibration factor.

m = Mass of NO_X as NO_2 in gas sample, μg.

P_f = Final absolute pressure of flask, mm Hg (in. Hg).

P_i = Initial absolute pressure of flask, mm Hg (in. Hg).

P_{std} = Standard absolute pressure, 760 mm Hg (29.92 in. Hg).

T_f = Final absolute temperature of flask, $^\circ$K ($^\circ$R).

T_i = Initial absolute temperature of flask, $^\circ$K ($^\circ$R).

T_{std} = Standard absolute temperature, 293°K (528°R).

V_{sc} = Sample volume at standard conditions (dry basis), ml.

V_f = Volume of flask and valve, ml.

V_a = Volume of absorbing solution, 25 ml.

2 = 50/25, the aliquot factor. (If other than a 25-ml aliquot was used for analysis, the corresponding factor must be substituted).

6.2   Sample volume, dry basis, corrected to standard conditions.

$$V_{sc} = \frac{T_{std}}{P_{std}} (V_f - V_a) \left[\frac{P_f}{T_f} - \frac{P_i}{T_i} \right]$$

$$= K_1 (V_f - 25 \text{ ml}) \left[\frac{P_f}{T_f} - \frac{P_i}{T_i} \right]$$

Equation 7-2

where:

K_1 = 0.3858 ($^\circ$K/mm Hg) for metric units.

= 17.64 ($^\circ$R/mm Hg) for English units.

6.3   Total μg NO_2 per sample.

$$m = 2 K_c A F$$

Equation 7-3

NOTE—If other than a 25-ml aliquot is used for analysis, the factor 2 must be replaced by a corresponding factor.

6.4   Sample concentration, dry basis, corrected to standard conditions.

$$C = K_2 \frac{m}{V_{sc}}$$

Equation 7-4

where:

K_2 = 10^3 (mg/m^3)/(μg/ml) for metric units

= 6.243 x 10^{-5} (lb/scf)/(μg/ml) for English units.

7. *Bibliography*

1. Standard Methods of Chemical Analysis. 6th ed. New York, D. Van Nostrand Co., Inc. 1962. Vol. 1, pp.329-330.

2. Standard Method of Test for Oxides of Nitrogen in Gaseous Combustion Products (phenoldisulfonic Acid Procedure). In: 1968 Book of ASTM Standards, Part 26. Philadelphia, Pa. 1968. ASTM Designation D-1608-60, pp.725-729.

3. Jacob, M. B. The Chemical Analysis of Air Pollutants. New York. Interscience Publishers, Inc. 1960. Vol. 10, pp.351-356.

4. Beatty, R. L., L. B. Berger and H. H. Schrenk. Determination of Oxides of Nitrogen by the Phenoldisulfonic Acid Method. Bureau of Mines, U.S. Department of Interior. R.I. 3687. February 1943.

5. Hamil, H. F. and D. E. Camann. Collaborative Study of Method for the Determination of Nitrogen Oxide Emissions from Stationary Sources (Fossil Fuel-Fired Steam Generators). Southwest Research Institute report for Environmental Protection Agency. Research Triangle Park, N.C. October 5, 1973.

6. Hamil, H. F. and R. E. Thomas. Collaborative Study of Method for the Determination of Nitrogen Oxide Emissions from Stationary Sources (Nitric Acid Plants). Southwest Research Institute report for Environmental Protection Agency. Research Triangle Park, N.C. May 8, 1974.

METHOD 8
DETERMINATION OF SULFURIC ACID MIST AND SULFUR DIOXIDE EMISSIONS FROM STATIONARY SOURCES

1. *Principle and Applicability*

1.1   Principle. A gas sample is extracted isokinetically from the stack. The sulfuric acid mist (including sulfur trioxide) and the sulfur dioxide are separated, and both fractions are measured separately by the barium-thorin titration method.

1.2   Applicability. This method is applicable for the determination of sulfuric acid mist (including sulfur trioxide, and in the absence of other partiuclate matter) and sulfur dioxide emissions from stationary sources. Collaborative tests have shown that the minimum detectable limits of the method are 0.05 milligrams/cubic meter (0.03 x 10^{-7} pounds/cubic foot) for sulfur trioxide and 1.2 mg/m^3 (0.74 x 10^{-7} lb/ft^3) for sulfur dioxide. No upper limits have been established. Based on theoretical calculations for 200 milliliters of 3 percent hydrogen peroxide solution, the upper concentration limit for sulfur dioxide in a 1.0 m^3 (35.3 ft^3) gas sample is about 12,500 mg/m^3 (7.7 x 10^{-4} lb/ft^3). The upper limit can be extended by increasing the quantity of peroxide solution in the impingers.

Possible interfering agents of this method are fluorides, free ammonia, and dimethyl aniline. If any of these interfering agents are present (this can be determined by knowledge of the process), alternative methods, subject to the approval of the Administrator, are required.

Filterable particulate matter may be determined along with SO$_3$ and SO$_2$ (subject to the approval of the Administrator); however, the procedure used for particulate matter must be consistent with the specifications and procedures given in Method 5.

2. *Apparatus*

2.1   Sampling. A schematic of the sampling train used in this method is shown in Figure 8-1. It is similar to the Method 5 train except that the filter position is different and the filter holder does not have to be heated. Commercial models of this train are

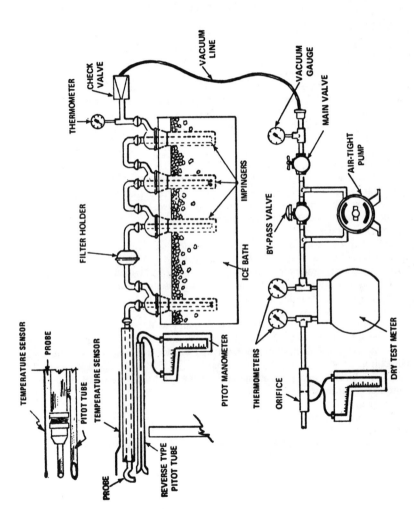

Figure 8-1. Sulfuric acid mist sampling train.

available. For those who desire to build their own, however, complete construction details are described in APTD-0581. Changes from the APTD-0581 document and allowable modifications to Figure 8-1 are discussed in the following subsections.

The operating and maintenance procedures for the sampling train are described in ATPD-0576. Since correct usage is important in obtaining valid results, all users should read the APTD-0576 document and adopt the operating and maintenance procedures outlined in it, unless otherwise specified herein. Further details and guidelines on operation and maintenance are given in Method 5 and should be read and followed whenever they are applicable.

2.1.1 Probe Nozzle. Same as Method 5, Section 2.1.1.

2.1.2 Probe Liner. Borosilicate or quartz glass, with a heating system to prevent visible condensation during sampling. Do not use metal probe liners.

2.1.3 Pitot Tube. Same as Method 5, Section 2.1.3.

2.1.4 Differential Pressure Gauge. Same as Method 5, Section 2.1.4.

2.1.5 Filter Holder. Borosilicate glass, with a glass frit filter support and a silicone rubber gasket. Other gasket materials, e.g., Teflon or Viton, may be used subject to the approval of the Administrator. The holder design shall provide a positive seal against leakage from the outside or around the filter. The filter holder shall be placed between the first and second impingers. Note: Do not heat the filter holder.

2.1.6 Impingers—Four, as shown in Figure 8-1. The first and third shall be of the Greenburg-Smith design with standard tips. The second and fourth shall be of the Greenburg-Smith design, modified by replacing the insert with an approximately 13 millimeter (0.5 in) ID glass tube, having an unconstricted tip located 13 mm (0.5 in.) from the bottom of the flask. Similar collection systems, which have been approved by the Administrator, may be used.

2.1.7 Metering System. Same as Method 5, Section 2.1.8.

2.1.8 Barometer. Same as Method 5, Section 2.1.9.

2.1.9 Gas Density Determination Equipment. Same as Method 5, Section 2.1.10.

2.1.10 Temperature Gauge. Thermometer, or equivalent, to measure the temperature of the gas leaving the impinger train to within $1^\circ C$ ($2^\circ F$).

2.2 Sample Recovery.

2.1.1 Wash Bottles. Polyethylene or glass, 500 ml (two).

2.2.2 Graduated Cylinders, 250 ml, 1 liter. (Volumetric flasks may also be used.)

2.2.3 Storage Bottles. Leak-free polyethylene bottles, 1000 ml size (two for each sampling run).

2.2.4 Trip Balance. 500-gram capacity, to measure to ± 0.5 g (necessary only if a moisture content analysis is to be done).

2.3 Analysis.

2.3.1 Pipettes. Volumetric 25 ml, 100 ml.

2.3.2 Burrette. 50 ml.

2.3.3 Erlenmeyer Flask. 250 ml (one for each sample blank and standard).

2.3.4 Graduated Cylinder. 100 ml.

2.3.5 Trip Balance. 500 g capacity, to measure to ± 0.5 g.

2.3.6 Dropping Bottle. To add indicator solution, 125-ml size.

3. *Reagents*

Unless otherwise indicated, all reagents are to conform to the specifications established by the Committee on Analytical Reagents of the American Chemical Society, where such specifications are available. Otherwise, use the best available grade.

3.1 Sampling.

3.1.1 Filters. Same as Method 5, Section 3.1.1.

3.1.2 Silica Gel. Same as Method 5, Section 3.1.2.

3.1.3 Water. Deionized, distilled to confrom to ASTM specification D1193-74, Type 3. At the option of the analyst, the $KMnO_4$ test for oxidizable organic matter may be omitted when high concentrations of organic matter are not expected to be present.

3.1.4 Isopropanol, 80 Percent. Mix 800 ml of isopropanol with 200 ml of deionized, distilled water.

NOTE—Experience has shown that only ACS grade isopropanol is satisfactory. Tests have shown that isopropanol obtained from commercial sources occasionally has peroxide impurities that will cause erroneously high sulfuric acid mist measurement. Use the following test for detecting peroxides in each lot of isopropanol: Shake 10 ml of the isopropanol with 10 ml of freshly prepared 10 percent potassium iodide solution. Prepare a blank by similarly treating 10 ml of distilled water. After 1 minute, read the absorbance on a spectrophotometer at 352 nanometers. If the absorbance exceeds 0.1, the isopropanol shall not be used. Peroxides may be removed from isopropanol by redistilling, or by passage through a column of activated alumina. However, reagent-grade isopropanol with suitably low peroxide levels is readily available from commercial sources; therefore, rejection of contaminated lots may be more efficient than following the peroxide removal procedure.

3.1.5 Hydrogen Peroxide, 3 Percent. Dilute 100 ml of 30 percent hydrogen peroxide to 1 liter with deionized, distilled water. Prepare fresh daily.

3.1.6 Crushed Ice.

3.2 Sample Recovery.

3.2.1 Water. Same as 3.1.3.

3.2.2 Isopropanol, 80 Percent. Same as 3.1.4.

3.3 Analysis.

3.3.1 Water. Same as 3.1.3.

3.3.2 Isopropanol, 100 Percent.

3.3.3 Thorin Indicator. 1-(o-arsonophenylazo)-2-naphthol-3, 6-disulfonic acid, disodium salt, or equivalent. Dissolve 0.20 g in 100 ml of deionized, distilled water.

3.3.4 Barium Perchlorate (0.0100 Normal). Dissolve 1.95 g of barium perchlorate trihydrate $[(Ba(Cl)_4)_2 \cdot 3H_2O]$ with isopropanol; 1.22 g of barium chloride dihydrate ($BaCl_2 \cdot 2H_2O$) may be used instead of the barium perchlorate. Standardize with sulfuric acid as in Section 5.2. This solution must be protected against evaporation at all times.

3.3.5 Sulfuric Acid Standard (0.0100 N). Purchase or standardize to $\pm$ 0.0002 N against 0.0100 N NaOH that has previously been standardized against primary standard potassium acid phthalate.

4. *Procedure*

4.1 Sampling.

4.1.1 Pretest Preparation. Follow the procedure outlined in Method 5, Section 4.1.1; filters should be inspected, but need not be desiccated, weighed, or identified. If the eflfuent gas can be considered dry, i.e., moisture free, the silica gel need not be weighed.

4.1.2 Preliminary Determinations. Follow the procedure outlined in Method 5, Section 4.1.2.

4.1.3 Preparation of Collection Train. Follow the procedure outlined in Method 5, Section 4.1.3 (except for the second paragraph and other obviously inapplicable parts) and use Figure 8-1 instead of Figure 5-1. Replace the second paragraph with: Place 100 ml of 80 percent isopropanol in the first impinger, 100 ml of 3 percent hydrogen peroxide in both the second and their impingers; retain a portion of each reagent for use as a blank solution. Place about 200 g of silica gel in the fourth impinger.

NOTE—If moisture content is to be determined by impinger analysis, weigh each of the first three impingers (plus absorbing solution) to the nearest 0.5 g and record these weights. The weight of the silica gel (or silica gel plus container) must also be determined to the nearest 0.5 g and recorded.

4.1.4 Pretest Leak-Check Procedure. Follow the basic procedure outlined in Method 5, Section 4.1.4.1, noting that the probe heater shall be adjusted to the minimum temperature required to prevent condensation, and also that verbage such as ". . . plugging the inlet to the filter holder . . . ," shall be replaced by, ". . . plugging the inlet to the first impinger . . . ." The pretest leak-check is optional.

4.1.5 Train Operation. Follow the basic procedures outlined in Method 5, Section 4.1.5, in conjunction with the following special instructions. Data shall be recorded on a sheet similar to the one in Figure 8-2. The sampling rate shall not exceed 0.030 m^3/min (1.0 cfm) during the run. Periodically during the test, observe the connecting line between the probe and the first impinger for signs of condensation. If it does occur, adjust the probe heater setting upward to the minimum temperature required to prevent condensation. If component changes become necessary during a run, a leak-check shall be done immediately before each change, according to the procedure outlined in Section 4.1.4.2 of Method 5 (with appropriate modifications, as mentioned in Section 4.1.4 of this method); record all leak rates. If the leakage rate(s) exceed the specified rate, the tester shall either void the run or shall plan to correct the sample volume as outlined in Section 6.3 of Method 5. Immediately after component changes, leak-checks are optional. If these leak-checks are done, the procedure outlined in Section 4.1.4.1 of Method 5 (with appropriate modifications) shall be used.

After turning off the pump and recording the final readings at the conclusion of each run, remove the probe from the stack. Conduct a post-test (mandatory) leak-check as in Section 4.1.4.3 of Method 5 (with appropriate modification) and record the leak rate. If the post-test leakage exceeds the specified acceptable rate, the tester shall either correct the sample volume, as outlined in Section 6.3 of Method 5, or shall void the run.

Drain the ice bath and, with the probe disconnected, purge the remaining part of the train, by drawing clean ambient air through the system for 15 minutes at the average flow rate used for sampling.

NOTE—Clean ambient air can be provided by passing air through a charcoal filter. At the option of the tester, ambient air (without cleaning) may be used.

4.1.6 Calculation of Percent Isokinetic. Follow the procedure outlined in Method 5, Section 4.1.6.

4.2 Sample Recovery.

4.2.1 Container No. 1. If a moisture content analysis is to be done, weigh the first impinger plus contents to the nearest 0.5 g and record this weight.

Transfer the contents of the first impinger to a 250-ml graduated cylinder. Rinse the probe, first impinger, all connecting glassware before the filter, and the front half of the filter holder with 80 percent isopropanol. Add the rinse solution to the cylinder. Dilute to 250 ml with 80 percent isopropanol. Add the filter to the solution, mix, and transfer to the storage container. Protect the solution against evaporation. Mark the level of liquid on the container and identify the sample container.

4.2.2 Container No. 2. If a moisture content analysis is to be done, weigh the second and third impingers (plus contents) to the nearest 0.5 g and record these weights. Also, weigh the spent silica gel (or silica gel plus impinger) to the nearest 0.5 g.

Transfer the solutions from the second and third impingers to a 100-ml graduated cylinder. Rinse all connecting glassware (including back half of filter holder) between the filter and silica gel impinger with deionized, distilled water and add this rinse water

PLANT _____
LOCATION _____
OPERATOR _____
DATE _____
RUN NO. _____
SAMPLE BOX NO. _____
METER BOX NO. _____
METER ΔH@ _____
C FACTOR _____
PITOT TUBE COEFFICIENT, Cp _____

STATIC PRESSURE, mm Hg (in. Hg) _____
AMBIENT TEMPERATURE _____
BAROMETRIC PRESSURE _____
ASSUMED MOISTURE, % _____
PROBE LENGTH, m (ft) _____
NOZZLE IDENTIFICATION NO. _____
AVERAGE CALIBRATED NOZZLE DIAMETER, cm (in.) _____
PROBE HEATER SETTING _____
LEAK RATE, m³/min, (cfm) _____
PROBE LINER MATERIAL _____
FILTER NO. _____

SCHEMATIC OF STACK CROSS SECTION

TRAVERSE POINT NUMBER	SAMPLING TIME (θ), min.	VACUUM mm Hg (in. Hg)	STACK TEMPERATURE (Ts), °C (°F)	VELOCITY HEAD (ΔPs), mm H2O (in. H2O)	PRESSURE DIFFERENTIAL ACROSS ORIFICE METER, mm H2O (in. H2O)	GAS SAMPLE VOLUME, m³ (ft³)	GAS SAMPLE TEMPERATURE AT DRY GAS METER		TEMPERATURE OF GAS LEAVING CONDENSER OR LAST IMPINGER, °C (°F)
							INLET, °C (°F)	OUTLET, °C (°F)	
TOTAL							Avg	Avg	
AVERAGE							Avg		

Figure 8-2. Field data.

to the cylinder. Dilute to a volume of 1000 ml with deionized, distilled water. Transfer the solution to a storage container. Mark the level of liquid on the container. Seal and identify the sample container.

4.3   Analysis.

Note the level of liquid in containers 1 and 2, and confirm whether or not any sample was lost during shipment; note this on the analytical data sheet. If a noticeable amount of leakage has occurred, either void the sample or use methods, subject to the approval of the Administrator, to correct the final results.

4.3.1   Container No. 1. Shake the container holding the isopropanol solution and the filter. If the filter breaks up, allow the fragments to settle for a few minutes before removing a sample. Pipette a 100-ml aliquot of this solution into a 250-ml Erlenmeyer flask, add 2 to 4 drops of thorin indicator, and titrate to a pink endpoint using 0.0100 N barium perchlorate. Repeat the titration with a second aliquot of sample and average the titration values. Replicate titrations must agree within 1 percent or 0.2 ml, whichever is greater.

4.3.2   Container No. 2. Thoroughly mix the solution in the container holding the contents of the second and third impingers. Pipette a 10-ml aliquot of sample into a 250-ml Erlenmeyer flask. Add ml of isopropanol, 2 to 4 drops of thorin indicator, and titrate to a pink endpoint using 0.0100 N barium perchlorate. Repeat the titration with a second aliquot of sample and average the titration values. Replicate titrations must agree within 1 percent or 0.2 ml, whichever is greater.

4.3.3   Blanks. Prepare blanks by adding 2 to 4 drops of thorin indicator to 100 ml of 80 percent isopropanol. Titrate the blanks in the same manner as the samples.

5. *Calibration*

5.1   Calibrate equipment using the procedures specified in the following sections of Method 5: Section 5.3 (metering system); Section 5.5 (temperature gauge); Section 5.7 (barometer). Note that the recommended leak-check of the metering system, described in Section 5.6 of Method 5, also applies to this method.

5.2   Standardized barium perchlorate solution with 25 ml of standard sulfuric acid, to which 100 ml of 100 percent isopropanol has been added.

6. *Calculations*

NOTE—Carryout calculations retaining at least one extra decimal figure beyond that of the acquired data. Round off figures after final calculation.

6.1   Nomenclature.

A_n = Cross-sectional area of nozzle, m^2 (ft^2).
B_{ws} = Water vapor in the gas stream, proportion by volume.
$C_{H_2SO_4}$ = Sulfuric acid (including SO_2) concentration, g/dscm (lb/dscf).
C_{SO_2} = Sulfur dioxide concentration, g/dscm (lb/dscf).
I = Percent of isokinetic sampling.
N = Normality of barium perchlorate titrant, equivalents/liter.
P_{bar} = Barometric pressure at the sampling site, mm Hg (in. Hg).
P_s = Absolute stack gas pressure, mm Hg (in. Hg).
P_{std} = Standard absolute pressure, 760 mm Hg (29.92 in. Hg).
T_m = Average absolute dry gas meter temperature (see Figure 8-2), $^\circ K$ ($^\circ R$).
T_s = Average absolute stack gas temperature (see Figure 8-2), $^\circ K$ ($^\circ R$).
T_{std} = Standard absolute temperature, $293^\circ K$ ($528^\circ R$).
V_a = Volume of sample aliquot titrated, 100 ml for H_2SO_4 and 10 ml for SO_2.

V_{lc} = Total volume of liquid collected in impingers and silica gel, ml.

V_m = Volume of gas sample as measured by dry gas meter, dcm (dcf).

$V_{m(std)}$ = Volume of gas sample measured by the dry gas meter corrected to standard conditions, dscm (dscf).

v_s = Average stack gas velocity, calculated by Method 2, Equation 2-9, using data obtained from Method 8, m/sec (ft/sec).

V_{soln} = Total volume of solution in which the sulfuric acid or sulfur dioxide sample is contained, 250 ml or 1,000 ml, respectively.

V_t = Volume of barium perchlorate titrant used for the sample, ml.

V_{tb} = Volume of barium perchlorate titrant used for the blank, ml.

Y = Dry gas meter calibration factor.

ΔH = Average pressure drop across orifice meter, mm (in.) H_2O.

Θ = Total sampling time, min.

13.6 = Specific gravity of mercury.

60 = sec/min.

100 = Conversion to percent.

6.2   Average dry gas meter temperature and average orifice pressure drop. See data sheet (Figure 8-2).

6.3   Dry Gas Volume. Correct the sample volume measured by the dry gas meter to standard conditions ($20°C$ and 760 mm Hg or $68°F$ and 29.92 in. Hg) by using Equation 8-1.

$$V_{m(std)} = V_m \, Y \left(\frac{T_{std}}{T_m}\right) \frac{P_{bar} + \left(\dfrac{\Delta H}{13.6}\right)}{P_{std}}$$

$$= K_1 \, V_m \, Y \, \frac{P_{bar} + (\Delta H/13.6)}{T_m}$$

Equation 8-1

where:

K_1 = $0.3858°K/mm$ Hg for metric units.

= $17.64°R/in.$ Hg for English units.

NOTE—If the leak rate observed during any mandatory leak-checks exceeds the specified acceptable rate, the tester shall either correct the value of V_m in Equation 8-1 (as described in Section 6.3 of Method 5), or shall invalidate the test run.

6.4   Volume of Water Vapor and Moisture Content. Calculate the volume of water vapor using Equation 5-2 of Method 5; the weight of water collected in the impingers and silica gel can be directly converted to milliliters (the specific gravity of water is 1 g/ml). Calculate the moisture content of the stack gas, using Equation 5-3 of Method 5. The "Note" in Section 6.5 of Method 5 also applies to this method. Note that if the effluent gas stream can be considered dry, the volume of water vapor and moisture content need not be calculated.

6.5   Sulfuric acid mist (including SO_3) concentration.

$$C_{H_2SO_4} = K_2 \, \frac{N(V_t - V_{tb}) \left(\dfrac{V_{soln}}{V_a}\right)}{V_{m(std)}}$$

8-2

Equation 8-2

where:
K_1 = 0.04904 g/milliequivalent for metric units
    = 1.081 x 10^{-4} lb/meq for English units.
6.6   Sulfur dioxide concentration.

$$C_{SO_2} = K_3 \frac{N(V_t - V_{tb})\left(\dfrac{V_{soln}}{V_a}\right)}{V_{m(std)}}$$

Equation 8-3

where:
K_3 = 0.03203 g/meq for metric units
    = 7.061 x 10^{-5} lb/meq for English units.
6.7   Isokinetic Variation.
6.7.1 Calculation from raw data.

$$I = \frac{100\, T_s\, [K_4\, V_{lc} + (V_m/T_m)\, P_{bar} + \Delta H/13.6)]}{60\, \Theta\, V_s\, P_s\, A_n}$$

Equation 8-4

where:
K_4 = 0.003464 mm Hg-m^3/ml-$^{\circ}$K for metric units.
    = 0.002676 in. Hg-ft^3/ml-$^{\circ}$R for English units.
6.7.2 Calculation from intermediate values.

$$I = \frac{T_s\, V_{m(std)}\, P_{std}\, 100}{T_{std}\, v_s \Theta A_n\, P_s\, 60(1 - B_{ws})}$$

$$= K_5 \frac{T_s\, V_{m(std)}}{P_s\, v_s\, A_n\, \Theta\,(1 - B_{ws})}$$

Equation 8-5

where:
K_5 = 4.320 for metric units
    = 0.09450 for English units.
6.8   Acceptable Results. If 90 percent $\leqslant I \leqslant$ 110 percent, the results are acceptable. If the results are low in comparison to the standards and I is beyond the acceptable range, the Administrator may opt to accept the results. Use Citation 4 in the Bibliography of Method 5 to make judgments. Otherwise, reject the results and repeat the test.

7. *Bibliography*

1. Atmospheric Emissions from Sulfuric Acid Manufacturing Processes. U.S. DHEW, PHS, Division of Air Pollution. Public Health Service Publication No. 99-AP-13. Cincinnati, Ohio. 1965.

2. Corbett, P. F. The Determination of SO_2 and SO_3 in Flue Gases. Journal of the Institute of Fuel. 24:237-243. 1961.

3. Martin, Robert M. Construction Details of Isokinetic Source Sampling Equipment Environmental Protection Agency. Research Triangle Park, N.C. Air Pollution Control Office Publication No. APTD-0581. April, 1971.

4. Patton, W. F. and J. A. Brink, Jr. New Equipment and Techniques for Sampling Chemical Process Gases. Journal of Air Pollution Control Association. 13:162. 1963.

5. Rom, J. J. Maintenance, Calibration and Operation of Isokinetic Source-Sampling Equipment. Office of Air Programs, Environmental Protection Agency. Research Triangle Park, N.C. APTD-0576. March 1972.

6. Hamil, H. F. and D. E. Camann. Collaborative Study of Method for Determination of Sulfur Dioxide Emissions from Stationary Sources (Fossil Fuel-Fired Steam Generators). Environmental Protection Agency. Research Triangle Park, N.C. EPA-350/4-74-024. December, 1973.

7. Annual Book of ASTM Standards. Part 31. Water, Atmospheric Analysis. pp.40-42. American Society for Testing and Materials. Philadelphia, Pa. 1974.

* * * * * * * * * * * * * *

(Secs. 111,114,301(a), Clean Air Act, sec. 4(a) of Pub. L. 91-604, 84 Stat. 1683; sec. 4(a) of Pub. L. 91-604, 84 Stat. 1687; sec. 2 of Pub. L. 90-148, 81 Stat. 504 [42 U.S.C. 1857c-6, 1875c-9, 1857g(a)].)

[FR Doc.77-13608 Filed 8-17-77; 8:45 am]